Date Due

Revolutions in Physics

Revolutions

by Barry M. Casper
and Richard J. Noer
CARLETON COLLEGE

in Physics

W · W · NORTON & COMPANY · INC ·
NEW YORK

FIRST EDITION

Library of Congress Cataloging in Publication Data

Casper, Barry M.
Revolutions in physics.

Bibliography: p.
1. Motion. 2. Relativity (Physics). 3. Particles (Nuclear physics). I. Noer, Richard J., 1937– joint author. II. Title.
QC125.2.C36 531'.11 75-162931
ISBN 0-393-09992-X

PRINTED IN THE UNITED STATES OF AMERICA

1 2 3 4 5 6 7 8 9 0

To Myra and Raymonde

CONTENTS

PREFACE

This book has grown out of our one-semester course for students who do not intend to specialize in a physical science. Such students, we find, are initially often indifferent or even hostile toward physics, which they picture as a sterile collection of facts and laws uncovered by systematic application of the scientific method. Many feel, with Alexander Pope, that "the proper study of Mankind is Man." In our judgment, the usual survey course is both inappropriate and ineffective for them. An attempt to deal with all the principal topics of physics in one semester is bound to result in a superficial treatment (replete with too many instances of "it can be shown . . .") with little insight into man's role in creating physics. Furthermore, it is not necessary to cover all of physics to communicate the essence of the discipline. By concentrating on a few topics, we can examine them in depth and hence provide a clearer conception of the physicist's enterprise.

Accordingly, we focus our attention upon two related topics—Newton's theory of motion and Einstein's conception of space and time—and the developments that lay behind them. From early star myths to the theory of relativity, we follow the changing course of man's ideas about motion in the heavens and on the earth. We examine how different men in different times, viewing the same phenomena, have seen different things and have explained what they saw in vastly different ways. In the process we pay particular attention to the times of conceptual revolution, when men cast off one world view in favor of another. As we study the evolution of ideas about motion we expose the interplay between experiment and theory in physics and examine the sense in which a physical theory can be said to explain observations of nature. The final chapter is devoted to elementary particle physics, where a revolution may be in progress today. Here the reader can compare the conception of physics he has gleaned from our historical study of motion with a problem of contemporary research.

Little mathematical proficiency is required of the reader; elementary algebra and geometry will suffice. Our purpose is not to produce accomplished problem solvers, but rather to communicate the spirit of physics. Questions at the end of each chapter are aimed at testing the understanding of concepts rather than sharpening the student's puzzle-solving ability. In addition, a few problems are included in the body of the text; we strongly encourage the reader to attempt these exercises as he encounters them. The mathematical level gradually increases as the book progresses, from very low in the first four chapters to rather extensive algebra in the treatment of the Lorentz transformation in Chapter 14. The more mathematical sections of Chapter 14 can easily be omitted. Chapter 15, while relatively non-mathematical, contains a number of rather difficult ideas, and may also be omitted, especially if the class is running short of time.

We acknowledge our debt to the many people who have contributed to this book. Our philosophical position has been strongly influenced by the work of Thomas S. Kuhn and Stephen Toulmin, whose books we regularly employ as supplements to our course. Donald Ferlen, Tom Videen, Finley Bishop, Mark Williams, and Orlen Fjelsted all performed invaluable services in helping us to prepare various editions of the manuscript for classroom use. Sharon Prink deserves special thanks for her excellent typing throughout the development of the book. We also wish to thank our colleagues, Gary Iseminger, David Sipfle, and Bruce Thomas, and the students of Physics 20, particularly David Loy and Richard Weiner, for their helpful advice. Davis Taylor of the Carleton College English Department and Mary Pell at Norton have provided skillful assistance with the style of the manuscript. We feel especially fortunate to have had Kenneth B. Demaree as our editor. Finally we must thank our wives, Myra and Raymonde, for reading the manuscript with such care and for the numerous other contributions that made this book possible.

BARRY M. CASPER
RICHARD J. NOER

Northfield, Minnesota
November, 1971

Revolutions in Physics

INTRODUCTION

> The mind of the most rational among us may be compared to a stormy ocean of passionate convictions based upon desire, upon which float perilously a few tiny boats carrying a cargo of scientifically tested beliefs. Nor is this to be altogether deplored: life has to be lived, and there is no time to test rationally all the beliefs by which our conduct is regulated. Without a certain wholesome rashness, no one could long survive. Scientific method, therefore, must in its very nature, be confined to the more solemn and official of our opinions.[1]
>
> —Bertrand Russell
> *The Scientific Outlook*

What is this "scientific method"? Is there really a way of thinking that clearly distinguishes science from the other pursuits of man?

CHAPTER ONE

The "Scientific Method"?

There is a view of science that is remarkably prevalent today. In this view, science is distinguished from other areas of human thought principally by its use of the so-called scientific method, which consists of a well-ordered series of steps apparently first set forth by Francis Bacon in the early seventeenth century. A modern statement has been given by Bertrand Russell, one of the most eminent philosophers and mathematicians of the present century. In *The Scientific Outlook* Russell describes the scientific process as composed of

> three main stages; the first consists in observing the significant facts; the second in arriving at a hypothesis which, if it is true, would account for these facts; the third in deducing from this hypothesis [further] consequences which can be tested by observation . . .[1]

If these new predictions are not borne out by experiment, the hypothesis must be rejected and a new one devised. The process, we are led to believe, is repeated again and again until finally a satisfactory hypothesis is discovered.

Many of us who are practicing scientists, however, find this picture not so much wrong as inadequate and even misleading. If science were as dry and ordered as this, it would have the intellectual and emotional appeal of a jigsaw puzzle. Yet many people of unusual intellect and sensitivity have found in science an outlet for their creative energy—an outlet as satisfying to them as is writing to the poet or painting to the artist.

So perhaps the common view of science is misleading. What then is science really about? What constitutes a physical theory and how is it obtained? What kinds of questions can science answer and what kinds are beyond its domain?

It is with inquiries of this sort that we shall deal in this book. Our approach, however, will be indirect: we will search for answers from the inside by immersing ourselves in science. We will not make any new discoveries, of course, but rather will experience vicariously the creative work of others. Our attention will be concentrated on two related topics: the nature of motion and the nature of space and time. We will examine the development of men's ideas concerning these topics in order to suggest answers to our questions about the nature of science.

The choice of physics as the subject of our investigation is a natural one. Physics is the most highly developed and most mature of the sciences, and its approach is often taken as a model by the others. This is not because of any inherent superiority, but rather because of the particularly simple and fundamental problems with which physics deals. This simplicity provides a sharp focus for the questions we shall ask.

The inadequacy of the traditional view of the scientific method is shown by the questions it leaves unanswered, questions that lie at the heart of the scientific process. In this chapter, we shall raise some of these questions and suggest another point of view about the way science operates.

Let us examine the above quotation. Consider the first two of Russell's three stages: "observing the significant facts" and "arriving at a hypothesis." One gets the impression of a sequential process. First the facts are observed and then a hypothesis is devised. But how does one know *a priori* which facts are significant?

The difficulty arises because any physical phenomenon has associated with it a vast number of facts. For example, one phenomenon we shall consider is the motion of the planets. Their wandering was a puzzle to early watchers of the heavens. Many facts were known, but which were significant? The number of planets? Their variations in brightness? Their distances from the sun? The changes in their speed as they traverse their paths? The nature of their paths relative to the stars? Or relative to the earth? Or to the sun?

If the task of physics is to find some order in natural phenomena, to devise a theoretical framework in which the observations fit, then clearly some observations will be more significant than others in the sense that they will contribute more fruitfully to the conception of a theory. The problem is that in the early stages of a scientific inquiry one does not yet know what the theoretical framework will be. In particular, one does not even know what questions about the phenomena the theory will answer. But which questions are asked determines which data are relevant or, to put it another way, which facts are significant.

The history of men's ideas about the planets contains many examples of this ambiguity. Should a theory of the planets explain their number and the distances between them? As we shall discuss, Johannes Kepler developed such a theory in the seventeenth century; for Kepler, the relevant data were the number of planets and their distances from the sun. Or should a theory explain the motions of the planets? In this case the relevant data depend on what kind of explanation will suffice. Ancient theories, which today we term mythological, tried to explain the motion of the planets as due to the influence of the constellations through which they pass. For proponents of these theories, the wandering of the planets relative to the field of stars constituted the relevant data. Late Greek and Roman astronomers explained the planetary motions as resulting from simple combinations of circular motions around the earth; the relevant data for them were the paths of the planets relative to the earth.

Today we have come to accept the Newtonian explanation of the motion of the planets, for which the relevant facts are the shapes of the orbits relative to the sun and the accelerations of the planets as they traverse these orbits. Thus, while the number of planets, their distances from the sun, and their motion relative to the stars and to the earth cannot be dismissed *a priori*, they are not directly relevant to our present understanding of the planetary problem.

The dilemma is this: The "significant" facts are significant because they are useful in suggesting a theoretical framework. However, until one finds a theoretical framework, one cannot know what the appropriate observations, the significant facts, will be. This does not mean that science is impossible—only that it is not the point-to-point process that the conventional description of the scientific method seems to imply. What we are suggesting is that the development of a physical theory involves a great deal of guesswork guided by such supposedly nonscientific factors as intuition and aesthetic and philosophical prejudices. In short, it requires a creative insight, an insight glossed over in the usual presentation of the scientific method.

Another aspect of Russell's second stage is also worth mentioning. It suggests that one seeks a hypothesis which "would account for these facts." There is an undeserved air of finality about the term "account for."

For example, consider once again the planetary problem. Suppose we believe that the question, "How can the observed motion of the planets be explained?" is a fruitful inquiry. We must then ask what sort of response constitutes a complete explanation. Is it sufficient to display the data in a succinct and ordered manner, perhaps condensing it in a mathematical formula? As we shall discuss, the heliocentric model proposed by Copernicus and modified by Kepler describes a solar system in which the planets move about the sun in elliptical orbits. The time it takes each planet to complete a full orbit and the variations in its speed during the orbit are related to its distance from the sun by simple mathematical expressions. Thus the apparently complex motion of the planets and the sun as observed from the earth can be deduced from this appealingly simple model.

Surely there is a sense in which this model explains the observed wandering of the planets about the sky. But is this explanation sufficient? One might argue that to explain the planetary observations we must show them to be a consequence of more fundamental laws. For example, Newton provided such an explanation when he demonstrated that the regularities in the planetary orbits discovered by Kepler follow directly from the assumption of a gravitational force of attraction between all objects.

But is even this sufficient? Perhaps an adequate explanation requires the answering of deeper "why" questions. An obvious one is: Why is there a gravitational force? What is the mechanism responsible for this force of attraction? Newton admitted that he had no answer to this. Yet his scheme, which indicates the appropriate questions to ask and the nature of satisfactory answers, has been so successful that it has provided a framework for inquiry in physics for three centuries.

Surely the sequence of "why" questions must end at some point. But how does the scientific process decide where?

The third stage of the scientific method, the crucial experimental test of a theory, would also seem to be a substantial oversimplification of what happens in practice. The notion is that after a theory is devised, further predictions that it makes are subjected to experimental test. If the predictions are not borne out by experiment, the theory is rejected.

This description rightly emphasizes the empirical basis of physical theories. A genuine discrepancy between theoretical predictions and experimental data certainly does call for reevaluation of the theory on some level. While we have emphasized the element of creativity in the construction of physical theories, this creativity is severely restricted by the unalterable characteristics of the real world with which the theories deal. But the implication that a whole theory must be rejected on the basis of a single bit of experimental evidence is an extreme oversimplification. The history of physics is filled with cases in which evidence did not quite fit and yet the basic theoretical framework was so compelling on other grounds that it was not rejected out of hand. Newton put forth his theory of gravitation, for example, as an explanation of the motion of all the bodies in the solar system, and yet for sixty years everyone who tried to apply it to the motion of the moon arrived at results that strikingly disagreed with observation. Nevertheless, the Newtonian theory was such an elegant and powerful way of looking at nature that it was simply assumed that the discrepancy would eventually be reconciled. And indeed it was.

The point is that a theory can be—and the more far-reaching theories are—more than simply a device for making predictions. In an important sense, a theory is an organizing principle, a prescription for making sense out of an otherwise disconnected jumble of observations. One does not easily reject that which provides one's *gestalt*, one's frame of reference for comprehending the world. Where the physical theory conflicts with an observation of nature, one makes minor modifications, perhaps at the expense of elegance in the theory. Repeated conflicts and the need for frequent modifications may give rise to doubts about the fundamental validity of the theory and even the world view on which it depends. Individuals may begin to seek radical new approaches. But it is generally true that the scientific community as a whole will not reject a previously accepted theory until an adequate, or at least promising, alternative has been devised. To do otherwise—to reject the old frame of reference without first seeing a new one—would be to give up whatever understanding one has achieved and face again a chaotic world. We are almost forced to cling to the accepted viewpoint, hoping, even believing, that the apparent discrepancy is only an illusion or that it will be resolved by a minor modification. Usually it is; only at particularly critical junctures is a fresh creative act needed to liberate us from the established point of view.

It is on some of these highly creative acts—theories which re-orient the whole thinking of a discipline—that we will center our attention. We do not wish to mislead, however; all of science does not require this order of creativity. As in any field, the efforts of men in science range from the highly imaginative to the less creative, with the balance weighted toward the latter end. For every Newton, whose conceptual scheme re-orients the thinking of the world, there are generations of scientists asking and answering the myriads of detailed questions raised by such a scheme.

For example, Newton explained the main features of planetary motion as resulting from the action of gravity between the planets and the sun. But his theory also implied a gravitational force between the planets themselves, leading to slight changes in their orbits. In the mid-nineteenth century, when the planet Uranus was observed to deviate from its predicted orbit, the astronomers Adams and Leverrier tried to explain this deviation by postulating the existence of another, as yet unknown, planet. Their success led to the discovery of Neptune. The problem was a challenging one, the solution immensely satisfying, and yet both question and answer were well defined within the context of the Newtonian theory. The work of Newton, and the imagination it required, were of a different order from that of Adams and Leverrier.

Similarly, in the twentieth century, the theory of quantum mechanics established a theoretical framework for understanding the behavior of atoms and nuclei; its creation was another of the high points in the history of physics. Now thousands of physicists are at work on experimental and theoretical investigations of the details of atomic and nuclear physics—using the quantum mechanical approach to study the nature of nuclear forces, the structure of complex nuclei, and the properties of solids, liquids, and gases. We can thus distinguish between problem-solving within an established framework, which characterizes much legitimate and important physics, and such towering conceptual advances as Newton's theory of motion, Einstein's theory of relativity, and the theory of quantum mechanics.

The common view of science is often determined primarily from a consideration of its less creative aspects—the testing, the working out of details, the taming of the grand theories. For science on this level, we have probably been unduly hard on Russell's view of the scientific method. Nevertheless, it is the great theories that expose the heart of science and furnish the best perspective on the entire enterprise, and it is in dealing with them that the common view of science is most inadequate.

Suggested Reading

Kuhn, T. *The Structure of Scientific Revolutions.* Chicago: University of Chicago Press, 1962. This and the following book are stimulating and insightful discussions of scientific thought and its development. The influences of both can be seen throughout this book.

Toulmin, S. *Foresight and Understanding.* Bloomington: Indiana University Press, 1961.

Russell, B. *The Scientific Outlook.* New York: W. W. Norton, 1931. For eloquent arguments in Russell's defense, see his own book.

THE PROBLEM OF CELESTIAL MOTION

To demonstrate that the appearances are saved by assuming the sun at the center and the earth in the heavens is not the same thing as to demonstrate that *in fact* the sun is in the center and the earth in the heavens. I believe that the first demonstration may exist, but I have very grave doubts about the second . . .[1]

—Cardinal Robert Bellarmine
April 4, 1615

Copernicus comprehended the true nature of the earth. He fully understood that it is a planet revolving around the sun in the company of the other planets.[2]

—*Encyclopedia Americana*
(1969 edition), vol. 7,
"Copernicus"

Although the sun is seen to move across the sky each day, you probably believe that this is due to the earth's really rotating on its axis rather than to the sun's really revolving around the earth. Can you cite any evidence to support this belief?

CHAPTER TWO

The Heavens around Man

One needs no historical expertise to realize that man's interest in the heavens must be nearly as old as man himself. Our primitive predecessors spent much of their lives outdoors, close to nature and largely under her dominion. Their survival and security depended on a knowledge of nature and an ability to anticipate her whims. A predictable world could be dealt with; a capricious one threatened destruction. The heavens seemed protective when the constellations moved across them night after night in familiar ways. Those same heavens seemed terrifying when the sun was suddenly extinguished in an eclipse.

Today man is relatively free from such fears, liberated by the understanding of the heavens he has attained. To begin our study of the development of this understanding, let us review some of the phenomena of the heavens that must have been well known to any primitive shepherd but are no longer familiar to most of us. It is difficult to say just how much was known so long ago, but some observations are so striking that we can assume primitive cultures were aware of them.

To a man standing in the open countryside, the earth appears to be a basically flat surface, bounded at its limits by the horizon. Extending up from the horizon all around are the heavens, illuminated by day by the sun, dark at night except for the stars and moon. The stars are scattered across the sky in patterns that imaginative minds have seen as all manner of objects and creatures—the Big Dipper, the hunter Orion, the flying horse Pegasus, and countless others.

Diurnal Motion

Once individual stars can be distinguished and followed by virtue of their positions in the constellations, we can see that the stars are not fixed in the sky; they appear to move together and in a particularly simple way. One star, the North Star, does seem motionless;* the others all appear to move around it in arcs of circles during the course of the night. (See Figure 2–1.) It is as if the stars were attached to a dome of the heavens that rotates on a tilted axis passing through the North Star, as depicted in Figure 2–2. On successive nights, the stars appear at about the same place at about the same time, but all night long, each night, they travel on their circular paths. Some stars near the North Star are visible throughout the night; others, near the southern horizon, rise in the east and set in the west during the course of the night as the dome lifts them briefly above the horizon.

Likewise the sun and moon. The sun seems to travel across the sky each day, rising in the east and setting in the west, following a course very similar to the path at night of those stars that rise and set. The moon moves across the sky in a

* Throughout this chapter, we take the point of view of an observer in the Northern Hemisphere, somewhere between the Tropic of Cancer and the Arctic Circle.

Figure 2–1. A time exposure photograph of the night sky as viewed from the middle latitudes of the Northern Hemisphere. (Yerkes Observatory photograph, University of Chicago.)

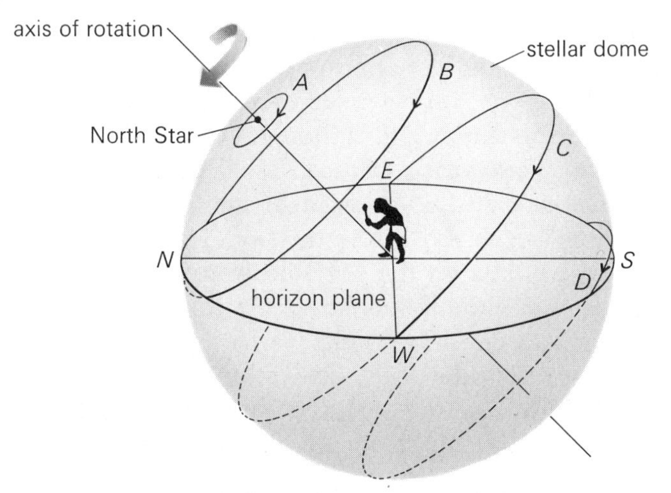

Figure 2–2. Diurnal motion as observed from a point in the middle latitudes of the Northern Hemisphere. As the stellar dome rotates through the night, stars are carried along arcs of circular paths such as *A*, *B*, *C*, and *D*. The horizon plane can be thought of as the extension to the stellar dome of the flat field on which the observer is standing. During the night, when objects on the dome are above the horizon plane, they are visible to the observer; when they are below the horizon plane they are not visible.

similar fashion. In fact, every object in the heavens seems to partake of this daily (or *diurnal*) motion.

Annual Motion

But there is more to it than that. Watching day after day for weeks and months, one becomes conscious of longer-term changes in the heavens. The most obvious are those of the moon, whose cycle of phases from full to new and changing times of rising and setting mark out the period known accordingly as the month. The long-term motion of the stars is slower, corresponding to a yearly cycle. Each night the constellations move around the sky with their diurnal motion, but any given star reaches a particular point in the sky about four minutes earlier than it did the previous night. It is as if the dome is rotating too fast, by about four minutes each day. Thus in a year the dome makes a full extra revolution and the stars return to their original schedule.* And so the sky appears radically different in different months and seasons: on a winter evening, Orion is prominent in the sky; in the summer it cannot be seen at all.

The sun too has a yearly motion. The length of the day varies with the seasons, reflecting a gradual change in the path of the sun as it moves each day across the sky. The sun's path and, in particular, the points on the horizon where it rises and sets are more northerly in the summer and more southerly in the winter, as shown in Figure 2–3. At the vernal equinox—the time in spring when the lengths of the day and night are exactly equal—the sun rises due east and sets due west, and its path is an arc between these points passing not directly overhead, but with its highest point somewhat to the south. From day to day the entire path shifts northward, until the summer solstice. This is the longest day of the year, and the sun's path is most northerly. At noon it appears most nearly overhead, and at sunrise and sunset it appears most north of due east and west. Then the

* 3 minutes, 56 seconds advance per day—a more precise figure—times 365¼ days per year equals 24 hours advance per year.

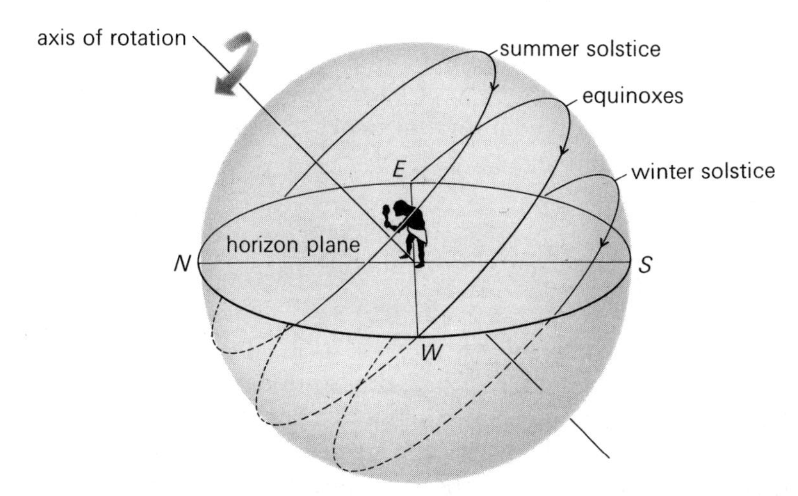

Figure 2–3. The diurnal motion of the sun at different times of the year.

motion reverses; at the autumnal equinox the path of the sun has returned to that of the vernal equinox, rising and setting due east and west. From this time on, the path becomes more and more southerly and the days grow shorter and shorter until the winter solstice, when the points of sunrise and sunset are most nearly south from due east and west.

A few "stars," however, behave in a different fashion from all the others. They participate in the diurnal motion, but as we watch from night to night, we see them slowly changing their positions relative to all the other stars. It is as if they are not fixed to the rotating dome like the other stars, but rather move slowly about as they are carried around each night. Furthermore, the path along which they move is not a simple one; the motion seems somewhat erratic, now forward, now backward, now stopping, having no obvious connection with the seasons or the months. In addition, the brightness of these objects changes with time—one month a particular one may be among the brightest stars in the sky, while the next it may be so faint as to be hardly visible. In antiquity, these bodies were given the name of planets (from the Greek word for wanderers) and they numbered only five—Mercury, Venus, Mars, Jupiter, and Saturn—among the myriads of well-behaved stars. Five relatively insignificant objects, which we mention here almost as an afterthought—and yet their unruly behavior was to intrigue and puzzle men for millennia until it forced a revolutionary change in man's view of himself and his universe.

Models versus Data

Let us pause for a bit of introspection about the preceding discussion. It claimed to be a description of the phenomena observable by primitive man, but it included statements that went beyond this. "The stars appear to move in circles about the North Star" describes an observed phenomenon, but "the stars appear fixed to a dome rotating on an axis passing through the North Star" adds a new element. The second statement implies the first, but not vice versa. The second statement describes the data in terms of a model.

At this point we are concerned with describing celestial observations. The dome model is a convenient way of remembering the primitive celestial data. Whether this model is anything more than a useful memory device, we leave for later discussion.

Primitive Explanations

Having reviewed the observations available to primitive man, let us consider some early explanations for them. The earliest of these were invariably of a character we would today call mythological. The sun, the most prominent and active heavenly body, demanded explanation—it was borne in a chariot pulled by gods across the sky, or it was a living god itself, riding in a boat through the heavens. The stars, on the other hand, seemed more neutral—mere tiny lamps suspended from a fixed vault in the heavens. Some common myths even ignored the stars' diurnal motion, which is readily apparent to anyone who watches all night.

Such explanations, based as they are on human (or human-like) will, are largely immune to questioning. *Why* do the gods pull the sun across the sky? Because they want to, for some reason of their own, or even for no reason at all. The questioner, realizing the arbitrariness of many of his own actions, sees the futility of further inquiry.

Beginnings of Technical Astronomy

The growth and development of primitive societies occasionally brought about a new kind of concern with the heavens. Men were no longer satisfied with approximate, qualitative observations; they began to watch the skies more carefully, and to make and record quantitative measurements. A striking prehistoric step in this direction is the monument of Stonehenge (Figure 2–4), built by the Stone Age people of southern England between about 1900 and 1600 B.C. It consists of several concentric circles, up to about 100 feet in diameter, formed from stone blocks, the largest of which have been cut into rectangular shapes several times the height of a man and surmounted by large flat lintels. The main axis of the structure extends through an obvious entrance, aligned roughly in a southwest-to-northeast

Figure 2–4. The ruins of Stonehenge. (British Crown Copyright.)

direction in just such a way that it singles out a particular point on the northeast horizon. This is the northernmost point of the rising of the sun at the latitude of Stonehenge. As we have noted, this northernmost rising occurs on the summer solstice. It is clear that the stones have been laid out precisely so that on that day, and only on that day, an observer at the center of the stone circles will see the sun rise directly between the entrance stones and immediately above a specially placed distant reference stone. Other similar special sighting lines allow one to determine the winter solstice and the vernal and autumnal equinoxes.

The most impressive use of Stonehenge, however, was apparently as an eclipse predictor. Many years of observation of eclipses of the moon enable one to discover certain regularities. The moon must be full, the sun and moon must rise and set at certain points on the horizon, and a particular number of years must have passed since the previous eclipse under the same circumstances. The last requirement is the most difficult to discover, for a complete cycle of seemingly irregular intervals between eclipses repeats itself only after 56 years. Recent measurements and calculations have shown convincingly that certain otherwise unexplained marker stones at Stonehenge indicate those points on the horizon from which the rising of the full moon may involve an eclipse. An outer circle of 56 holes was apparently used to keep track of the years. A Stone Age astronomer could consult the markers in the 56 holes and, if an eclipse year was indicated, follow the risings and settings of the sun and moon until the special stones indicated that a critical alignment was at hand. He could then announce with some confidence that an eclipse was imminent.*

Stonehenge thus functioned as a gigantic astronomical instrument. Its importance to the people who built it can be seen in the immense effort involved in its construction. Many of the huge stones were hauled over 200 miles, partly by water on rafts, partly by land on log rollers, using only human motive power. Far more impressive is the accumulated experience of

* In fact, the system seems to have been even more elaborate and impressive than we have been able to relate in this short space. Interested readers are referred to a most entertaining and readable account in *Stonehenge Decoded*, by G. S. Hawkins, the astronomer who made these discoveries.

astronomical observation that the structure represents. Even to be aware that the seemingly random eclipses occur with some regularity, especially when the cycle extends over a period of many years, must have required generations of observation, with the results being passed from person to person by word of mouth.

Stonehenge was unique, but the interest in quantitative observation that it represents was not. Technical astronomy was a central concern of other early civilizations, and it seems likely that in each case there were two primary motivations: religious rites and calendar regulation. The former involved an obvious extension of the earlier mythological explanations of heavenly events. If these events were controlled by deities, then a study of the heavens was a clear duty of the priesthood. From a knowledge of the movement of heavenly bodies one could discern the will of the gods and thus discover the propitious times for religious festivals. Perhaps one could even foretell the acts of the gods that precipitate such disasters as earthquakes and eclipses.

The calendar application was also important. Many practical operations, in particular agricultural ones, depended on an accurate knowledge of the progression of the seasons, and the heavens provided the only reliable timekeeping device. The planting season in Egypt depended on the flooding of the Nile, which occurred each year at about the time of the first appearance of the star Sirius above the horizon. As early as the third millennium B.C., Egyptians were using this event to mark the beginning of each new year.* Care in observing and attempting to predict the rising of Sirius must soon have led to the realization that a full year contains not 365 but 365¼ days, though the adoption of the leap year came only much later, in Roman times.

In Mesopotamia, the art of astronomical observation was developed for the same religious and calendrical reasons, but to an even higher order. By 1200 B.C., the Babylonians were using sophisticated instruments for measuring the positions of the sun and stars. With these instruments, they made elaborate tables, detailing the times of rising and setting of the planets and

* The calendar had previously been based on the periods of the moon. However, since the lunar month is incommensurable with the year, this proved awkward.

of a large number of stars. This information was used to forecast all manner of events affecting the life of man. An inscription dating from the twentieth century B.C. reads:

> If on the 21st of Ab, Venus disappeared in the east, remaining absent in the sky for two months and 11 days, and in the month Arkhsamna on the 2nd day Venus was seen in the west; there will be rains in the land; desolation will be wrought.[1]

The astronomy of the eastern Mediterranean civilizations and of the Stonehenge people shared one feature of special importance to our study. The emphasis was entirely on empirical observation, with no apparent attempts at explanations other than those of a mythological character. By observing eclipses over long periods, the Stonehenge people could discover regularities in their occurrence and thus accurately predict future ones. But they could not be said to have acquired any understanding of why eclipses occur when they do other than to say, "That's what happened the last time." The Babylonians made extensive tables of conjunctions of planets with various constellations but seemed unconcerned with what went on between these conjunctions. We have no evidence that these people tried to invent models of the heavens or to picture what is actually going on between observations. One could predict the eclipse and announce the planting time; that was sufficient. The emphasis was on technology rather than science. The practical outweighed the theoretical.

Early Greece

We now shift our attention forward in time and westward in place to the scattered city-states of Greece in the seventh century B.C. Intellectually, this world was quite different from that of the Egyptians and Babylonians, and for the first time we can associate ideas with individual men. We find that these men are asking new kinds of questions and giving new kinds of answers. The astronomer-priests and technical astronomy have vanished, and in their place we find philosophers. Man has changed from a technician to a speculative intellectual.

The earliest of these men—in fact, probably the first Greek philosopher whose name we know—was Thales of Miletus, a city of Ionia on the west coast of what is now Turkey. Thales lived during the late seventh and early sixth centuries B.C. He seems to have been a practical man, working in the olive oil business and dabbling in politics. In no sense of the word was he an astronomer; like Greeks before his time, he apparently made no detailed observations or studies of the skies, though a trip to Egypt during his youth had made him aware of the extensive Egyptian data. At some point in his life, however, he turned to philosophical matters, and his questions are indicative of the new directions astronomy was soon to take.

Thales must have been impressed with the Egyptians' discoveries of regularities in the heavens, of a hidden order behind even such apparently random occurrences as eclipses. But while the Egyptians simply accepted this order at face value, Thales reacted to it in what was to be a typically Greek way: he questioned its meaning. Might not the hidden order of the eclipses point to an order throughout all of nature? Perhaps the apparent complexity of the world is only an illusion. How can one look at the world to see this order? Perhaps there is some organizing principle in light of which the universe can be seen not as chaos, but rather as a simple unity.

The old mythological pictures of the world, dependent as they were on capricious and arbitrary gods, could not readily supply such an organizing principle. Thales' own approach was to try to understand the universe in terms of some basic substance or matter that underlies it and gives it form. That is, his organizing principle was: to find a unity in the world, view the world as a manifestation of the properties and character of this basic element.

Thales' attempt to carry out such a program seems almost anticlimactic in its naïveté. He postulated Water as the basic substance of the world. It appears all around us as liquid, mist, ice, and rain, and is the substance necessary for all life. The earth, he said, must have been condensed out of water, and it floated on an infinite sea. The heavens were water; the sun and moon floated up overhead from east to west through this water, and returned again to the east through the waters around the earth. Taken as an attempt to describe the physical world realistically, this view now seems crude at best. Perhaps it was meant only as a series of poetic images. The important thing is that the attempt was made, that the need was felt to see the universe as an ordered unity, expressed in natural as opposed to supernatural terms.

Other early Greek philosophers were also concerned with a search for the same sort of unity, though the terms in which they expressed it varied. The details of their teachings do not matter; the approach, and particularly the contrast with that of the Egyptians and Babylonians, is what interests us. To the Egyptians and Babylonians, an empirical description was the ultimate goal. There is no record of their having asked anything more. The early Greeks, on the other hand, began a search that extended intermittently through centuries of Western thought—a search for unifying principles to explain the diverse phenomena of nature.

Pythagoras

Another approach to celestial motion was initiated by the Greek philosopher and mathematician Pythagoras, who was born about 590 B.C. As with other early Greek philosophers, nothing written by Pythagoras himself has survived. We can only attempt to piece together his ideas from fragments written by his followers and from the commentaries of later Greeks, notably Aristotle.

Pythagoras had an almost mystical view of the relation between numbers and natural phenomena. Apparently his knowledge of the properties of musical instruments was an important influence in the development of this point of view. For example, he is supposed to have made the following observation: Suppose we stretch two strings so that they sound the same note when plucked. If we then press one of the strings at its center, as indicated in Figure 2–5(a), and pluck both strings together, the sound is harmonious—an octave. The ratio of lengths is 2:1. If we press the same string a third of the way

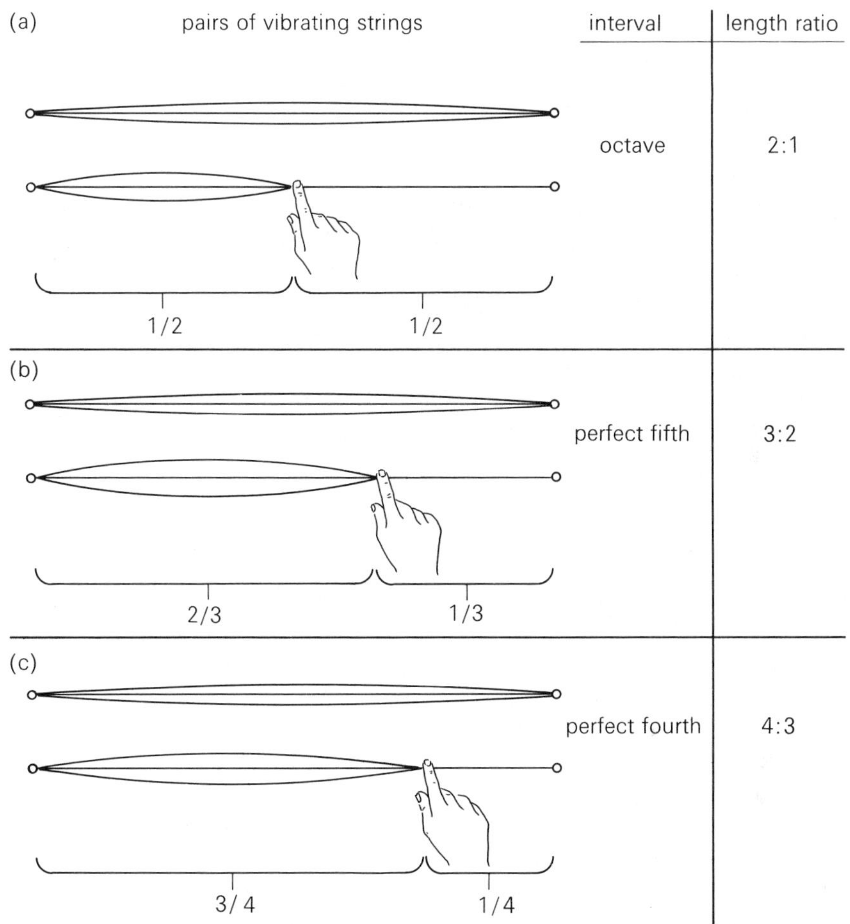

Figure 2–5. The basic musical intervals and their relation to the integers.

from the end [Figure 2–5(b)] and pluck both together, again the sound is harmonious—what musicians call a perfect fifth. The ratio of lengths is 3:2. And if we press a quarter of the way from the end [Figure 2–5(c)], we have a ratio of 4:3, and a pleasing perfect fourth. These three basic intervals, common to both Greek and present-day music, involve the simple ratios of 4:3, 3:2, and 2:1.

Properties of Numbers

To Pythagoras and his followers, this result represented a fundamental truth about the physical world—that there is a deep inherent relation between the simple abstract numbers and the universe. To understand the world, one had first to understand numbers. It was to this task that the Pythagoreans turned. Numerical notation as we know it had not been developed in early Greece. Instead, numbers were represented by geometric arrangements of pebbles or marks on the ground. Of particular interest were the so-called triangular and square numbers, some of whose properties are indicated in Figures 2–6 and 2–7. The regularities made apparent by such arrangements were intriguing then and remain so today. Men have often derived a certain pleasure from contemplating the way the simple integers—odds, evens, squares—fit together.

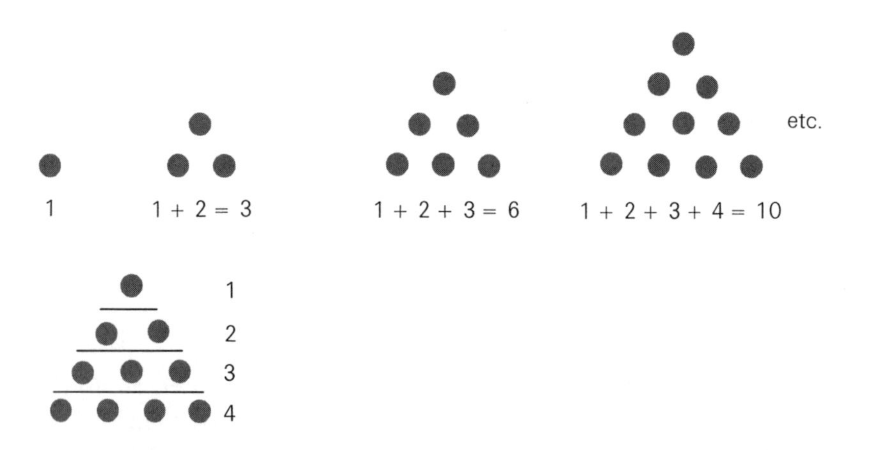

Figure 2–6. The triangular numbers. Each triangular number represents a sum of successive integers. Each can be derived from the previous one by adding the next integer.

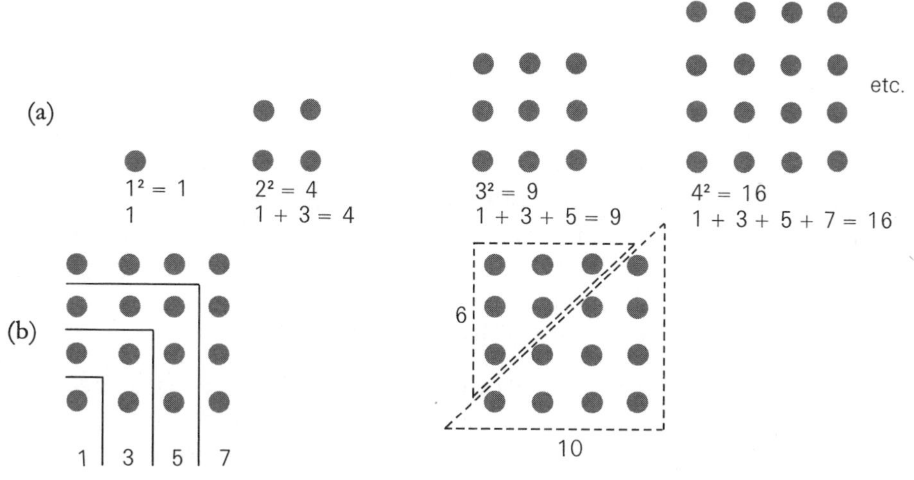

Figure 2–7. The square numbers. (a) shows that each square number represents a sum of successive odd integers; (b) shows that each is also the sum of two triangular numbers.

Similar geometrical constructions may have led Pythagoras to the discovery of the 3-4-5 right triangle (Figure 2–8), though there is doubt as to whether either he or his followers actually discovered the familiar "Pythagorean theorem" traditionally attributed to them.

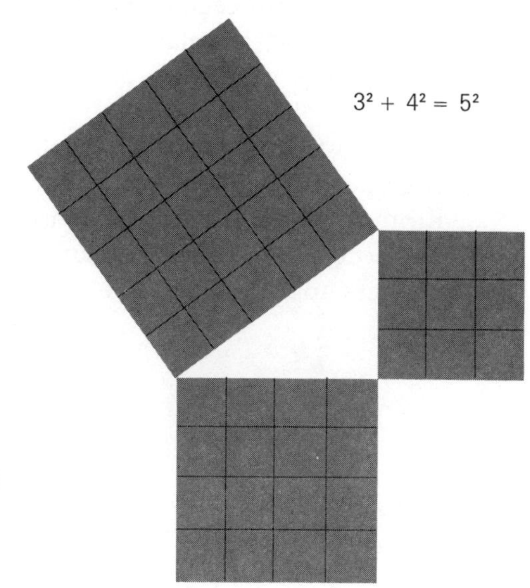

Figure 2–8. The 3-4-5 right triangle.

The Nature of Things Is Number

The integers that seemed to have the closest connection with the world were the first four, both directly (as 1, 2, 3, and 4) and in their combination as the fourth triangular number, 10 (see Figure 2–6). The number 10 was so ubiquitous that it assumed a special significance as a sacred number, the Decad. We note several examples of its importance:

1. The elements of geometry. The point, the basic geometric element, can be associated with the number 1. A line, generated by moving a point, corresponds to the number 2. A surface, generated by moving a line, is 3. Finally, a volume, obtained by moving a surface, is 4. Thus the four Decad integers are related to the four elements of geometry.
2. The basic musical harmonies. We have already seen how these correspond to the ratios 4:3, 3:2, and 2:1.
3. The constituents of matter. A belief prevalent in early Greece was that all matter is composed of four basic elements: Earth, Water, Air, and Fire. One could be associated with each of the Decad integers.
4. The simplest regular solid, the tetrahedron [Figure 2–9(a)], has four faces and six edges; $4 + 6 = 10$, the Decad.
5. The most familiar regular solid, the cube, has twelve edges, eight corners, and six faces [Figure 2–9(b)]. From these we can form the ratios 8:6, 12:8, and 12:6—or, more simply, 4:3, 3:2, and 2:1. Again, the first four integers appear.

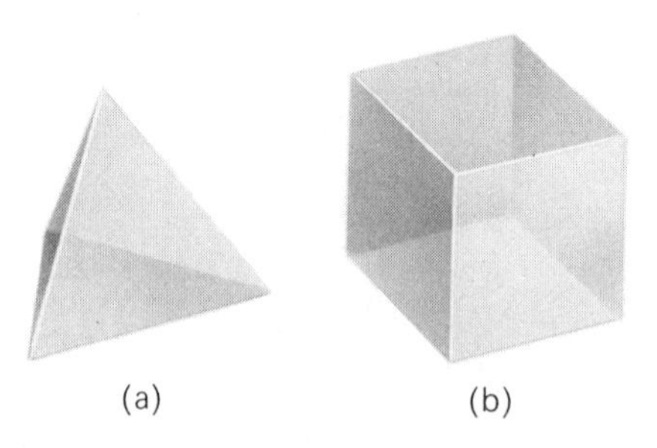

Figure 2–9. The simplest regular solids: (a) the tetrahedron; (b) the cube.

Connections like these between numbers and the world—some of them seeming rather farfetched to us—were seen by Pythagoras and his followers as evidence for their belief that numbers provide not simply a description of the world but, in addition, a reason for its being the way it is. The nature of things is number, they said, and they saw, for example, the combination of the first four integers in the Decad as an explanation for the existence of four and only four elements. The interpretation of the world in terms of number was their organizing principle.

Philolaus, a follower of Pythagoras, wrote:

> Consider the effects and the nature of number according to the power that resides in the Decad. It is great, all-powerful, all-sufficing, the first principle and the guide in life of Gods, of Heaven, of Men. Without it all is without limit, obscure, indiscernible. The nature of number is to be a standard of reference, of guidance, and of instruction in every doubt and difficulty. Were it not for number and its nature, nothing that exists would be clear to anybody either in itself or in its relation to other things . . . One can observe the power of number exercising itself not only in the affairs of demons and of gods, but in all the acts and the thoughts of men, in all handicrafts and in music.[2]

There is a great deal of mysticism in this. In fact, Pythagoras' observations became the foundation for a religious cult that flourished for two centuries after his death. Number and the Decad were the regulators of the universe and the relations between the integers we have just examined were among the secrets of the initiates. Thus the properties of numbers were studied by the Pythagoreans in much the same spirit in which Christian monks later studied the Scriptures.*

The notion that numbers control nature, that their properties provide an explanation of natural phenomena, might seem so primitive that it is of purely historical interest. One is tempted to ask what possible relevance such concepts could have to the

* The "irrational" numbers like $\sqrt{2}$, which cannot be expressed exactly by integers or ratios of integers, were discovered by the Pythagoreans and kept as one of their most closely guarded secrets. The existence of such numbers threatened their whole system.

sophisticated explanations given by the "exact science" of modern times. And yet, as we follow man's attempts to understand the heavens, it will be worth bearing in mind the Pythagorean viewpoint. Perhaps it will turn out not to be so primitive after all.

The Pythagorean Universe

The outlook of the Pythagoreans is exemplified by their ideas of the earth and heavens. Reports by travelers of differences in the skies as they journeyed north and south indicated that the earth could not be flat. (Figure 2–10.) If it is curved, the Pythagoreans argued, then it must be a sphere, for the sphere is the most perfect curved solid. By the same argument, the surface on which the stars are located must also be spherical. Furthermore, the simplest and most symmetrical arrangement of two spheres is concentric. Thus, the Pythagoreans concluded, the center of the earth must coincide with the center of the sphere of the stars.

In like manner they argued that the objects in the sky that move separately from the stars—the sun, the moon, and the planets—must move in circles about the center of the earth, each at a uniform speed. Not only were the shapes and motions of the orbits governed by geometrical simplicity, but the very sizes of these orbits were seen to result from the properties of the integers—or the musical harmonies, which were the same thing. The Pythagoreans reasoned that (as Aristotle later reported):

> bodies so great must inevitably produce a sound by their movements . . . Taking this as their hypothesis, and also that the speeds of the planets, judged by their distances, are in the ratios of the musical consonances, they affirm that the sound of the planets as they revolve is concordant. To meet the difficulty that none of us is aware of this sound, they account for it by saying that the sound is with us right from birth and has thus no contrasting silence to show it up; for voice and silence are perceived by contrast with each other.[3]

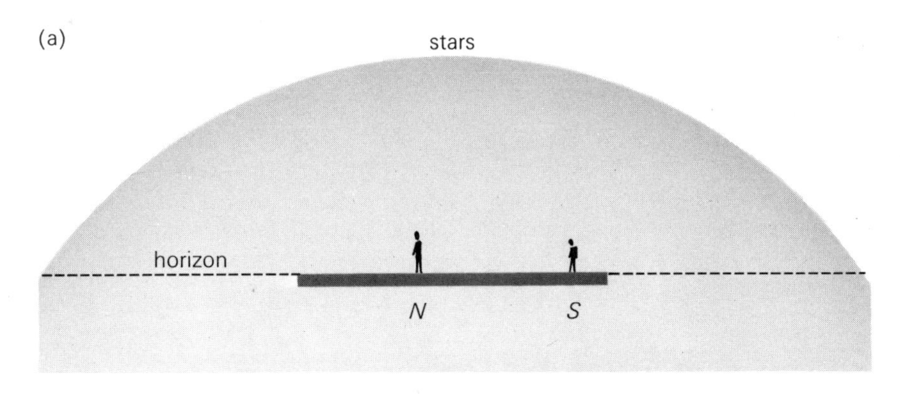

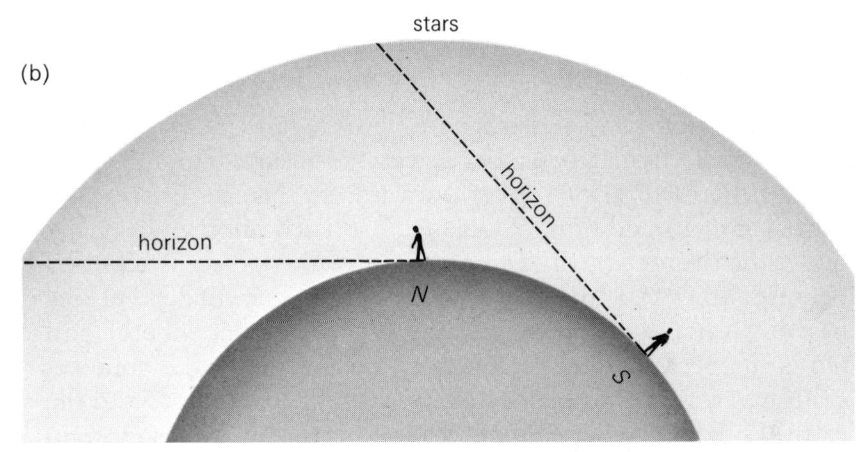

Figure 2–10. Changes in the night sky as one travels north and south imply that the earth is not flat. (a) Flat earth. The same portion of the sky is visible above the horizon to observers *N* and *S*. (*N* is taken to be farther north than *S*.) (b) Curved earth. Different parts of the sky are visible to observers *N* and *S*.

One can thus view the integer ratios (musical consonances) as an explanation of why the planets move as they do.

Though the geocentric (earth-centered) picture of the universe outlined above was the one generally advanced by the Pythagoreans, the somewhat different view of Philolaus, a Pythagorean of the early fifth century, is worth noting. His model is shown in Figure 2–11. He imagined what he called a central fire at the

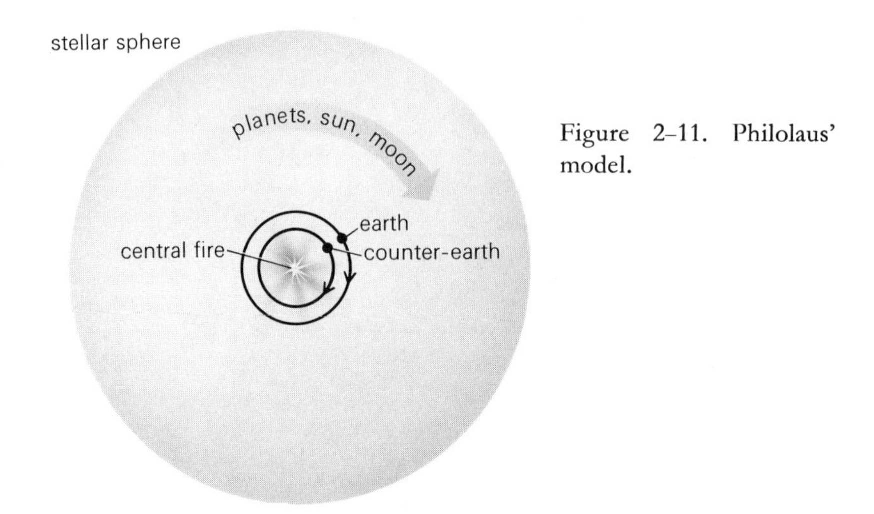

Figure 2–11. Philolaus' model.

center of a motionless stellar sphere, with the earth moving around it in a small circular orbit once every 24 hours. The inhabited part of the earth always faces outward, so men do not see the central fire. In addition, Philolaus postulated another heavenly body, the counter-earth. The existence of this object is required because otherwise there would be only nine bodies in the heavens (earth, sun, moon, five planets, and stellar sphere), and ten is the perfect number. Since we do not see it, the counter-earth was assumed to be located between the central fire and the earth. Philolaus' scheme is interesting because it comes close to a modern idea that the diurnal motion of the stars is only apparent, resulting from the actual motion of the earth. His scheme, however, was not widely accepted. The more conventional view of a stationary earth seemed not only simpler but more rational. After all, doesn't the assumption of a moving earth violate all common sense?

The Basic Geocentric Model

In the preceding discussion we met what we shall call the *basic geocentric model*, the model of the universe put forth by the Pythagorean cult: a spherical earth at the center surrounded by a rotating spherical shell in which the stars are imbedded. In addition, the sun, moon, and planets move in circular paths about the earth. This model is so important to our story that we shall examine it and its consequences in some detail.

Let us first consider the earth, the sun, and the stars. The basic observations about the motion of the sun and stars as seen from the earth were discussed earlier in this chapter: The stars travel on circular paths across the sky each night, as if attached to a sphere (we called it a dome before). This sphere rotates about an axis passing through the earth's center and the North Star, a complete rotation taking 23 hours and 56 minutes. The sun travels on a circular arc across the sky each day, as if rotating about the same axis as the stars; only it takes 24 hours to make a full rotation. (This should not be surprising—the day, and thus the hour, have been defined according to the sun's motion.) In addition, the position of the sun's diurnal path changes from northerly in the summer to southerly in the winter.

The basic geocentric model encompasses these observations as well as a number of others not yet discussed. Consider Figure 2–12, which is virtually the same as Figure 2–2. We shall give it a slight reinterpretation to allow it to represent our model accurately. Imagine the (rather oversized) man at the center to be standing on a spherical earth whose diameter is very small

Figure 2–12. In the basic geocentric model, the only stars that can be seen from the position on the earth where the man is standing are those on the hemisphere above the horizon plane.

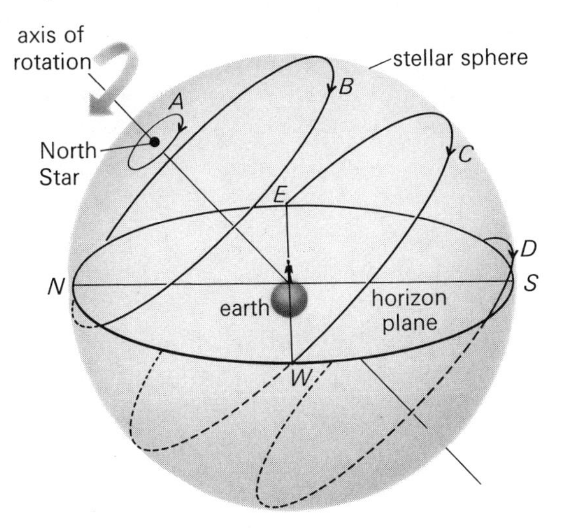

compared to that of the stellar sphere, and yet large with respect to the man. The stellar sphere rotates about the axis once every 23 hours and 56 minutes. Since the stars are rigidly attached to the stellar sphere, they always appear in the same relative positions, thus maintaining the patterns we have come to know as constellations.

The flat plane on which the man appears to be standing is the *horizon plane*—an imaginary plane tangent to the earth's surface at the man's feet. The horizon plane indicates the man's field of view. That is, all stars on the hemisphere above the horizon plane are potentially visible from the vantage point of the man, though of course if it is daytime they will not actually be seen. This simple combination of a rotating stellar sphere and a fixed horizon plane can tell us anything we want to know about the motions of any of the stars in the sky. For example, a star on circle *A* is *circumpolar*—it is always above the horizon. On the other hand, a star on circle *B*, *C*, or *D* rises and sets. (Of course the risings or settings may not be visible if they occur during the daytime.)

And now notice what else Figure 2–12 can predict. We can see how the night sky changes as an observer moves north or south. Let our man in the figure walk north—that is, along the earth's surface in the direction indicated by the North Star. His view of the heavens after a certain journey is indicated in Figure 2–13(a). Because of the curvature of the earth, the North Star appears higher overhead. At the same time, the horizon plane has become more nearly perpendicular to the rotation axis. His view of the stars has changed: the stars on circle *B* no longer rise and set, but become circumpolar stars like those on circle *A*; stars on *C* still rise and set due east and west but never rise as high in the sky; and stars on *D* are never seen at all.

When the man reaches the North Pole, his orientation is shown in Figure 2–13(b). The North Star is directly overhead. The only stars visible are those in the half of the stellar sphere above circle *C* (which for obvious reasons is called the *celestial equator*), and all of them are circumpolar. At the North Pole, all of the stars travel about the sky in circular paths parallel to the horizon. On the other hand, if the man travels south to the earth's equator [Figure 2–13(c)], his horizon plane there will

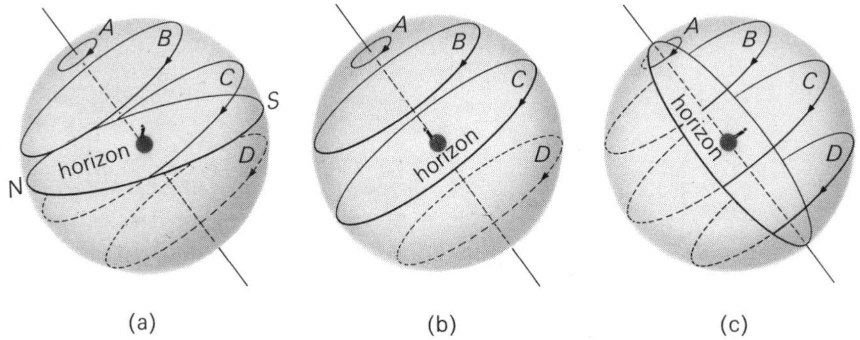

Figure 2–13. Changes in the sky at different latitudes. The visible portion of the stellar sphere is indicated for an observer located (a) in the Northern Hemisphere near the Arctic Circle, (b) at the North Pole, and (c) at the equator.

contain the rotation axis. All stars will rise and set; stars on the celestial equator (circle *C*) will pass directly overhead; and many new stars will be visible to the south.

These changes in the skies on traveling north and south were alluded to before as evidence against the earth's being flat. In fact, detailed observations of the stars show that this simple model of the earth and stars accounts exactly for what is seen from any vantage point on the earth. For example, compare these predictions with Figure 2–14, which shows time exposure photographs of the night sky taken at the North Pole and at the equator.

Figure 2–14. Time exposure photographs of the night sky (a) at the North Pole and (b) at the equator. (Yerkes Observatory photographs, University of Chicago.)

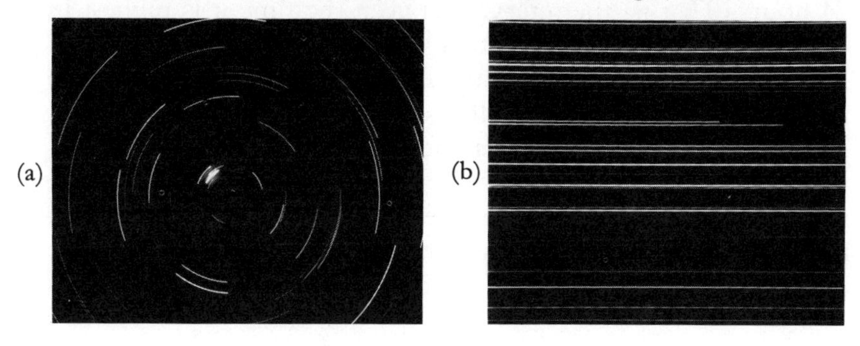

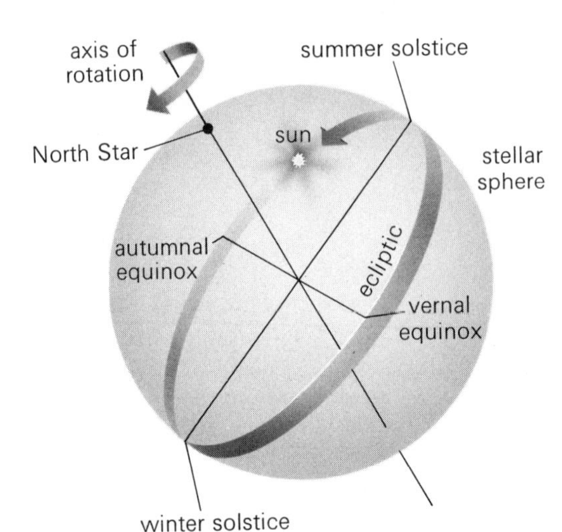

Figure 2–15. The sun in the basic geocentric model.

So much for the stars. How is the sun to be included in the basic geocentric model? As we have remarked, the sun is observed to move relative to the stars along a path known as the *ecliptic*. It travels very slowly along the ecliptic from west to east, returning to its starting point after exactly a year and then beginning the journey again. Suppose we draw this path of the sun through the stars on the stellar sphere. As indicated in Figure 2–15, the ecliptic is a very simple curve, a great circle of the sphere.

It should now be easy to guess how the sun can be incorporated in the basic geocentric model. It is located on the stellar sphere. It moves from west to east along the ecliptic, completing a full circuit in one year. This simple addition to the model accounts for all the observed features of the sun's motion.

For example, the sun has a diurnal motion whose period is 24 hours. That is, the interval between the time it is highest in the sky on one day and the time it is highest in the sky on the next is 24 hours. This is exactly what the model predicts. In the model, the sun is on the stellar sphere, which completes a full revolution in 23 hours 56 minutes. But as the stellar sphere is revolving, carrying the stars and the sun from east to west, the sun moves slowly in the opposite direction along the ecliptic—from west to east. As a result of this additional motion, the time between two successive appearances of the sun at its

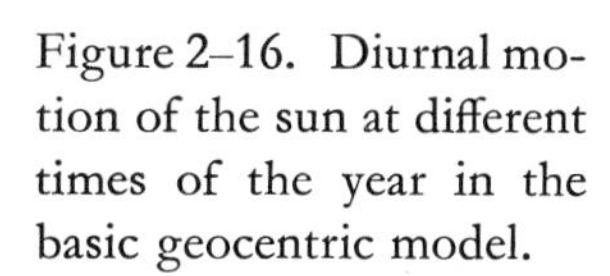

Figure 2–16. Diurnal motion of the sun at different times of the year in the basic geocentric model.

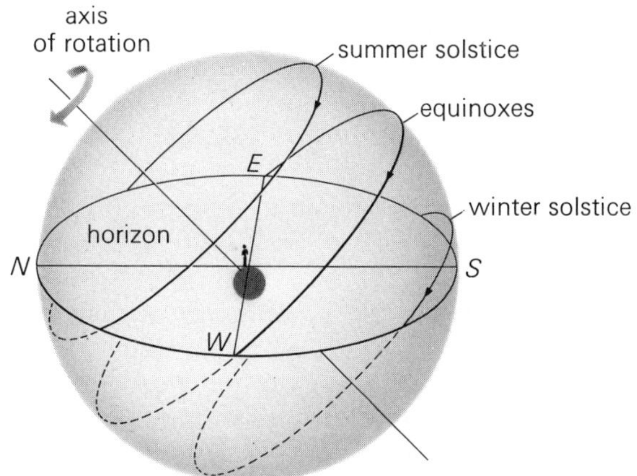

highest point in the sky is four minutes longer than the 23 hours 56 minutes it takes the stellar sphere to revolve. Thus the model predicts a diurnal period of 24 hours for the sun and 23 hours 56 minutes for the stars.

Another feature of the sun's motion explained by the basic geocentric model is the seasonal variation of its path. At the winter solstice, the sun appears in the southern part of the sky. The path gradually moves northward until the summer solstice, when it is most directly overhead. It then reverses itself and begins to move toward the south (Figure 2–16).

Figure 2–17 shows how this behavior is built into the basic geocentric model. At the winter solstice the sun is at the position on the ecliptic indicated in Figure 2–17(a). The rotation of the stellar sphere then leads to the diurnal motion shown in Figure 2–17(b). Similarly, at the summer solstice the sun has moved along the ecliptic to the position shown in Figure 2–17(c), leading to the diurnal motion of Figure 2–17(d).

> Show how the basic geocentric model explains the phenomenon of six months of continuous daylight and then six months of continuous darkness at the North Pole.

Note the essential simplicity of this model. It consists of a spherical earth surrounded by a rotating sphere of stars, and a

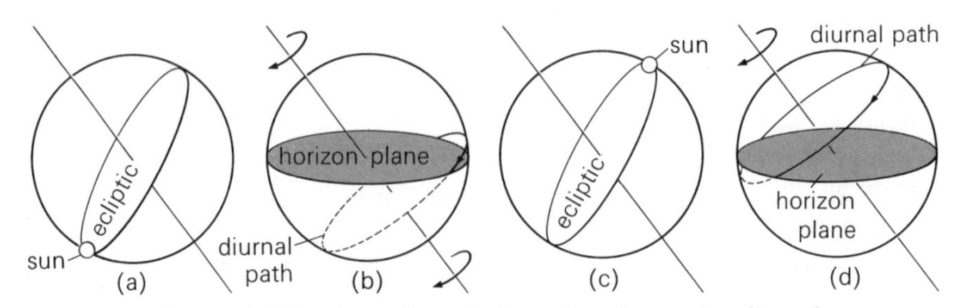

Figure 2–17. Annual variations in the sun's diurnal motion according to the basic geocentric model. (a) Position of the sun on the ecliptic at the winter solstice. (b) Diurnal motion of the sun at the winter solstice. (c) Position of the sun on the ecliptic at the summer solstice. (d) Diurnal motion of the sun at the summer solstice.

sun, on the sphere, which rotates diurnally with the stars and at the same time slowly proceeds along the ecliptic. Although simple, the model predicts successfully all of man's primitive observations concerning the motion of the stars and the sun, including

1. the positions of the stars and sun as viewed at all points on earth at all times; in particular
2. the times and horizon positions of rising and setting of the stars and the sun at all times of the year and from all points on earth; and
3. the times and locations of any other special positions of the sun and stars (e.g., a *culmination*—the highest point in the sky of a star's path).

This basic geocentric model is so accurate that it is still used today for navigation, and yet we do not believe in it—we say it is not the way the universe really is. Why do we not accept the model? To answer this question, we must turn to the small group of ostensibly insignificant objects that do not fit so readily into this simple scheme—the wandering planets.

Suggested Reading

Kuhn, T. *The Copernican Revolution.* New York: Random House, 1959. This and the following reference are excellent, more detailed treatments of the problem of celestial motion.

Toulmin, S., and Goodfield, J. *The Fabric of the Heavens.* New York: Harper & Row, 1961.

de Santillana, G. *The Origins of Scientific Thought.* Chicago: University of Chicago Press, 1961. A discussion of early Greek philosophy and its bearing on scientific ideas.

Sarton, G. *A History of Science: Ancient Science Through the Golden Age of Science.* Cambridge: Harvard University Press, 1952. A good reference for the astronomy of Egypt and Mesopotamia, as well as for Greek science and philosophy in general.

Hawkins, G. S. *Stonehenge Decoded.* Garden City, New York: Doubleday, 1965. An extremely readable account of Stonehenge and its function as an eclipse predictor.

Questions

1. Astrology attempts to use observations of the heavenly bodies to predict the course of human affairs. Can astrology legitimately be called a science?
2. The Pythagoreans also studied rectangular numbers, the first two of which can be written as $\circ\ \circ$ and $\begin{smallmatrix}\circ&\circ&\circ\\\circ&\circ&\circ\end{smallmatrix}$. Notice how the second is related to the first: $\begin{smallmatrix}\circ&\circ&\circ\\\circ&\circ&\circ\end{smallmatrix}$. Construct the third and fourth rectangular numbers, and show that, in general, each rectangular number represents a sum of successive even integers. Can you find a simple numerical relationship between the rectangular and triangular numbers?
3. "If the earth is curved, then it must be in the shape of a sphere, for the sphere is the most perfect curved solid." This might be termed an aesthetic argument. Does such an argument have any place in science?
4. Which of these observations is more significant? Why?
   (a) Seven stars seem to cluster in the shape of a water dipper.

(b) The object Mercury, which otherwise appears to be a star, moves relative to the other stars in an erratic fashion.

5. The basic geocentric model can be made exceedingly accurate and yet we do not believe in it. Is the model wrong? Can a model be wrong if accurate predictions can be made from it?

6. Use the basic geocentric model to answer the following questions about the sun:
   (a) At the equator, how does the length of the day vary with the seasons?
   (b) In what region of the earth does the sun appear directly overhead at least once each year?
   (c) In what regions of the earth is there at least one day each year when the sun never appears at all?
   (d) Use the result of part (b) or (c) along with a globe of the world to determine the angle between the plane of the ecliptic and the axis of rotation of the stellar sphere in the basic geocentric model. (See Figure 2–15.)

7. At the latitude represented in Figure 2–16, Sirius rises just at dawn on a particular day in the spring. Indicate on that figure where Sirius and the sun might be located on that day.

8. How might the basic geocentric model explain seasonal climatic variations?

9. Show that in the basic geocentric model the time by which the sun lags the stars each day is four minutes. (Recall that in one full year, $365\frac{1}{4}$ days, the sun completes its journey around the ecliptic. Thus, in this time, the stars complete exactly one more revolution than the sun.)

10. If the moon is observed from night to night, it is seen to move relative to the stars. It moves along a path close to the ecliptic, completing a circuit each month. Can you suggest how the moon might be included in the basic geocentric model?

11. Another kind of evidence for the curvature of the earth comes from differences in the lengths of shadows cast by

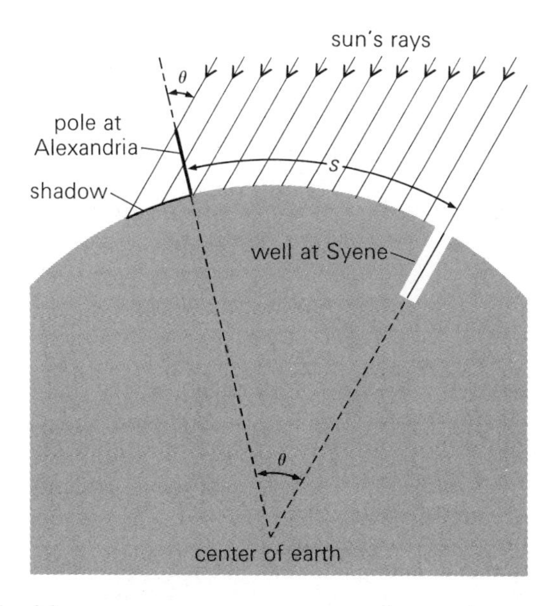

vertical objects at noon as one travels north or south. An early observation of such a difference was made by Eratosthenes in 230 B.C., and used to deduce the radius of the earth. At noon on the summer solstice, the sun at Syene could be seen reflected at the bottom of a deep well, implying that the sun's rays were coming down vertically, as shown in the sketch. (A vertical pole in Syene would cast no shadow.) In Alexandria, 490 miles due north, a pole did cast a shadow; by measuring the length of the pole and its shadow, Eratosthenes found the angle θ to be 7.2°. As shown in the sketch, this is the same as the angle formed at the center of the earth by radii drawn to Syene and Alexandria. Show by simple proportions that

$$\frac{s}{2\pi R} = \frac{\theta}{360^\circ}$$

and thus find Eratosthenes' estimate of the radius R of the earth. Compare with the modern value (3960 miles).

CHAPTER THREE

Stars that Wander

Later Greek civilization readily accepted the basic geocentric model, not only as a succinct way of summarizing observations of the sun and stars, but even more as a view of reality. By imagining the universe to be constructed in this simple fashion, one could *understand* what he saw in the heavens.

Yet there were certain objects in the sky whose motion did not seem consistent with the basic geocentric model. These were the planets. In contrast to the simple circular motions of the sun and stars, planetary motions appear far more complex.

For example, Figure 3–1 shows a typical motion of Mars relative to the stars. Its path, like those of the other planets, lies close to the ecliptic. The general motion of each planet is from west to east, like that of the sun, but by no stretch of the imagination is it circular. Reversals of direction (termed *retrogressions*), like that shown in Figure 3–1, occur at apparently irregular intervals. In addition, Mercury and Venus are never seen far from the sun (they are always "morning stars" or "evening

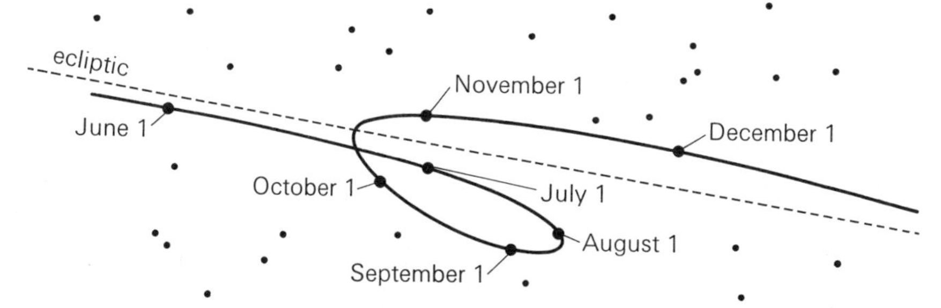

Figure 3–1. A typical motion of Mars relative to the stars, as seen from the earth.

stars"), while the positions of Mars, Jupiter, and Saturn bear no apparent relation to that of the sun.

How were the motions of these objects to be explained? Did their existence require discarding the basic geocentric model, or could the model be modified to accommodate them? It is only natural that the latter possibility was pursued first. For one thing, a man-centered (and hence earth-centered) universe was an underlying assumption of Greek thought. For another, the basic geocentric model provided an appealingly simple explanation for the multitude of heavenly bodies other than the planets.

To appreciate the Greek reaction to this planetary problem, let us briefly examine the intellectual and philosophical climate of the day. Plato (427–347 B.C.) was the most influential philosopher of his time and his views clearly influenced later Greek inquiries into the nature of planetary motion.

Plato—A Digression

Plato's philosophical approach is illustrated in his famous allegory of the cave. Our view of reality, he says, is like that of prisoners who pass their entire lives chained at the entrance to a cave in such a way that they must always face the interior wall, unable to look back at the outside world. A large fire burns outside the entrance so that people and objects in the area between the fire and the cave cast shadows on the rear wall.

The shadows are illusions giving only distorted hints of the reality outside.

So it is with our existence, Plato claims. The physical world around us and all the activity within it are merely illusions, like the shadows in the cave. Truth lies beyond our immediate view. It is the philosopher's task to find truth, but he cannot do this by extrapolating directly from everyday observations, just as the prisoners in the cave cannot arrive directly at the idea of colors from the black shadows they have always seen. The method must be deduction from abstract principles, not induction from experience; one must go from the general to the specific, not the other way around.

According to Plato, the philosopher must begin with abstractions. These Forms and Ideas, as he calls them, are eternal, but their manifestations in our world are merely ephemeral shadows. For example, the Idea of man is timeless, while any individual man is soon departed. These eternal abstractions constitute the Good, which exists over and above the transitory world.

One might rightly guess from this that Plato had no particular interest in what we call science, for he believed that attempts to deal with a physical world that is merely illusion were a waste of effort at best. Mathematics, on the other hand, was an important discipline for Plato and his followers, and a moment's reflection on the difference between mathematics and science will make clear the reason for this. Mathematics is really nothing more than a system of codified logic; it needs no objects as a base. $2 + 2 = 4$ is true whether it refers to apples or pears or philosophers, or even if it refers to nothing in particular. Numbers are abstractions of the purest sort, and it is not surprising that Plato found them compatible with his outlook. The same is true for other branches of mathematics. "All points on a circle are equally distant from its center" is true without reference to *what* circle; it is simply a matter of definition. Applied mathematics—mathematics in the service of science—is another matter. "The moon is always a fixed distance from the earth" may or may not be true, depending upon whether the orbit of the moon is in fact a circle. This is a matter of observation, not logic. But pure mathematics deals only in abstractions, and

Plato thought it to be of the utmost importance in the training of a philosopher.

Astronomy occupied a somewhat ambivalent position in Plato's views. While it was a physical science, he still found reasons for taking it rather more seriously than other sciences. He accepted the Pythagorean view that the universe had two different domains: that of the earth, where imperfection, chaos, and decay were everywhere evident; and that of the heavens, where all was order and perfection. Perhaps the heavens were less an illusion than the earth. Perhaps they might permit a glimpse of the Good, the ideal abstraction.

In fact, there was much to be learned along these lines from the Pythagoreans. Their view, which posited a spherical earth at the center of the universe with the heavenly bodies moving uniformly in circles about it, derived from prior ideas of the symmetry of the universe and was thus compatible with Plato's philosophy. As we have noted, the basic Platonic approach was deductive. The ideas of symmetry and uniform circular motion seemed so simple and natural that Plato accepted them without question as the basis for his deductions about the heavenly domain. To put it another way, he believed that uniform circular motion was so basic to the existence of the heavens that it needed no explanation. That heavenly bodies should move in this way was the unwritten assumption and starting point not only for Plato but for everyone of his time who tried to understand the physical world. Today we may find it a strange and arbitrary assumption, but perhaps our present ideas about the physical world are based on similar assumptions about what is natural, assumptions with no better *a priori* justification.

Eudoxus and the Homocentric Spheres

The first model of planetary motion is generally attributed to Eudoxus of Cnidus (ca. 408–355 B.C.), a student of Plato and one of the foremost Greek mathematicians. Eudoxus, like any man, was significantly constrained by the philosophical climate of his time. It was therefore natural that his model should place the earth at rest at the center of the universe, since this is a necessary condition for man to be at the center of the universe. In the basic geocentric model, the motions of the stars and the sun had been understood in terms of the ideal of uniform circular motion. Plato's influence extended this requirement to the explanation of the planets. For Eudoxus, the problem of the planets could be stated: "In what way can the observed motion of the planets be understood in terms of uniform circular motion?"

Within this framework, Eudoxus developed the model of planetary motion that has come to be called the *homocentric sphere model*. We can understand it most easily by examining first its application to the sun. Recall that in Eudoxus' time the sun's motion was viewed as the combination of two effects: a diurnal motion, circular and uniform, and a slow annual motion along the ecliptic, also circular and uniform. In Eudoxus' model, each of these motions is produced by the uniform rotation of one of two homocentric (i.e., concentric) spheres whose common center is the center of the earth. (See Figure 3–2.) The sun itself is imagined to be attached to a point on the equator of the inner sphere, whose axis is tilted so that this equator coincides with the ecliptic. This sphere rotates once each year, carrying the sun around the ecliptic. The ends of this sphere's axis are embedded in the second, outer sphere. This outer sphere rotates about an axis through the North Star once every 23 hours, 56 minutes, carrying the slowly rotating inner sphere around with it, and thus producing the diurnal part of the sun's motion.

The moon's motion can be fitted in a similar way, except that a third sphere is necessary. The outer sphere accounts for the diurnal part of the motion; a second sphere inside the first accounts for the moon's monthly motion around the ecliptic; and a third sphere, the innermost, accounts for a further feature of the moon's motion that differs markedly from that of the sun. If one observes the moon's path through the stars, he finds that it does not exactly follow the ecliptic. In any given month, it is parallel to the ecliptic and close to it but usually slightly to one side or the other. Over the years, however, this path appears to drift across the ecliptic and back again. An entire cycle takes 18.6 years. In order to account for this motion the third homo-

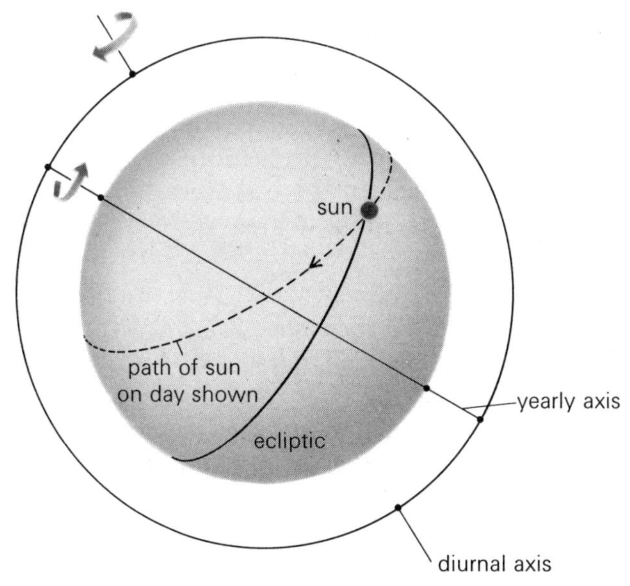

Figure 3–2. Homocentric spheres for the sun. The inner sphere with sun attached rotates once each year about the yearly axis. The outer sphere rotates once each day about the diurnal axis, carrying the inner sphere and sun along with it.

centric sphere is very slightly tilted with respect to the second, and rotates once every 18.6 years.

The more complex motions of the planets—in particular their retrogressions—require a more complex arrangement of spheres, in which the skill of Eudoxus is most clearly seen. For a given planet, he began with two spheres like those mentioned for the sun—one for the diurnal motion, a second for the motion along the ecliptic. To reproduce the retrograde motions, he used a third and a fourth sphere, the planet being on the equator of the fourth, innermost one. The net effect of the uniform rotation of the third and fourth spheres was to give the planet a figure-eight movement. When this effect was combined with the motion of the second (ecliptic) sphere, the net motion was predominantly forward movement plus occasional retrogressions.

With four spheres for each of five planets, plus three spheres for the moon, two for the sun, and one more for the stars,

Eudoxus needed a total of twenty-six homocentric spheres to account for the observed motions of the heavenly bodies. All things considered, it was a remarkably simple system, at least in conception. Of course, unraveling the details was no small task. Eudoxus had to decide on the orientation and rotation speed of each sphere for each planet to give the best fit to the available data, which required a high degree of geometric skill and calculational ability.

It is not clear how much reality was attributed to the spheres. Eudoxus may have viewed them simply as mathematical devices, a way of looking at the planets' motions that allows one to make sense of them. However, later investigators, including the philosopher Aristotle, most clearly did interpret the spheres quite literally as part of an almost mechanical system.

Refinements of the Homocentric Sphere Model

We have examined Eudoxus' model and remarked on its qualitative success. Quantitatively, it was less satisfactory. Perhaps we should be lenient with it, considering that it was the first attempt by anyone in Western civilization to invent a model of planetary motion specific enough to allow numerical comparison with the physical world. In addition, we are not sure how much data was available to Eudoxus when he worked out the details of his model. At any rate, some years later his pupil, Calippus, found it necessary to add a fifth sphere for Mercury, Venus, and Mars, and two more each for the sun and the moon.

Calippus' alterations were the last mathematical improvements on the homocentric sphere model, but the philosopher Aristotle (384–322 B.C.) was responsible for a final conceptual change. He viewed the universe as a connected, interrelated system, and thus felt that a purely mathematical conception of the spheres was not enough. The spheres had to be made physically more concrete—that is, if they existed, one ought to be able to understand their mechanical interconnection. And so Aristotle devised a way of linking together all the spheres of Eudoxus and Calippus with additional spheres and made of the heavens a gigantic clock-like system.

According to Calippus' modification of Eudoxus' scheme there were four spheres for the outermost planet, Saturn, with the planet attached to the inner sphere. Aristotle added three extra spheres inside the one carrying Saturn, identical to the outer ones but moving in opposite directions. The outer sphere of the next inner planet, Jupiter, was then attached to the innermost of these countermoving spheres, and so forth. The countermoving spheres both linked the planets mechanically and insured that the particular motion of a planet was not transmitted to the next inner planet. Thus the mathematical, conceptual scheme of Eudoxus, with an independent set of spheres for each planet, was replaced by Aristotle with a mechanical, clockwork-like set of nested spheres, fifty-five in all. The outermost sphere, containing the stars, drives the next inner one, that one drives the next, and so on down through all the planets, sun, and moon. In turn, the motion of the moon's inner sphere supplies the motive power for the apparently chaotic motion of the elements on earth.

The Epicycle Theory

Aristotle's incorporation of homocentric spheres into his system of the heavens insured the widespread acceptance of this model among his followers, who dominated the intellectual world through the Middle Ages. We find references to the spheres in the writings of philosophers, scholars, and poets for more than fifteen hundred years, as this conceptually simple and appealing model of the heavens captured men's imaginations. As late as the seventeenth century, John Donne wrote:

> The spheres have music, but they have no tongue,
> Their harmony is rather danc'd than sung.[1]

To men concerned with the details of astronomical observation, however, a number of defects threatened the homocentric sphere solution to the planetary problem. The first written reference to them came about a generation after Aristotle's death, but it is hard to imagine that Aristotle was not aware of them.

The most glaring defect involved the planets' obvious changes in brightness when undergoing retrogression. Mars, for example, is one of the brightest objects in the heavens when it is in retrograde motion; at other times it is still easily visible but hardly striking. The simplest explanation for the changes in brightness would involve changes in the planet's distance from the earth. The homocentric sphere model, however, requires that each planet remain at a constant distance from the earth since the planet is fixed to the surface of a sphere whose center is at the center of the earth. If the homocentric sphere model is accepted, changes in brightness cannot be caused by changes in the planet–earth distance. They would have to come from changes in the planet itself, an idea incompatible with the universally accepted view of the permanence and stability of the heavens.

Another difficulty arose from changes in the apparent size of both the sun and the moon at different times of the year. These changes, while not large, are easily measurable. Here again the simplest explanation would involve variations in the distances from the earth, changes not allowed by the homocentric sphere model.

Further progress seemed to demand a radical alteration in the model or even an entirely new way of describing the motions of the heavenly bodies.

Eccentrics and Epicycles

Eudoxus was the first in a line of technical astronomers whose primary interest was observing and recording phenomena of the heavens and devising detailed mathematical models to account for the data. Philosophical astronomers had been concerned with astronomy primarily as a part of a more general conception of existence, a system of the world. The difference between them was apparent in their attitudes toward the interaction of observation and theory. If the data didn't quite fit the theory—if the planets couldn't be described by a model involving uniform circular motion—something had to give. The new technical astronomer was willing to relax his philosophical demands if by doing so he could more successfully account for the data. The central question—"In what way can the observed motion of the planets be understood in terms of uniform circular motion?"—must still be answered, but to arrive at an answer, one might be forced to compromise a bit on the nature of the uniform circular motion.

To visualize more clearly what was known about the planets, let us consider the planet Mercury. Whenever Mercury undergoes one of its intermittent reversals of direction, it increases significantly in brightness. If one attributes the increase in brightness to the planet's coming closer to the earth, then the path of Mercury relative to the earth would have to be something like that shown in Figure 3–3. While the net motion of the planet is counterclockwise, when it dips in close to the earth (and hence appears brighter), it briefly reverses its direction. This was how the later Greeks interpreted their observations of Mercury. In fact, Figure 3–3 also represents what modern measurements indicate to be the path of Mercury relative to the earth (excluding, of course, its diurnal motion).

The inability of the homocentric sphere model to account for variations in earth–planet distance of the sort shown in Figure 3–3 led technical astronomers to search for other mathematical devices to explain planetary motion. One of the first such devices was the *epicycle*, put forth by the mathematician Apollonius late in the third century B.C.

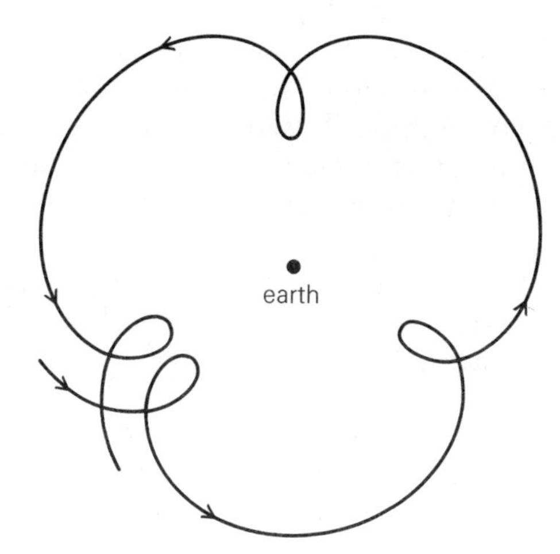

Figure 3–3. The orbit of Mercury relative to the earth.

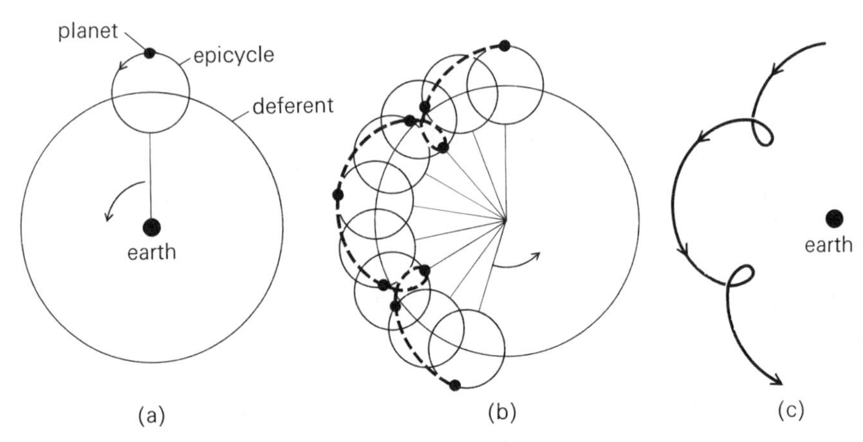

Figure 3–4. A retrograde epicycle-deferent. (a) shows the basic components; (b) is a multiple-exposure view of the motion; and (c) shows the net motion alone. Note that the direction of motion reverses.

Figure 3–4 illustrates the idea. The planet moves at a uniform rate around a small circle, which is called an epicycle. The epicycle itself travels, again at a uniform rate, around another circle, the *deferent*, centered on the earth. If the motion on the deferent is in the same direction as that on the epicycle (counter-clockwise in the figure), and if the planet's motion around the epicycle is fast enough, retrograde motion will result—and the retrogression will occur when the planet is closest to the earth and thus appears brightest, just as is actually observed. By adjusting both the relative size of epicycle and deferent and the speeds of the two motions, one can generate many different types of planetary motion. Figure 3–5 shows a typical example. We call this type of epicycle-deferent combination *retrograde* because it produces a retrograde motion of the planet.

> On Figure 3–5, illustrate the net motion for the second year, showing that there will be no retrogression as stated.

A somewhat different use of the epicycle was made by Hipparchus of Rhodes about a century later. It had been known for

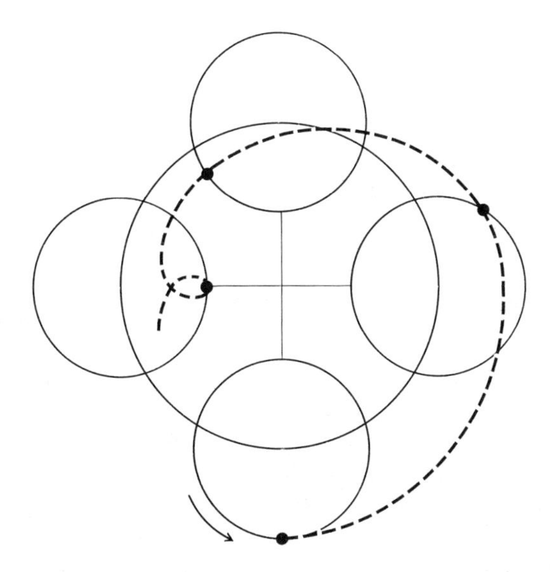

Figure 3–5. Complex planetary motion resulting from a retrograde epicycle-deferent. The period of the planet's motion on the epicycle is $1\frac{1}{2}$ times the period of the motion of the epicycle around the deferent. Note the irregularity of the net motion (the dashed path): in the first year the planet retrogresses once; in the second year (not shown) the motion is purely forward. This is typical of the observed motion of Venus.

some time that the sun does not move around the ecliptic at a precisely uniform rate. The time from the vernal to the autumnal equinox is about seven days longer than the time from the autumnal to the vernal equinox. In fact, the two extra sun spheres that Calippus added to Eudoxus' model were needed to account for this observation.

Hipparchus found he could fit the same motion with the epicycle scheme illustrated in Figure 3–6. In this case, the epicycle motion is opposite to the deferent motion. The sun moves once around the epicycle in one year, the same time in which the epicycle center moves once around the deferent. The net motion of the sun is thus a circle whose center is displaced from the earth. Since the motion has no retrogression, we shall call this device a *direct* epicycle-deferent. Such a direct epicycle-deferent gives a net motion which is precisely that of the *eccentric*, another

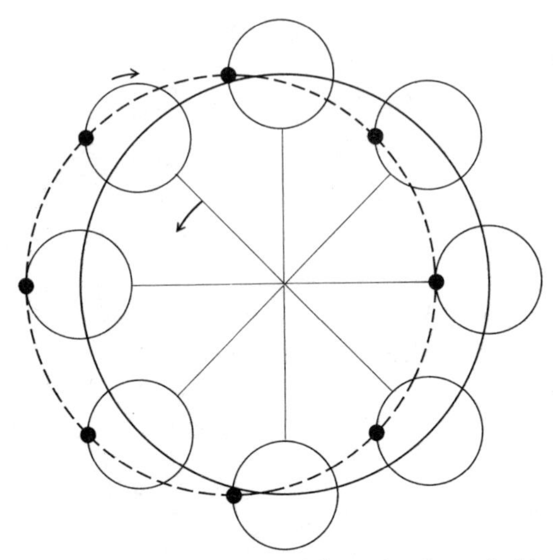

Figure 3–6. A direct epicycle. The epicycle period is equal to that of the deferent. This multiple-exposure view shows successive positions, while the dotted line indicates the resulting net motion.

device in general use at the time (see Figure 3–7). In this variant, the sun moves uniformly about a circle whose center is displaced from the center of the earth. The epicycle approach, however, was generally preferred. It made the motion seem symmetrical about the earth, while the eccentric was disturbingly centered on a seemingly arbitrary point out in space.

Interaction between Model and Observation

In one aspect of Hipparchus' work we find a particularly illuminating example of the relationship that inevitably exists between theory and observation. In his examination of the motion of the moon, Hipparchus could fit the available data by using a direct epicycle-deferent scheme similar to the one he had previously used for the sun, except that the period of the moon's motion around the epicycle was no longer quite the same as that for the motion of the epicycle around the deferent. Hipparchus'

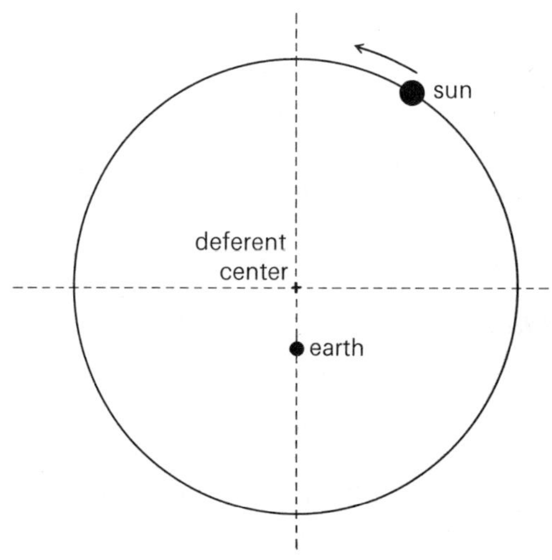

Figure 3–7. An eccentric. The sun moves uniformly on a circle whose center is displaced from the earth. The net motion is the same as that in Figure 3–6. Note that as seen from the earth, the sun appears to speed up when it is closest to the earth. The effect is very similar to looking out the window of a moving car; telephone poles nearby appear to move by faster than distant trees.

original data, which came from Babylon, was limited because the Babylonians were interested primarily in eclipses and therefore focused on the moon's position at these times and not on its full path. They had data, therefore, only for the position of the moon during lunar eclipses, when the moon is full, and during solar eclipses, when the moon is new. Hipparchus used these data to devise an epicycle scheme for the moon and then asked whether his model would work properly for other points on the moon's path. To answer this question, he made further measurements of his own—measurements which in fact proved disastrous for his lunar solution. Discrepancies were immediately apparent; they were resolved only much later by Ptolemy.

The point is that men tend to determine the significance of data in the context of a theoretical framework. Physical measurements are frequently difficult and tedious. For the Babylonians there was no point in measuring the positions of the half-moon, since such measurements had no obvious connection with eclipses. Only after Hipparchus had found a specific theory that allowed specific comparisons did such measurements become significant. True, the measurements when made showed the theory to be wrong; but this in itself was a gain. It is often said in this respect that a wrong theory is better than no theory at all.

The Ptolemaic System

Hipparchus could never obtain a completely satisfactory model describing the motion of the planets, and for over two centuries the problem lay dormant. Rome had come to rule the world, and the Romans tended to concentrate on practical matters. The Greek and Hellenistic tradition survived in the eastern part of the Roman world, but only as a weak vestige of what had gone before. The speculative philosopher, the creative mathematician, the searcher after a system of the heavens: these were typical of the Greek world and out of place in the Roman. The planetary problem had no practical significance and no longer commanded much attention.

In the later days of the Roman Empire, however, one astronomer emerged to climax man's attempt to organize the universe around a central earth. Claudius Ptolemaeus lived in Alexandria, in the southern part of the Roman world, during the first half of the second century A.D. We do not know how Ptolemy, as he is known, came to be interested in the planetary problem because our only source of his thoughts is his writings, and they are purely technical. The greatest of his works was called the *Almagest* by the Arab scholars who later preserved and studied it. For fourteen hundred years, this book was the Bible for anyone interested in the heavens. In it, Ptolemy surveyed the work of his predecessors, including Apollonius and Hipparchus, and then set forth the details of his presumably complete solution of the planetary problem. Let us examine some of these details.

Ptolemy accepted Hipparchus' solution for the motion of the sun, consisting of a direct epicycle and deferent. However, Hipparchus had never found a satisfactory solution for the moon, and Ptolemy could find a sufficiently accurate one only by

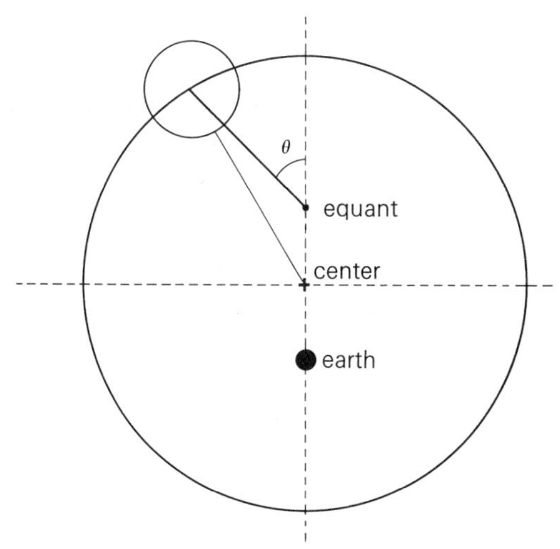

Figure 3–8. The equant. The angle θ changes uniformly. This has the effect of moving the epicycle faster at the bottom of the figure than at the top. Ptolemy ordinarily took the deferent center to be halfway between the earth and the equant.

combining with a direct epicycle-deferent two other devices, one old and one new. Figure 3–8 shows the scheme. The old device is the eccentric, in which the center of the deferent does not coincide with the earth but is located at a point in space some distance away. The new device is the *equant*, a further point in space that controls the speed with which the epicycle center travels along the deferent. Instead of moving in a uniform manner with respect to the center of the deferent, the epicycle now moves uniformly with respect to the equant—that is, the angle θ in the figure changes at a constant rate.

Ptolemy's success, however, was most striking in his solutions for the planets. Hipparchus had recognized two major anomalies in the motions of these bodies. First and most obvious were the retrograde motions. To account for these, the retrograde epicycle-deferent combinations had been proposed.

The second anomaly was more difficult to explain. The retrograde epicycle-deferent device predicted that the times between successive retrogressions of a planet should always be the same. However, planetary observations revealed small but measurable variations. The epicycle of a planet thus seemed to move along the deferent at a non-uniform rate. To represent this using only uniform circular motion again necessitated the use of an eccentric deferent and an equant, just as with the moon. The *coup de grâce* came in the case of Mercury, one of the most troublesome of the planets. In Ptolemy's final solution, the center of the deferent is no longer the earth, or even a fixed eccentric point in space away from the earth; instead this center itself moves in a small circle, as shown in Figure 3–9. Complicating matters still further, it was necessary to assume an equant that governed the motion of the epicycle center around the deferent.

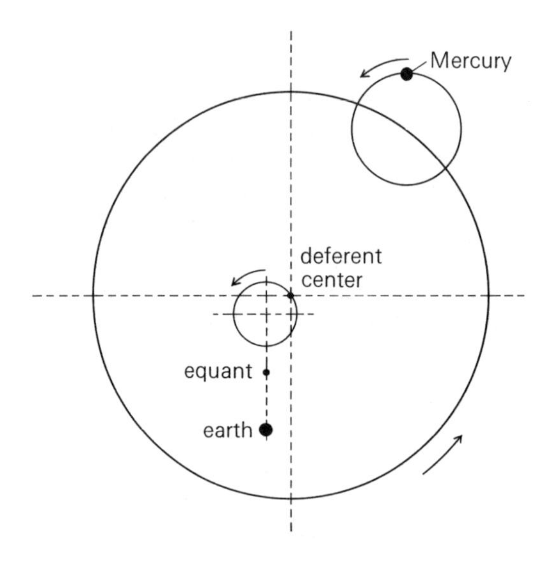

Figure 3–9. Ptolemy's scheme for Mercury. The deferent center moves in a small circular path. The motion of the epicycle is governed by the equant.

Another feature of the planets, important in view of later developments, is the apparent influence of the sun on their movements. The inner planets, Mercury and Venus, are always near the sun. To account for this, Ptolemy's model fixed the centers of the epicycles of Mercury and Venus on a line from the earth to the sun, as shown in Figure 3–10.

For the outer planets, Mars, Jupiter, and Saturn, retrogressions occur when, and only when, the sun is diametrically opposite in the sky from the planet—that is, when the planet is highest in the sky (or culminates) at midnight. To account for

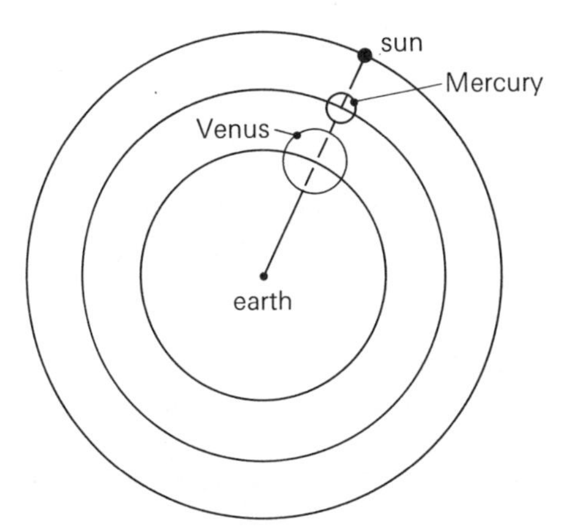

Figure 3–10. In the Ptolemaic model, the epicycles of Mercury and Venus were postulated to lie on the line between the earth and the sun. This ensured that Mercury and Venus would never stray far from the sun.

this, the motion of the planet's epicycle had to be related in a definite though rather complex way to the motion of the sun. These relationships were intriguing, but the Ptolemaic system had nothing very illuminating to say about them; they simply had to be appended to the model in an *ad hoc* fashion.

As Ptolemy left it, his system of the universe was undoubtedly complex, but it worked. It fitted the available observations of the heavens with remarkable accuracy. There seems to be no question that Ptolemy thought of his system as only a mathematical one, a way of accounting for the motions we observe. Any tangible realization would clearly be awkward at best, with all the epicycles and deferents and equants and eccentrics physically getting in each other's way. It was enough that one could accurately predict future positions of the heavenly bodies and that one could somehow explain the complex motions of these bodies without straying too far from simple Platonic principles.

In fact, some damage had been done to these principles of uniform circular motion. Since Plato clearly intended that the circles be centered on the earth, one can assume that he would have been bothered by the eccentric. He surely would have objected to the equant, in which the motion is circular about one point and uniform with respect to another. Ptolemy acknowledged these difficulties when, in the *Almagest*, he defended his approach as pragmatic:

> The astronomer must try his utmost to explain celestial motions by the simplest possible hypothesis; but if he fails to do so, he must choose whatever other hypotheses meet the case.[2]

What Ptolemy did in choosing these "other hypotheses" is typical of the way physical theories evolve. The first attempts to solve the problem of the planets using the purest of uniform circular motions failed. The homocentric sphere model could not explain what was seen as a significant phenomenon—the changes in brightness of the planets. Apollonius succeeded in explaining these changes with his retrograde epicycle-deferent scheme, but failed to make this device account for the detailed motions of the planets. Rather than overthrow the ideal of uniform circular motion in a geocentric universe, Hipparchus and Ptolemy made minor alterations, patching the model as necessary to fit the observations. The patching seemed successful; there was no need for a revolution.

A Footnote to Greek Astronomy: Heracleides and Aristarchus

While the work of Ptolemy followed the dominant line of late Greek astronomy, other approaches to a solution of the planetary problem were suggested from time to time. We digress briefly to consider two unorthodox astronomers whose ideas were not in the mainstream of ancient astronomy, but who did anticipate much later views of the universe.

Heracleides of Pontus was a contemporary of Aristotle. His writings have all been lost, but we can catch a bare glimpse of his ideas through the accounts of later men. He is said to have suggested that the motions of the stars could be accounted for by assuming that the earth turns on its axis once each day while the sphere of the stars, still centered on the earth, is at rest. He also tried to account for the fact that Mercury and Venus never appear far from the sun by assuming that these planets move in circular orbits around the sun itself as the sun travels around the earth.

Aristarchus of Samos lived three or four generations after Heracleides, during the first half of the third century B.C. It is not clear how much he knew of Heracleides' ideas, but the system he put forth was a natural extension of them. Although his writings on the subject have not survived, we may quote his younger contemporary, Archimedes:

> Aristarchus of Samos has published in outline certain hypotheses . . . He supposes that the fixed stars and the sun are immovable, but that the earth is carried round the sun in a circle . . .[3]

No more detailed accounts of these hypotheses exist, and we may infer that the scheme was little more than a suggestion, though, according to Archimedes, Aristarchus did go so far as to grapple with a possible argument against it: If the earth travels around the sun and if the stellar sphere is centered on the sun as suggested by Aristarchus, then at different times of the year the earth should be at different distances from any given region of the stars. These stars should therefore appear spread out when the earth is close and more densely concentrated when the earth is more distant. (See Figure 3–11.) Such an effect, called *parallax*, had not been observed; this forced Aristarchus to assume that the stellar sphere was so much larger than the earth's orbit that parallax would be too small to detect.

How did Aristarchus arrive at the idea of this modern-sounding heliocentric model? Perhaps a source can be found in his one surviving book, *On the Dimensions and Distances of the Sun and Moon*, which contains one of the earliest serious attempts to determine these quantities. From a rather elementary set of arguments, he concluded that the size of the sun is three hundred times that of the earth. Perhaps this conclusion led Aristarchus to postulate the massive sun as the fixed object, with the tiny earth as its satellite.

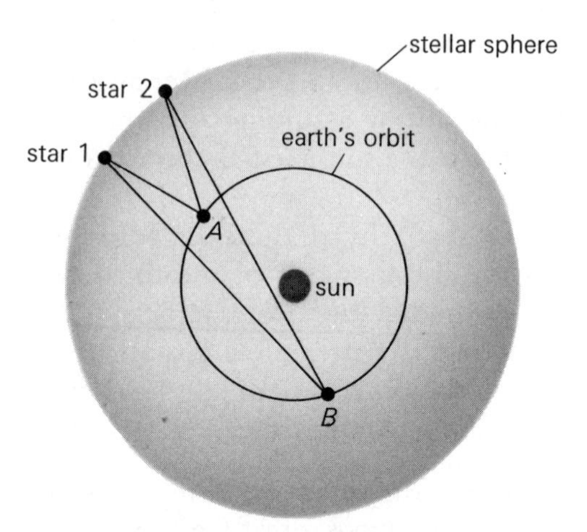

Figure 3–11. Parallax in a heliocentric model. 1 and 2 represent two stars; *A* and *B* are two positions of the earth at different times of the year. At *A* the earth is closer to 1 and 2; the angle 1*A*2 is large and the stars appear relatively far apart. At *B* the earth is farther away; angle 1*B*2 is small, and the stars appear closer together.

Such a thought was so much at odds with the general outlook of the age that it was unthinkable for all but the most adventurous of minds. It is interesting how little impact even the less strikingly heretical ideas of Heracleides had on the later views of both philosophers and astronomers. The idea of a spinning earth, much less one that moved through space away from the center of the universe, was so far from both sense experience (we certainly feel the earth to be solid and motionless) and the philosophical point of view of the era that no particular reason existed to take it seriously. Later astronomers explicitly rejected the heliocentric model. Their continued use of a geocentric system certainly cannot be attributed to their not having thought to try a heliocentric one.

Suggested Reading

Kuhn, T. *The Copernican Revolution*. New York: Random House, 1959. Chapter 2 gives a clear discussion of Greek planetary astronomy.

Dreyer, J. L. E. *A History of Astronomy from Thales to Kepler*. 2nd ed. New York: Dover, 1953. A complete and authoritative reference.

Heath, T. *Aristarchus of Samos*. Oxford, 1913. A classic, detailed discussion of Greek astronomy to the time of Aristarchus.

Questions

1. Distinguish between induction and deduction. Are these modes of reasoning peculiar to the physical sciences? If not, give examples of inductive and deductive reasoning in other areas of thought.
2. Eudoxus seems to have been the first person to attempt a quantitative solution to the planetary problem. What reason did he have to assume:
   (a) that the motions of the planets followed some sort of order? and
   (b) that he would be able to reduce this order to combinations of uniform circular motions?
3. In Figure 3–2, what is the angle between the directions of the diurnal and yearly axes? What observations can you cite to support your conclusion? (Cf. Question 6, Chapter 2.)
4. In Figure 3–4, how many retrogressions will occur as the planet moves once completely around the earth?
5. What would be the effect on the planetary motion of Figure 3–4 if both the radius of the epicycle and the radius of the deferent were doubled but the rotation rates were left unchanged? Would the change be apparent to an observer watching the planet from the earth? If not, how could Ptolemy's geocentric model determine the relative distances of the sun, moon, and planets from the earth?
6. To achieve a greater variation of speed of the planet in Figure 3–8, should the equant be moved toward or away from the center?

7. Why does the observation that Mercury and Venus are always "evening stars" or "morning stars" imply that these planets are never far from the sun?
8. Consider the change from Eudoxus' homocentric sphere model to the Ptolemaic epicycle model in the light of the discussion in Chapter 1. Did the significant facts change? Was there a change in the kinds of questions asked? Did the two models attempt different sorts of explanations? Did they work from different presuppositions?
9. In what sense had Ptolemy *explained* the observed motions of the planets? Does his explanation seem satisfactory to you? What questions about the planets do you think an explanation of planetary motion should answer?
10. Aristarchus of Samos made the first recorded attempt to determine the relative distances of the sun and moon from the earth. His method is shown in the figure below. When the

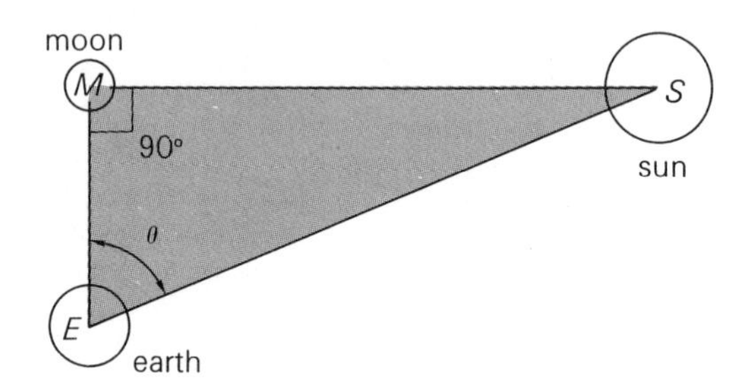

moon, seen from the earth, is exactly half illuminated, the angle between EM and MS must be exactly 90°. The angle θ can then be measured directly; Aristarchus found it to be 87°. Use Aristarchus' measurement to find the ratio ES/EM. (Students who know some trigonometry can simply look up the cosine of 87° and work with that; others may find it easiest to draw an appropriate triangle and measure the ratio of the sides directly.) We now believe the sun to be much farther away than your results indicate. The difficulty was not with the method devised by Aristarchus; it is correct. However, modern measurements show that the angle θ is actually closer to 89.8° than to the 87° Aristarchus estimated. Show that this small error in the measurement of θ leads to a very large error in the ratio ES/EM.

11. With the aid of the pattern provided, find a direct epicycle-deferent scheme that leads to a circular planetary orbit with the earth displaced from the center of the circle. Consult Figure 3–6 for more details concerning this construction.

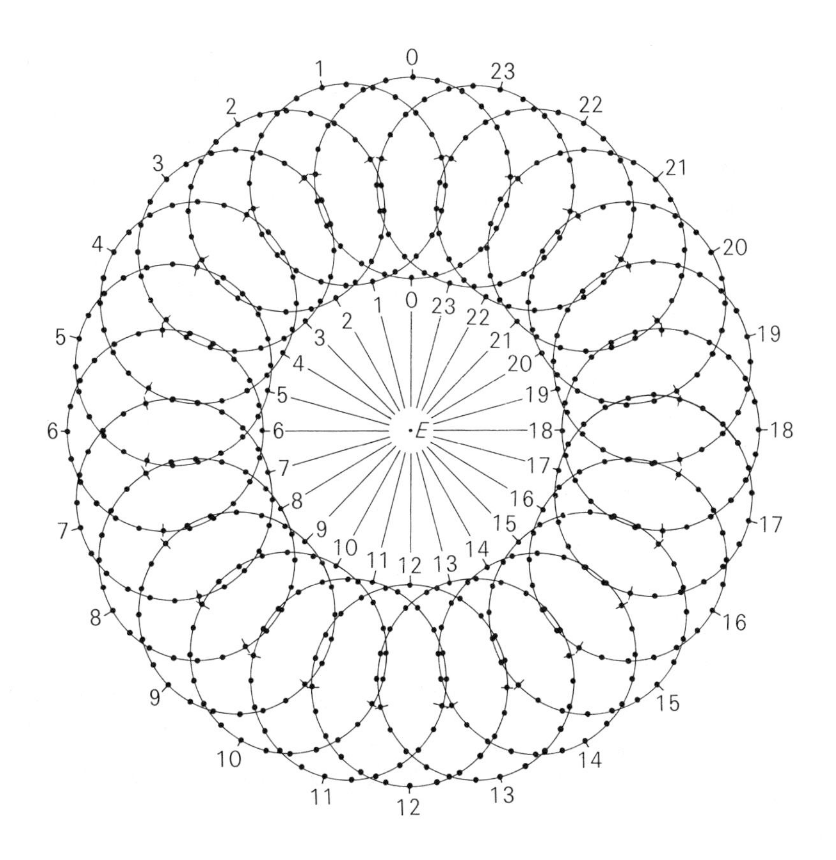

12. With the aid of the pattern on page 66, find epicycle-deferent schemes that result in the following planetary orbits about the earth:

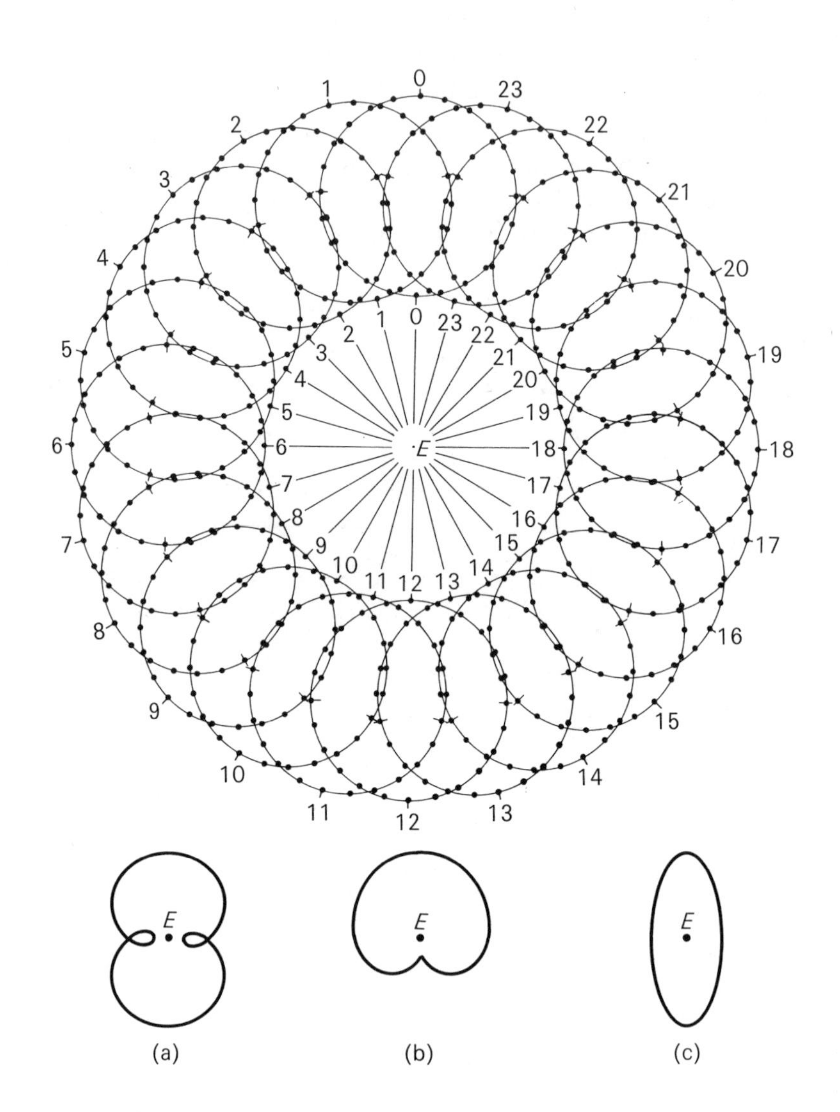
0
1
2
3
4
5
6
7
8
9
10
11
12
13
14
15
16
17
18
19
20
21
22
23
E
(a)
(b)
(c)

13. What sort of epicycle-deferent combination will reproduce the motion of Mercury shown in Figure 3–3? Construct this orbit with the aid of the pattern provided. Suggestion: First do the easier problem of constructing an orbit with three retrogressions that closes on itself as the sketch shows.

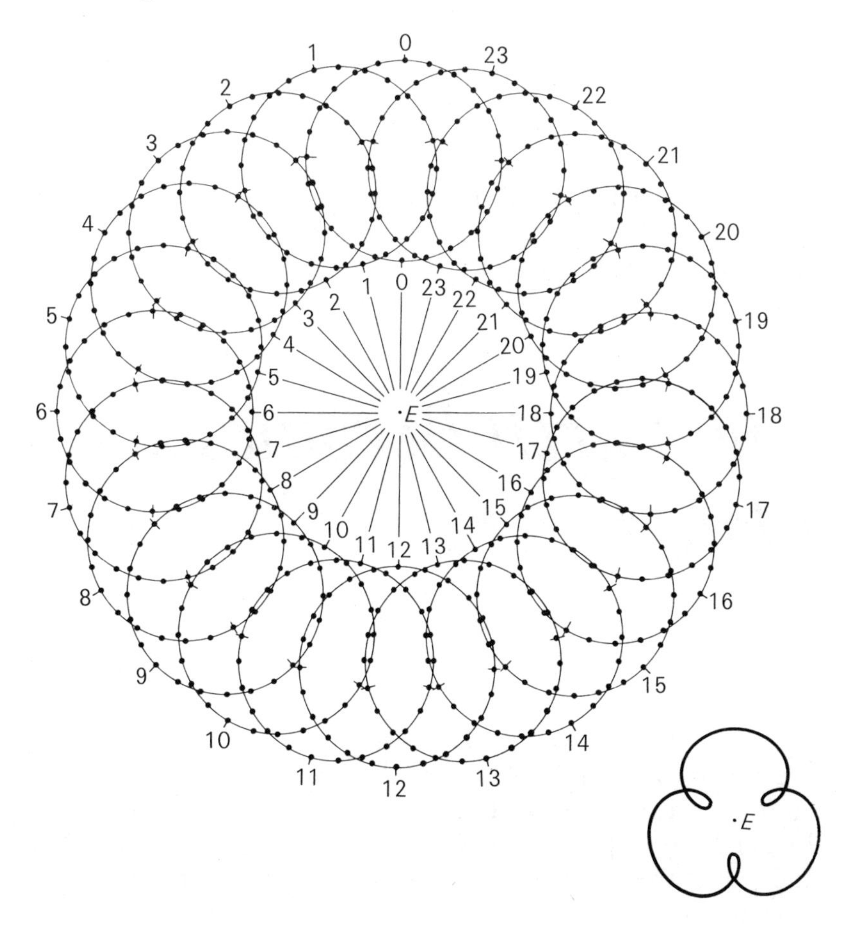

CHAPTER FOUR

The Earth Becomes a Planet

The work of Ptolemy was the high point of Greco-Roman science. Along with the Roman Empire itself, interest in astronomy and science in general went into first a slow and then a precipitous decline. Europe sank into the Dark Ages, and the accomplishments of the past were forgotten. Only in the East did there remain any interest in the Greek astronomical tradition. In the fourth century B.C., Alexander the Great had spread Greek ideas as far east as India, and contact with Western astronomy had continued through the time of Ptolemy. In the eighth century A.D., these ideas were introduced from India into the Arab world, and Arab astronomers sought out the Greek and Alexandrian originals, including the writings of Ptolemy, for study and translation.

The Arabs' interests were largely the practical ones of calendar regulation and astrological prediction. However, their many attempts to provide some physical basis for the Ptolemaic scheme demonstrated a sympathy with the Greeks' search for an understanding of the structure of the heavens. For example, they invented systems of counter-rotating crystal spheres that could drive epicycles physically the way balls are driven in ball bearings. Though all these attempts failed in the end, and other efforts to simplify or improve the Ptolemaic system achieved no significant success, the Arabs did keep alive not only the Ptolemaic approach but even the details of Ptolemy's solutions. In addition, their substitution of trigonometric methods of calculation for Ptolemy's geometric ones made the use of the system less cumbersome.

This legacy from the past was returned to a reawakening Europe in the eleventh and twelfth centuries, principally through Spain, where contact between Arabs and Europeans was extensive. Translations of the Arabic works spread to Italy and France, where they soon aroused the interest of scholars. The Church, once hostile to Greek astronomy because it was pagan in origin and because it seemed to conflict with the Bible, slowly came to encourage this interest.

At the universities in Bologna, Paris, and Oxford, scholars began to relearn Greek and to study the writings of the ancients in the original. Thomas Aquinas (1225–1274) interpreted the writings of the Greek philosophers, particularly Aristotle. His sympathetic views brought the works of Aristotle to the attention of educated men, and gradually Aristotle came to occupy a place in Christian teaching next to that of Saint Augustine and the Bible. Aristotle was studied in monasteries and universities all over Europe, and explication and interpretation of his views became a primary task for men of intellect.

By the sixteenth century, Europe had regained a knowledge of the Ptolemaic system of the heavens, a system that, despite the efforts of centuries of astronomers, remained essentially the same as Ptolemy had left it. There had been numerous attempts to improve the details—alterations in numbers and sizes of epicycles and the like—but none had significantly improved Ptolemy's fit of the observations. Ptolemy's solution, as we have seen, was good—remarkably good. However, minor discrepancies between calculation and observation did exist, and these were difficult to correct without vastly increasing the complexity of the system—by adding more and more epicycles, for example, or even by assuming epicycles on other epicycles, as some scholars proposed.

These discrepancies had to be dealt with. The Ptolemaic system, never simple, seemed destined to grow even more complex.

Copernicus

From our modern perspective, the first notable conceptual change in man's view of the heavens for seventeen hundred years resulted from the work of Nicholaus Copernicus (born Niklas Koppernigk in 1473). Copernicus was brought up in the north of Europe, in what is now Poland. He was educated at the universities of Cracow and Bologna, and at the age of thirty-three returned to Frauenburg, near his birthplace. Until his death in 1543 he held the ecclesiastical position of canon of the cathedral there and devoted most of his time to astronomy.

Copernicus seems to have been struck by the complexity and inelegance of the Ptolemaic system. Furthermore, it seemed to him that one element of that system, the equant, clashed philosophically with the principle of uniform circular motion. The equant did involve uniform circular motion, but the motion was uniform about one point and circular about another, with neither point at the center of the universe. "A system of this sort seemed neither sufficiently absolute nor sufficiently pleasing to the mind," Copernicus wrote.[1]

He saw that one could avoid the use of equants, but only at the cost of adding many more epicycles to a solution already overburdened with epicycles and eccentrics. Why couldn't someone find a combination of epicycles that would be markedly less complex and still adequately describe the motions of the heavenly bodies? Fourteen hundred years of fruitless attempts along these lines, however, were not encouraging.

Perhaps the geocentric approach had been carried as far as it could go; perhaps one should search for another approach that would allow a simpler view of the heavens.

The model Copernicus eventually proposed must have seemed preposterous to men of the sixteenth century. The sun replaced the earth at the center of the universe. Around the sun were placed the earth, now a planet, and the five other planets, in the order Mercury, Venus, Earth, Mars, Jupiter, and Saturn. Each planet was assumed to travel about the sun with uniform circular motion. The moon was put in orbit around the earth, again with uniform circular motion, while the earth itself was taken to rotate about its own axis once each day. Far beyond Saturn was the sphere of fixed stars, at rest.

We shall refer to this as the *basic heliocentric model*. How Copernicus was led to it we can only guess. He gives us a clue in the introduction to his major work, *De Revolutionibus*. After briefly alluding to Philolaus and Heracleides as having taught the motion of the earth, he writes:

> Taking advantage of this, I too began to think of the mobility of the Earth; and though the opinion seemed absurd, yet knowing now that others before me had been granted freedom to imagine such circles as they chose to explain the phenomena of the stars, I considered that I also might easily be allowed to try whether, by assuming some motion of the Earth, sounder explanations than theirs for the revolution of the celestial spheres might so be discovered.[2]

Let us examine the Copernican solution in some detail to try to understand both why Copernicus felt it was an improvement over that of Ptolemy and why, generally speaking, his contemporaries refused to accept it.

The Basic Heliocentric Model

The Earth: The motion of the earth in the basic heliocentric model is illustrated in Figure 4–1(a). In this model the earth rotates about its own axis once every 23 hours and 56 minutes.

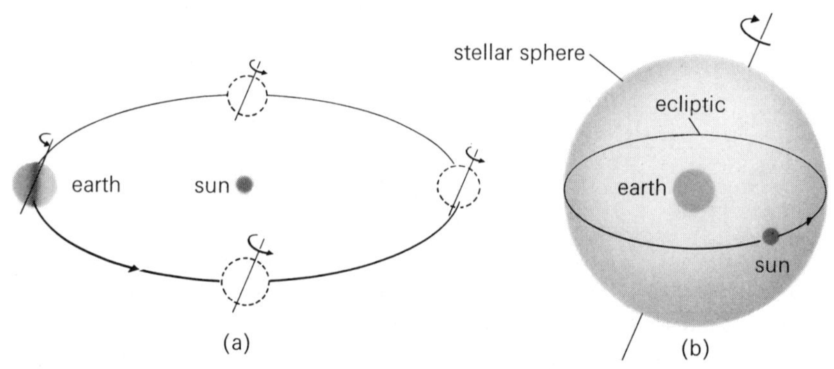

Figure 4–1. (a) The earth and sun in the basic heliocentric model. The earth moves about the sun in a circular path in one year while rotating about its axis daily. (b) The earth, sun, and stars in the basic geocentric model. The sun moves about the ecliptic on the stellar sphere in one year, while the stellar sphere rotates about its axis daily.

This axis is tilted with respect to the plane of the earth's orbit about the sun by just the angle that the axis of the stellar sphere makes with the plane of the ecliptic in the basic geocentric model [Figure 4–1(b)].

The earth moves about the sun in a circular path with constant speed, one full revolution taking precisely one year. Throughout this motion around the sun, the earth's axis of rotation continues to point in the same direction in space, toward the North Star.

Thus this model explains the diurnal motion of the heavens by the daily rotation of the earth about its axis. This leads to precisely the same predictions for the diurnal motion of the sun and stars as did the rotation of the stellar sphere about a fixed earth in the basic geocentric model. The annual motion of the earth in its circular orbit about the sun explains the apparent motion of the sun about the ecliptic. As the earth moves, it seems to an observer on earth that the sun changes its position with respect to the background of stars.

This is seen more clearly in Figure 4–2, which is a top view of Figure 4–1(a) with the stellar sphere added. Four arrows, 1, 2, 3, and 4, are drawn showing the distance and direction of the sun as seen from the earth at four different times of the year, t_1,

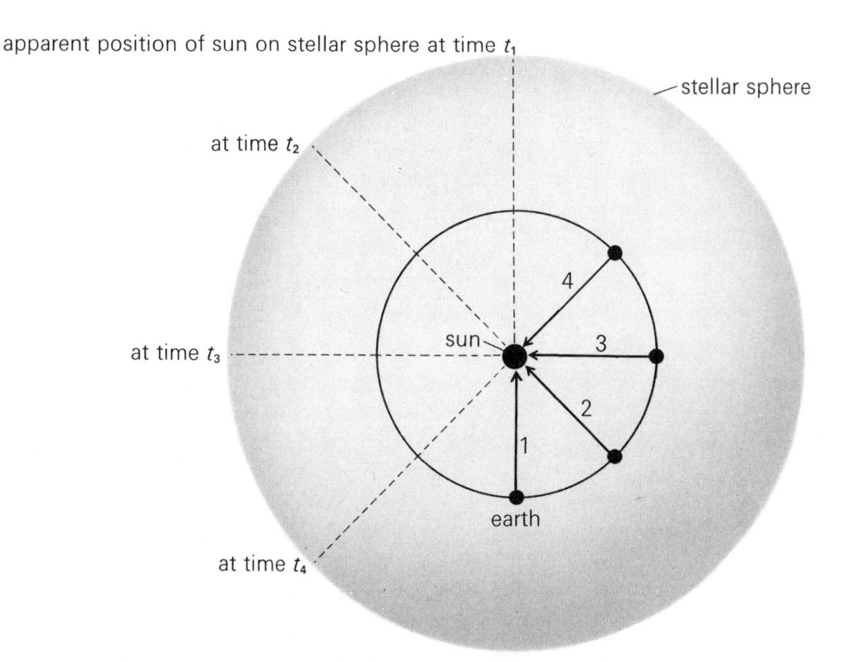

Figure 4–2. The apparent motion of the sun on the stellar sphere in the basic heliocentric model.

t_2, t_3, t_4. The dotted extensions of these arrows indicate the points on the stellar sphere (i.e., the neighborhoods of stars) where the sun will appear at the times in question. It is clear from the figure that the sun will seem to move in a counterclockwise direction through the stars.

One possible difficulty with this model can be seen in Figure 4–2. This is the problem of parallax, which we discussed in connection with Aristarchus' heliocentric model. If the earth moves about the sun, its distance from a given region of the stellar sphere will vary. This should lead to variations in the appearance of the stars, as illustrated in Figure 3–11. To account for the apparent absence of any such effect, Copernicus, like Aristarchus, assumed that the stellar sphere is immense compared to the size of the earth's orbit. Thus, even though the earth moves, its distance from a given point on the stellar sphere varies only negligibly.

The apparent motion of the sun can be explained in another way that will prove useful in our later discussion. To an observer on the earth, the earth *appears* to be at rest and the sun *appears* to be moving. Let us use Figure 4–2 to construct the apparent motion of the sun as viewed from the earth, according to the heliocentric model. Since the earth seems to an earthbound observer to be at rest, we represent it by a fixed point in Figure 4–3. Let us construct the motion of the sun as seen by this observer. For example, at time t_1 we note from Figure 4–2 that to see the sun he would have to look in the direction of arrow 1.

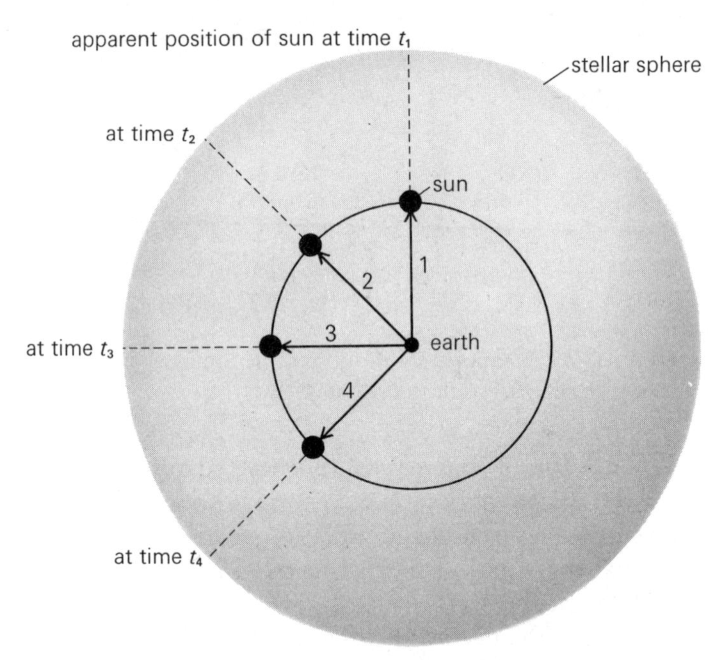

Figure 4–3. The motion of the sun relative to the earth as constructed from Figure 4–2. This is how the sun would appear to move to an earthbound observer if the sun were really at rest and the earth were moving as shown in Figure 4–2.

The length of arrow 1 indicates the distance from the earth to the sun at that time. We can therefore draw the same arrow in Figure 4–3 to show the position of the sun relative to the earth at time t_1. In the same way we indicate by arrows 2, 3, and 4 in Figure 4–3 the positions of the sun relative to the earth at times t_2, t_3, and t_4 respectively. Notice that each arrow has the same length and points in the same direction as the corresponding arrow in Figure 4–2.*

From Figure 4–3 we conclude that if the basic heliocentric model is valid, the sun, as seen by an observer on earth, will still appear to move in a circular path around the earth just as predicted by the basic geocentric model. Thus, the *relative* motion of the earth and the sun is the same in the heliocentric and geocentric models. Since all measurements of the sun from observatories on the earth give only the relative position of the sun with respect to the earth, on the basis of such measurements alone one simply cannot decide whether the earth or the sun is *really* at rest.

The Planets: For the earth and sun, both heliocentric and geocentric models seem equally simple—one body moves in a circle about the other. However, for the motion of the planets, the Copernican system promises a remarkable simplification. In the basic heliocentric model, the planets are assumed to orbit the sun in circular paths with constant speeds, as illustrated in Figure 4–4. These motions are conceptually much less complex than the

* At first glance there may seem to be an inconsistency between Figures 4–2 and 4–3. Parallax seems implied by 4–2 and not by 4–3. The reason is that these figures are not drawn to scale. The radius of the stellar sphere was taken by Copernicus to be so much larger than the distance from the earth to the sun that if the figures were drawn to scale, both the sun and the earth would appear extremely close to the center of the immense stellar sphere. Hence, to an excellent approximation, either the earth or the sun can be placed at the center.

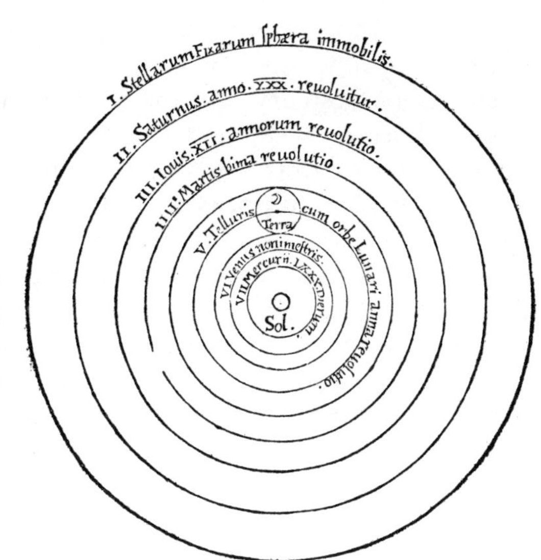

Figure 4–4. Copernicus' illustration of the basic heliocentric model, published in 1543. (Yerkes Observatory Collection.)

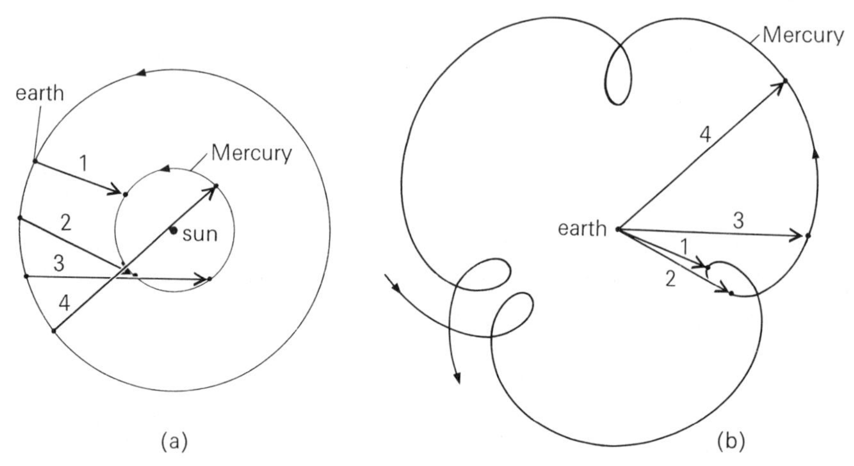

Figure 4–5. Two geometrically equivalent representations of the earth and Mercury in the basic heliocentric model. (a) shows the earth and Mercury moving in circular orbits about the sun. (b) indicates how Mercury would appear to move to an observer on the earth if the two planets were really moving as shown in (a). The *relative* motion of the two planets is the same in (a) and (b). (After Gerhart and Nussbaum, "Motion," University of Washington, 1966.)

intricate epicycles the planets were assumed to follow in the Ptolemaic scheme. Yet the simple planetary model of Figure 4–4 is at least qualitatively consistent with the observations of planetary motions from the earth, including retrogressions. This is shown in Figure 4–5(a) and (b). Part (a) depicts the uniform circular motions of the earth and one of the inner planets, Mercury, around the sun in the basic heliocentric model. The positions of Mercury and the earth at a succession of equally spaced times t_1, t_2, t_3, and t_4 are indicated. Part (b) shows the same motions redrawn with respect to the earth. In this figure, the earth is fixed in position while Mercury appears to move in an epicyclic path. But Figures 4–5(a) and (b) are completely equivalent, as we can see by comparing the solid arrows labeled 1, 2, 3, and 4 in part (a) with the corresponding arrows in (b), which have the same directions and lengths. The basic heliocentric model thus accounts for the retrograde motion of the planets as viewed from the earth without the use of epicycles.

The appearance of retrograde planetary motion in the basic heliocentric model can also be seen in another way. Figure

4–6(a) shows the relative motion of the earth and an inner planet, while Figure 4–6(b) shows the relative motion of the earth and an outer planet in the basic heliocentric model. The dots represent successive positions of the earth and planet at equal intervals of time. The straight lines connect corresponding positions, and the intersections of these lines with the stellar sphere show the apparent path of the planet with respect to the stars. Note that in both cases the retrograde motion occurs when the planet is closest to the earth and hence appears brightest. We have previously noted that this is precisely what is observed. Furthermore, in the case of the outer planet, retrogression occurs when the planet is in the opposite direction from the sun—when the planet culminates at midnight—again in agreement with observation.

Figure 4–6. Retrograde motion as incorporated in the basic heliocentric model for both (a) an inner planet and (b) an outer planet.

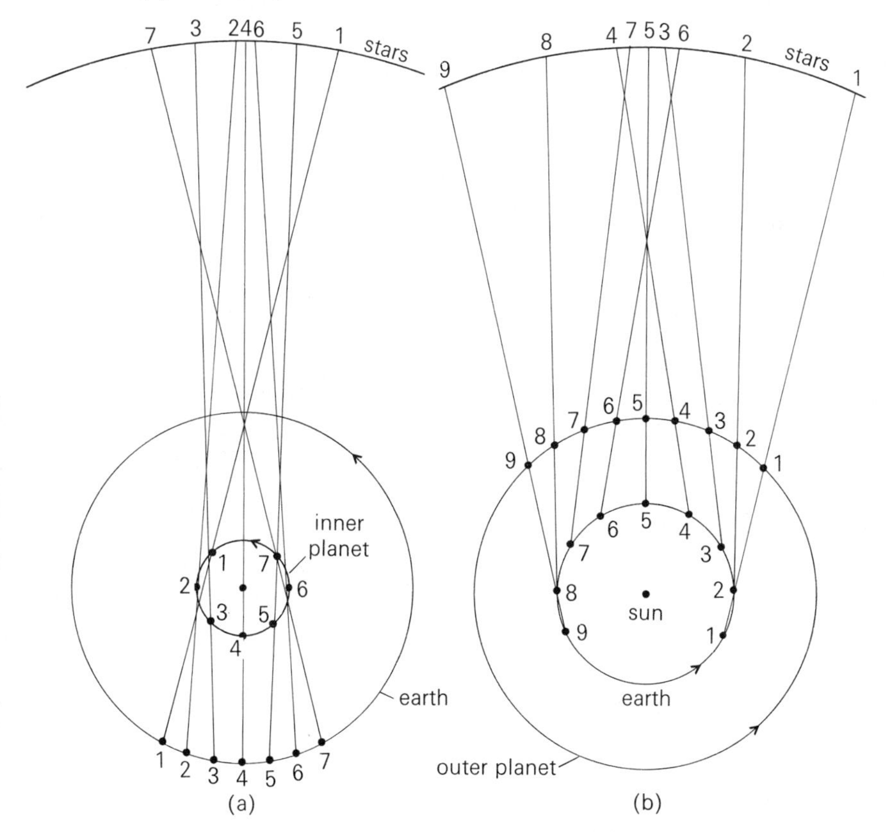

Thus the Copernican model can explain the special relations between the sun and the retrogressions of the planets in a natural way. This is one of the greatest triumphs of the system. To the question, "Why is the sun always diametrically opposite in the heavens when an outer planet retrogresses?" the Ptolemaic astronomer could only shrug his shoulders and say, "That's apparently the way it is," but the Copernican could answer, "It could not be otherwise."

In like manner, the Ptolemaic model could account for the observation that the inner planets always appear near the sun only somewhat artificially by connecting the deferents of the sun, Mercury, and Venus in a line with the earth (Figure 3–10). In the Copernican system, since the orbits of Mercury and Venus lie inside that of the earth, these planets *must* appear close to the sun.

Another attractive feature of the Copernican model is that it determines unambiguously the relative size and sequence of the planetary orbits. The Ptolemaic model determined the relative size of epicycle and deferent for each planet, but the sequence of the planets could be chosen arbitrarily. Figure 4–7 shows the situation for Copernicus' model: a simple measurement gives

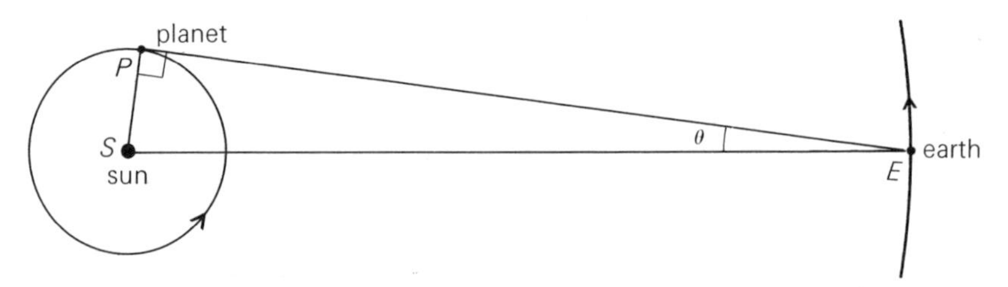

Figure 4–7. Determination of relative sizes of orbits for the earth and an inner planet. When the planet appears at its maximum distance from the sun, the angle EPS must be a right angle. Measuring θ allows us to calculate $SP = ES \sin \theta$. A similar though slightly more complicated relation holds for the outer planets. In this way the distance from each planet to the sun can be determined in terms of the earth-sun distance ES. Even if ES is not known, this method enables one to calculate the relative sizes of the planetary orbits.

the distance of each planet from the sun in terms of the earth-sun distance. Thus, if one believes the model, one can determine the order and relative size of the planetary orbits.

Drawbacks

However, there were difficulties, both technical and philosophical, with the basic heliocentric model. As we have noted, the problem of parallax was still present. The only solution consistent with the model seemed to be that of Aristarchus—to assume the stellar sphere to be so large with respect to the size of the earth's orbit that the effect would be too small to be seen. If one accepted the Copernican model, one would thus be forced to contemplate a universe far larger and emptier than that to which sixteenth-century men were accustomed.

An even more compelling argument against the Copernican system was the very absurdity of the notion of a moving earth. If the earth were moving, surely we would sense this motion just as we sense the motion of a cart or a ship on which we ride. Furthermore, wouldn't a moving earth leave the air behind and cause objects thrown directly upward to return to the ground far off?

Another serious difficulty became apparent as Copernicus tried to fit actual planetary observations to his system. Of course, what was known about the planets consisted of observations from the earth; the raw observational data was (and still is) earth-centered. In addition, the only planetary measurements possible until very recent times were angular positions in the sky. (Consider, for example, the positions of the sun and Mercury as measured from the earth. All that can be determined from simple visual observation is the angle between the directions of the sun and Mercury.) The actual distances to the planets were not known; only very crude estimates could be made from changes in apparent brightness.

Thus, for each planet, Copernicus had to go through a procedure very similar to that involved in constructing Figure 4–5. Assuming a model like that of Figure 4–5(a), with both earth and planet moving about the sun in circular paths with constant speed, he calculated their corresponding motions relative to the earth. That is, he constructed a figure like Figure 4–5(b). He could then compare the angular positions of the planet predicted by this construction with actual earthbound observations.

Unfortunately, the agreement between such a simple model and the observations could not be made completely satisfactory. Copernicus found significant discrepancies that could be resolved only by modifying his basic heliocentric model so that a planet did not move with uniform speed about the sun, but rather speeded up and slowed down in different parts of its orbit. The only way to account for these changes while maintaining the principle of uniform circular motion was to augment the model with direct epicycles, as shown in Figure 4–8. The use of these epicycles, however, forced Copernicus to destroy the simplicity of his model, the very simplicity that had made it so appealing to begin with. It would probably have been difficult to ignore a model that used only simple circular orbits about the sun and could account for earthbound observations of planetary retrogressions without the use of epicycles. But in the end, Copernicus had to specify more parameters (radii of deferents and epicycles, periods of deferent and epicycle motion) than the Ptolemaic solutions had used, and the resulting fit was no better. On the basis of simplicity, the final Copernican model had no clear-cut advantage over the Ptolemaic scheme.

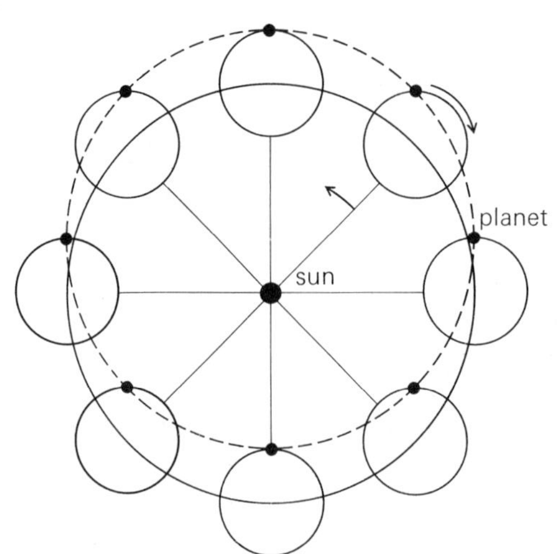

Figure 4–8. Epicycles as used in the final Copernican model.

The philosophical difficulties were at least as serious as the technical ones. The geocentric model had the whole Platonic-Aristotelian world view behind it, as well as the blessing of the Church. The heliocentric model had no such support.

With all its conceptual shortcomings and with neither the prevalent philosophical bias nor aesthetic simplicity behind it, Copernicus feared that his model would meet with a negative reception both from astronomers and from learned theologians. He therefore kept his work unpublished until very late in his life. When *De Revolutionibus* did appear, it was with a preface arbitrarily inserted by Osiander, the Lutheran theologian who supervised the printing. Osiander felt it necessary to represent the Copernican system simply as a new calculational device in the tradition of Ptolemy and the Greeks, although Copernicus indicated clearly in his text that he believed in the reality of his system.

In fact, Copernicus' approach was in many respects still basically conservative. Radical as was the sun-centered perspective, Copernicus was still bound to the ideal of uniform circular motion and the devices of Ptolemy. Something more was required to consummate a revolution.

The Coming of the Crisis

As the sixteenth century advanced, the new ideas of Copernicus spread and were much discussed. A few astronomers accepted them wholeheartedly, and a set of planetary tables determined on the basis of the Copernican model was widely adopted as the most accurate and complete compilation available. And yet, for the reasons discussed above, the intellectual community, including most astronomers, was reluctant to accept the heliocentric view except as a calculational method. The arguments against it were too fundamental to allow its acceptance as anything more.

The fate of Giordano Bruno was indicative of the spirit of the times. Although Bruno was a writer, not a professional astronomer, he espoused the Copernican system and carried it even further. If the earth rotates, but the stars do not, he reasoned, then perhaps the stellar sphere does not even exist. Why couldn't the stars be spread throughout an infinite universe, with some of them accompanied by their own earths inhabited by other peoples? Such ideas not only seemed outlandish in themselves but also presented a clear threat to the Church. They challenged the philosophical and theological view of man as the focal point of the universe. Other worlds, other men—how could they fit into teachings of original sin, of mankind as redeemed by the death of Christ? Bruno was burned at the stake as a heretic in 1600.

And so a silent geocentric majority faced a vocal heliocentric minority. The Aristotelians argued the simplicity of a spherical universe with a spherical earth at its center; the Copernicans countered with the simplicity of planetary orbits around the sun. The Aristotelians argued that a moving earth would leave the air behind; the Copernicans replied that the air must naturally be carried along with the earth. Isn't it absurd to imagine the huge, sluggish earth moving? But isn't it more absurd to imagine the motion of the even larger sun? Proofs abounded—the trouble was that different proofs rested upon different assumptions.

From time to time, new elements were added to the controversy. In 1572 the world was shaken by the appearance of an unexpected phenomenon: a new star. Day by day it grew brighter until it was visible even in broad daylight, and then, after eighteen months, it disappeared. What of the Aristotelian vision of the immutability and permanence of the heavens if such a thing could happen? In 1577 a comet appeared. Comets had previously been accepted as some sort of atmospheric phenomenon, belonging to the imperfect and impermanent sublunar region, but careful observations of this comet showed a lack of parallax indicating that it must be farther away from the earth than the moon. The Aristotelian foundations were weakening.

Tycho Brahe

During this period of argument and uncertainty, the major technical contributions were made by the Danish astronomer Tycho Brahe. At the height of his career, Tycho served as a sort

Figure 4–9. The great mural quadrant at Uraniborg, used by Tycho to map the heavens. From his *Astronomiae Instrumentae Mechanica*, 1598.

of court astrologer-astronomer to King Frederick II of Denmark, who in 1576 gave him the use of the island of Hveen and sufficient funds to construct Uraniborg, the most elaborate astronomical observatory the world had ever seen (Figure 4–9). Tycho was a gifted and dedicated observer. He concluded that astronomy's greatest need was better measurements, and he applied himself to this task for over twenty years.

Copernicus had not been much of an observer; he had generally used older measurements, many from Ptolemy himself. These measurements had often been extremely inaccurate, and at best had a precision of only 10 minutes of arc.* For example, a measured angle between Mars and Jupiter would have been stated as 30° 22′ ± 10′, indicating that the observing apparatus was not precise enough to pin down the value any more exactly than somewhere between 30° 12′ and 30° 32′.** Tycho at Uraniborg could follow the motions of the heavenly bodies with

* A minute is a measure of angle equaling 1/60 of a degree.

** 30° 12′ means 30 degrees, 12 minutes.

a precision of 4 minutes (approximately the angle subtended by the thickness of a dime held at arm's length—rather impressive for naked-eye measurements).

For twenty years Tycho measured the heavens at Uraniborg, accumulating the greatest treasure of precise planetary data that anyone had yet possessed. With it he intended to construct a definitive version of the motion of the planets, using his own scheme of the universe. Tycho was definitely not a Copernican. He argued that the heliocentric scheme was untenable on several grounds. It was against the Scriptures; it required too immense a stellar sphere; and it contradicted what every man could see and feel for himself—that the earth is at rest.

On the other hand, some features of the heliocentric model attracted Tycho. It accounted for the retrograde motion of the planets without retrograde epicycles. It also provided a natural explanation for the observed relations between the sun and planets, and enabled one to determine the sequence and relative sizes of the planetary orbits.

Tycho found a way to retain the best features of both the heliocentric and geocentric models. His scheme is illustrated in Figure 4–10. The earth remains at rest, the sun revolves around the earth, and the other planets revolve around the sun. This seemed to be the perfect model—the Copernican advantages are

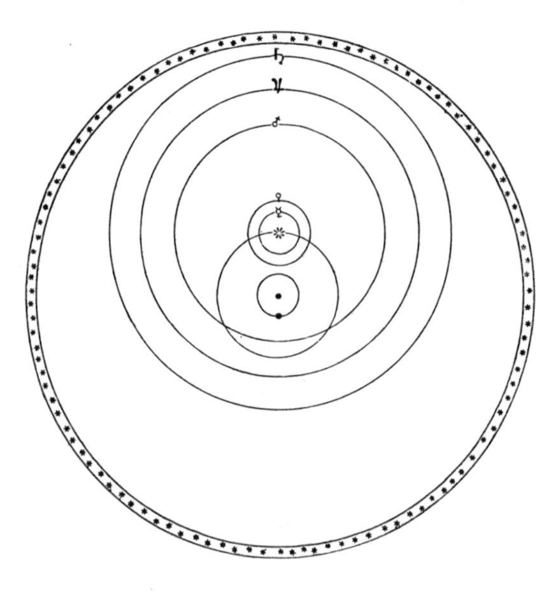

Figure 4–10. Tycho's illustration of his model, published in 1588. The sun and moon orbit a stationary earth, while the other planets orbit the sun.

retained, and yet the earth is not required to move. Tycho settled upon this model, using it as a framework for his data. Others followed his example, and soon it was almost universally adopted by astronomers.

But it was too late. It was becoming apparent that any one of several points of view—heliocentric, geocentric, or Tychonic—could encompass the observed motions of the heavens. Was there no way to choose between them on grounds more substantial than philosophical or aesthetic taste? Was there no way to determine how the heavens *really* move?

Kepler

In the end, progress was made not so much by accumulating new data as by asking new questions. In contrast with the sober reticence of Copernicus and the elegant precision of Tycho, Johannes Kepler (1571–1630) had a predilection for wide-ranging speculation and an attraction to mystical ideas. This led him to markedly different questions from those of Copernicus and Tycho, questions that would have a major impact on subsequent investigations of the planetary problem.

Kepler was born in Würtemberg, Germany, and studied at the University of Tübingen under Michael Mästlin, one of the foremost Copernican astronomers of the period. Having accepted the heliocentric model from his teacher, he published his first work, the *Mysterium Cosmographicum*, at the age of twenty-five. In this book he set forth and attempted to answer the question that was to underlie most of his life's work: Why is the universe the way it is and no other? In particular, why are there six and only six planets?

Of Kepler's approach, Mästlin wrote:

> The subject is new and has never before occurred to anybody. It is most ingenious and deserves in the highest degree to be made known to the world of learning. Who has ever dared before to think, and much less to try to expose and explain *a priori* and, so to speak, out of the hidden knowledge of the Creator, the number, order, magnitude and motion of the spheres?[3]

In fact, the Pythagoreans had raised the issue long before with their vision of a universe ordered by integers and the sacred Decad. Kepler's approach two thousand years later was remarkably similar. He sought to account for the number of planets by appealing to a simple geometrical theorem.

Plato's followers had proved that there are only five *perfect solids*—regular solids that can be constructed from regular polygons. They are shown in Figure 4–11: the tetrahedron (four triangles), the cube (six squares), the octahedron (eight triangles), the dodecahedron (twelve pentagons), and the icosahedron (twenty triangles). Kepler proposed that the number of planets is six *because* there are five and only five perfect solids.

Figure 4–11. The five perfect solids.

Figure 4–12 illustrates his idea. Let us imagine the planets as having circular orbits around the sun. We take Saturn's orbit to be the equator of a sphere. Inside this sphere we construct a cube so that the corners of the cube just touch the sphere. Then we construct another sphere inside the cube so that the faces of the cube just touch this inner sphere. The equator of this sphere would then be the orbit of Jupiter. Next we construct a tetrahedron, its corners just touching the sphere of Jupiter. Next comes the sphere of Mars, touching the faces of the tetrahedron, and so on, as shown in Figure 4–12. By choosing the proper order of the solids, Kepler hoped to find agreement with and thus explanation for the observed radii of the planets' orbits as well as the reason why there are six and only six planets. Since he began with Saturn's sphere on the outside and ended with Mercury's sphere on the inside, there was one more sphere than regular solid. Thus the fact that there are six planets follows from the fact that there are five perfect solids.

Of course, the planets did not move in perfect circles about the sun—Copernicus had had to use eccentrics or direct epicycles. So Kepler had to endow each sphere in his scheme with a thickness large enough to encompass the extremes of the correspond-

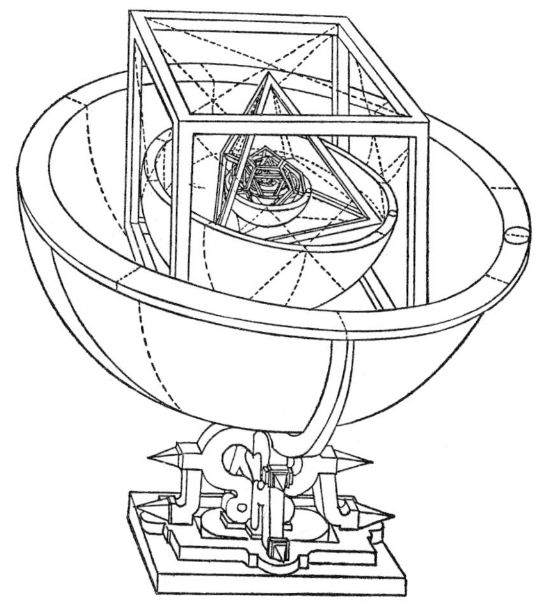

Figure 4–12. Kepler's perfect solids model of the solar system.

ing planet's orbit. After some trial and error, he arrived at an arrangement that was rather good—at least sufficiently good to allow him to feel continued enthusiasm for his scheme. Yet there were serious discrepancies. The spheres, chosen to fit the planetary orbits, could not be made to contact precisely the perfect solids. Perhaps the trouble lay in the orbits, which he had taken from the work of Copernicus.

In 1599 Kepler went in search of better data, which in that day meant to Tycho Brahe. For eighteen months they worked together, trying to reduce Tycho's measurements to accurate planetary orbits. Kepler's primary concern was Mars, the most intractable of the planets. After Tycho's death in 1601, Kepler continued the difficult task alone.

The model he eventually chose was heliocentric, but not quite the same as that adopted by Copernicus. The Copernican solution had two defects in Kepler's eyes. First, when Copernicus modified his original model (the basic heliocentric model) with direct epicycles, he calculated the motions of the planets, including Mars, with respect to the center of the earth's deferent, not the sun. But, Kepler argued, the sun was the largest of the heavenly bodies, and the source of all energy and thus life.

Shouldn't the deferents of the planets be related to the sun rather than the earth?

The second difficulty, Kepler felt, was Copernicus' arbitrary rejection of the use of the equant, a choice that complicated his solution immensely. Couldn't one find a simpler solution for Mars by using the simple equant device to control its motion?

To this end, Kepler chose four of the best of Tycho's measurements of Mars and searched for a solution of the form shown in Figure 4–13. The orbit is taken as a simple eccentric circle whose center C is displaced from the sun S; the motion of Mars is taken to be uniform with respect to the equant E. It soon became apparent that a solution could not be found using the Ptolemaic assumption that $CS = CE$. The equant position must be found by trial and error.

Five years later, after some nine hundred pages of calculation for seventy different trials, Kepler arrived at what first appeared to be a satisfactory solution. His orbit agreed with the four observations of Tycho to within two minutes of arc—excellent agreement considering that these data themselves were subject to uncertainties of about four minutes. Then he tested the solution against other data of Tycho's, and still the agreement was good—except for two particularly accurate points, which missed the calculated orbit. Kepler adjusted the orbit; he could get these points to fit, but then some of the formerly good points missed the curve. He found no solution for which at least some points did not miss the orbit by as much as eight minutes—twice the precision of the measurements. In the days of

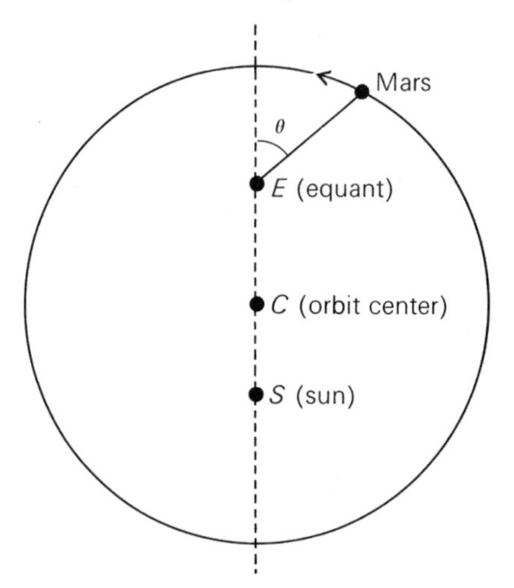

Figure 4–13. Kepler's original attempt to fit the motion of Mars, using both an equant and an eccentric.

Copernicus, this would have been a satisfactory fit, for the old data were only good to ten minutes. But Tycho had done better, and, as Kepler said years later:

> It is fitting that we should acknowledge this divine gift and put it to use . . . For, if I had believed that we could ignore these eight minutes, I would have patched up my hypothesis accordingly. But since it was not permissible to ignore them, those eight minutes point the road to a complete reformation of astronomy . . .[4]

Two possibilities existed: either to continue on the same general path with a new complication—letting the equant oscillate along the line *SCE*—or else to give up the idea of circular motion entirely. Kepler's historic choice was the latter, requiring a complete rejection of the previous five years' labor. He set about the fresh new task of simply determining the orbit of Mars with respect to the sun, abandoning any *a priori* assumption of uniform circular motion. His result seems almost anticlimactic, although it was of the utmost importance for further developments: the orbit of Mars could be represented by an ellipse. All Tycho's data fitted this shape in the sense that each data point was no further from the curve than four minutes—the uncertainty in the measurement. Kepler noted a further striking feature—the sun was located at a focus of the ellipse (Figure 4–14).

Figure 4–14. An ellipse. F_1 and F_2 are the foci. An ellipse is a figure such that for each of its points P, the sum of the distances to the two foci is the same; that is, $F_1P + F_2P$ is constant. It may be readily constructed by tacking the ends of a string at points F_1 and F_2, pulling the string taut with the point of a pencil (at P in the figure), and tracing out the curve. All points P on the curve satisfy the relation:

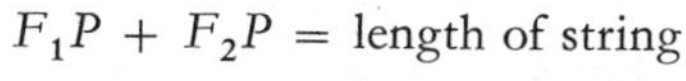

$$F_1P + F_2P = \text{length of string}$$

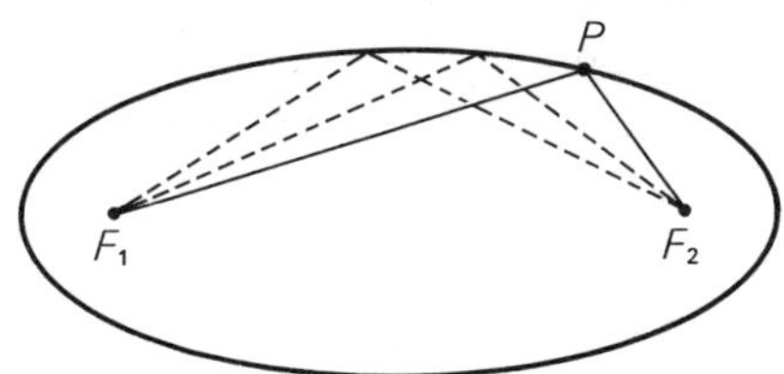

Kepler's Laws

Kepler followed the same approach with the measurements for the other planets and the earth. In each case he concluded that the orbit could be represented by an ellipse with the sun at a focus. This observation is known as Kepler's First Law: *The planets move in elliptical orbits around the sun, with the sun at one focus.*

Such a result must have been doubly satisfying to Kepler. Having given up trying to force the planets into circular motion, he found ellipses appearing almost by themselves; the new shape was still a relatively simple geometric one and fitted nicely with his Pythagorean predispositions. But to have the sun appear at such a special point in the ellipses was even better, for if the sun controlled the motions of the planets, what more fitting place for it?

One aspect of the planets' motion, however, remained unclear. Uniform circular motion had been given up and ellipses had replaced circles, but what about the speed of the motion? The data indicated that the speed varies as a planet moves around its orbit, and the sun is clearly involved—the planet moves most rapidly when closest to the sun. Eventually Kepler hit upon the simple result shown in Figure 4–15. It is now known as Kepler's Second Law: *The line drawn from the sun to a planet sweeps out equal areas in equal times.*

But why ellipses and why equal areas? Such questions had not been asked about planetary motion since Greek times. With the abandoning of uniform circular motion, they again came to the fore. Kepler felt the answers must have something to do with the sun. Perhaps the sun was somehow responsible for planetary motions; perhaps the sun actually made the planets move. Kepler thus raised an entirely new issue. If he could find out how the sun made the planets move, then possibly he could explain the ellipses and the equal areas.

Kepler's solution was to postulate that the sun possessed what he called an *anima motrix*, a kind of force that emanated

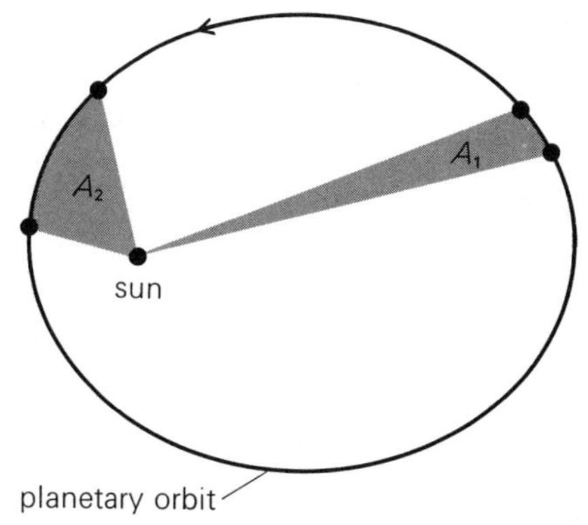

Figure 4–15. Kepler's Second Law. The areas A_1 and A_2 are swept out by the line from the planet to the sun in equal time intervals. Since Kepler's Second Law implies $A_1 = A_2$, the planet must move farther during the time interval when it is nearer the sun. Thus it moves faster when it is nearer the sun.

from it and pushed the planets around in their orbits. This idea fitted nicely with the equal areas law. If the force originated in the sun, then it presumably grew stronger when the planet approached the sun. The closer it came, the more rapidly it would be driven. By the same token, when the planet was farther from the sun, the force would be weaker and the planet would be driven more slowly. A possible candidate for such a force was magnetism, which was much discussed at the time because of the researches of William Gilbert in England. Magnetism seemed to act across large distances without needing physical contact like other forces, such as the pull of a spring or a rope. Furthermore, it was known that the earth itself had magnetic properties. Kepler became convinced that magnetism ran the solar system; he even put forth an explanation for the elliptical shape of the orbits in terms of that force. His specific arguments are no longer convincing today. What is significant for our study is that Kepler considered, for the first time, a physical explanation of a less mechanical sort than the clockwork model of Aristotle or the ball-bearing system of the Arabs.

In considering the influence of the sun on the planets, Kepler discerned yet another regularity in planetary motion. If the *anima motrix* propelling the planets emanated from the sun, shouldn't the planets closer to the sun receive more of a push than the ones farther out? Kepler noted that indeed the farther a planet is from the sun the longer it takes to complete a full orbit. Table 4–1 lists the average distance r from each planet to the sun, and the period T, the time the planet takes for one orbit.

Table 4–1. Modern data for the periods of the planets and their distances from the sun.

Planet	r (a.u.*)	T (years)
Mercury	.39	.241
Venus	.72	.615
Earth	1.00	1.00
Mars	1.52	1.88
Jupiter	5.20	11.68
Saturn	9.54	29.5

* r is measured in astronomical units (a.u.), where one a.u. is the average distance from the sun to the earth.

Could there be some simple numerical relationship between T and r? Kepler believed there was. Like the Pythagoreans long before him, he felt that there were mystical connections between mathematics and the universe. This intuition motivated his perfect solids model and his search for a simple relation between T and r.

Let us construct a graph of period T vs. average distance from the sun r, as shown in Figure 4–16. The data of Table 4–1 are indicated by crosses on the graph. It is clear from the table that as the distance increases, the period also increases. But is there a simple mathematical relation between them? The spirit of Kepler's approach was simply to try various ways T might depend on r. For example, two simple ways in which T might increase with increasing r are:

$$(1) \quad T = k_1 r$$

$$(2) \quad T = k_2 r^2$$

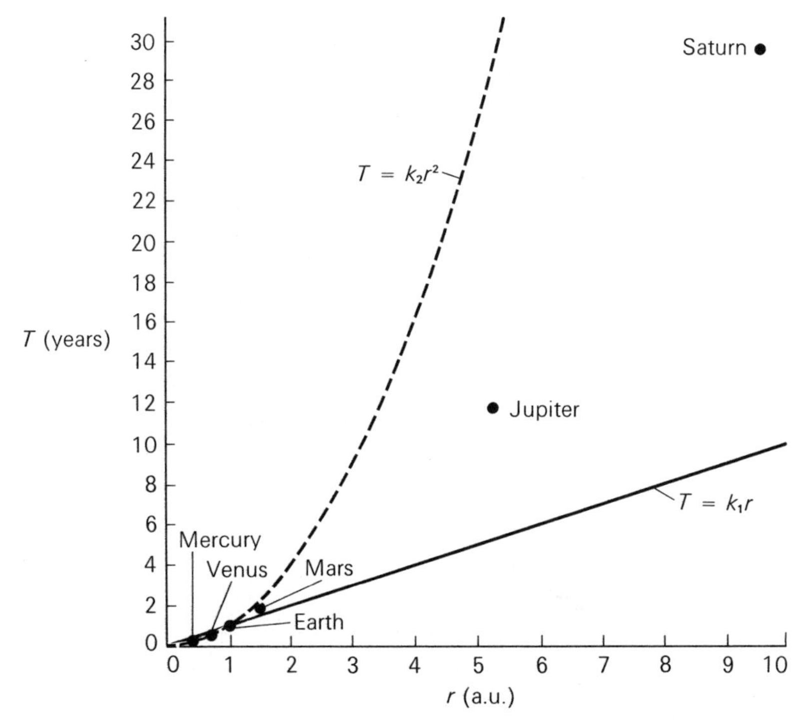

Figure 4–16. Two unsuccessful attempts to fit the planetary data with simple T vs. r relationships. The solid line assumes that T is proportional to r, while the dashed line assumes that T is proportional to r^2.

where k_1 and k_2 are constants to be determined from the data.

Figure 4–16 shows how these relations fit the planetary data.* Clearly, neither does very well. T proportional to r^2 rises too rapidly for increasing r, while T proportional to r does not rise rapidly enough. But Kepler persevered in the search. He tried a curve with a gentler increase than r^2, yet stronger than r, namely:

$$(3) \quad T = k_3 r^{3/2}**$$

This curve is shown in Figure 4–17. It fits the planetary data beautifully. For some reason there does seem to be a very

* We have chosen the constants k_1 and k_2 for the curves by requiring the curves to pass through the point representing the earth.

** $r^{3/2}$ means "take the square root of r and cube the result."

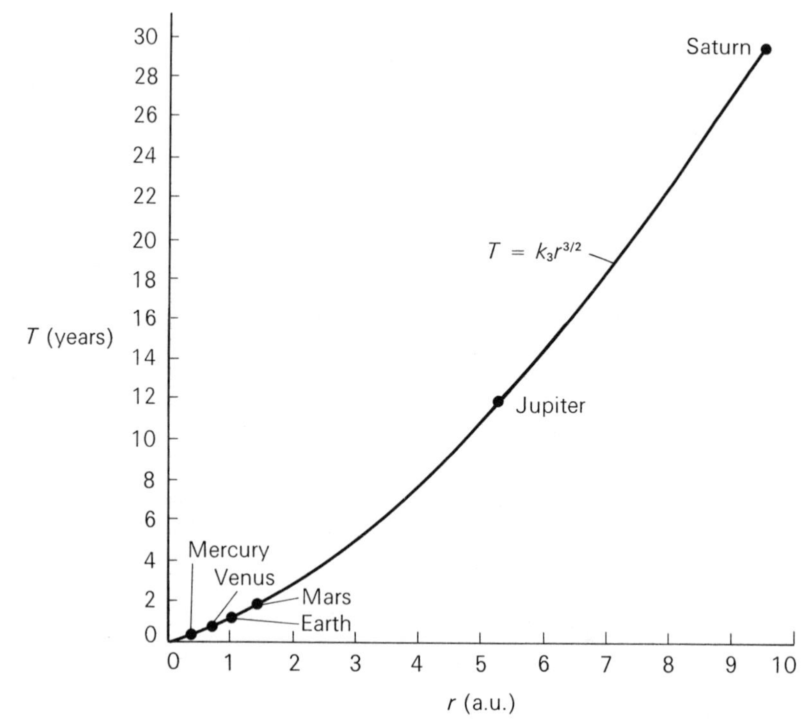

Figure 4–17. Kepler's successful T vs. r relationship for the planets. The solid line assumes that T is proportional to $r^{3/2}$.

simple relation between the period of a planet's motion and the size of its orbit. This relation is often written in the form:

$$T^2 = kr^3$$

It is Kepler's Third Law: *The square of the period of a planet is proportional to the cube of its average distance from the sun.* Although the planets differ in many physical properties, the periods of their orbits seem to depend on none of these properties. The periods depend only on the distances from the sun, and in a particularly simple manner.

The significance of Kepler's three laws warrants special emphasis. Not only did the laws describe the planetary orbits and motions in a particularly simple way, but in addition, each law applied in the same way to each planet. Kepler had uncovered a unity more fundamental than the general statement that each planet moves around the sun. Now, it appeared, each planet moved in the same way around the sun.

Despite these accomplishments, Kepler never succeeded in his attempt to understand the planets in terms of the five perfect solids. Yet, ironically, his pursuit of this goal was responsible for the discovery of a unity in the behavior of the planets, a unity that cried out for some kind of explanation. Indeed, as we shall see, Kepler's laws were a key element in the formulation of Newton's theory of planetary motion. They thus provided the empirical foundation for an understanding of the planets far different from the conception that motivated their discovery.

Descartes

A notably different explanation of planetary motion was advanced by the French philosopher and mathematician, René Descartes. Working with the same data as Kepler and assuming the same sun-centered planetary system, Descartes asked different questions and was directed by a different set of philosophical prejudices. The contrast between his theory of the planets and those of Kepler and Newton provides a good illustration of the inevitable ambiguity in the development of physical theories.

Descartes was born in France in 1596. After eight years at a Jesuit school in Anjou and one or two more at the University of Poitiers, he spent much of his early manhood wandering about Europe, joining an army here and there, moving from city to city in search of new experiences. Finally, at the age of thirty-two, he settled in Holland and began serious work on the questions that had long been bothering him, questions about the foundations of philosophy and mathematics.

In 1633, five years later, Descartes completed his first work, *Le Monde*, in which he presented his explanation of the motion of heavenly and terrestrial bodies. However, in that same year the Italian astronomer and physicist Galileo Galilei was condemned by the Church in Rome and his *Dialogues Concerning the Two Chief World Systems* was banned, primarily because it contained the heretical notion that the earth moves around the sun. On hearing of the fate of Galileo, Descartes decided to withhold *Le Monde* from publication rather than risk its suppression or modification by the Church. Consequently, his planetary theory did not actually appear in print until much later, with the publication in 1644 of his *Principia Philosophiae*.

For Descartes, the central question about the planets was: "What is the mechanism causing the planets to orbit the sun?" In his view a theory of the planets was a theory that would answer this question. Furthermore, he had very definite ideas about what constituted a satisfactory answer. He dismissed as appeals to the supernatural such explanations as Kepler's *anima motrix* or similar forces of attraction that had been proposed by others. Postulating such forces without describing the means by which they acted seemed no explanation at all to Descartes. He required a detailed description of the physical mechanism that causes the planets to orbit the sun.

Descartes' formulation of the planetary problem was strongly influenced by his conviction that there was no such thing as a void, a region of space in which there is no matter. He argued quite plausibly that the notion of a void is a contradiction in terms. Either there is something or there is nothing. It is difficult to conceive of what could be meant by a region of space containing nothing. To Descartes, the notions of extension (space) and substance (matter) were indivisible. One could not exist without the other.

Descartes believed that all of space is filled with matter, which he hypothesized to be of three different kinds. The first element, Fire, is the constituent of the stars. It consists of finely divided particles in rapid motion, which are the source of all light. The matter making up the bulk of the heavens is the second element, Air, composed of spherical, transparent particles through which light may be transmitted. Fine particles of the first element fill the interstices between the closely packed spheres so that the heavens are completely filled with matter. Large, odd-shaped, and slowly moving particles of the third element, Earth, are the constituents of the earth, the moon, and the other planets.

Within the framework of a universe filled with matter, Descartes constructed an intuitively plausible model of planetary motion. In this model, the sun is located at the center of a

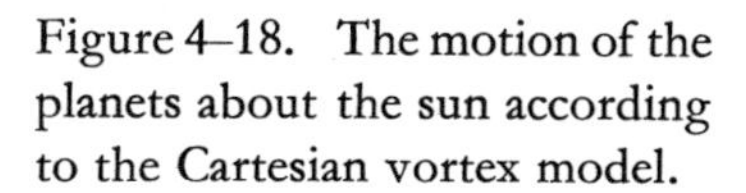

Figure 4–18. The motion of the planets about the sun according to the Cartesian vortex model.

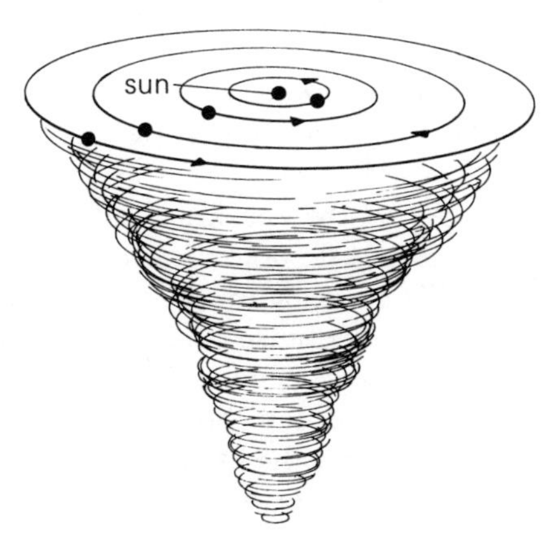

whirlpool of the second element, Air, as depicted in Figure 4–18. The earth and other planets are embedded in this matter and hence are swept around the sun by the vortex motion of the surrounding medium. Thus the motion of the planets can be understood in appealingly simple mechanical terms by anyone who has ever seen a whirlpool or water emptying down a drain.

Descartes pointed out that the vortex motion of whirlpools is rarely precisely circular. Consequently, the model is consistent with Kepler's conclusion that the paths of the planets are not perfect circles.

Perhaps more intriguing is his argument that his model was consistent with the Church's requirement that the earth be at rest. Descartes' universe was completely filled with the three elements, so that a body was always in contact with other particles of matter. In accordance with this conception, Descartes *defined* the motion of a body to be the transfer of that body from contact with one group of particles to contact with another group of particles. In the Cartesian model of the solar system, not only the planet but also its entire surrounding medium travels around the sun. Therefore, Descartes argued, the planet does not really move:

> No motion, in the proper sense of the word, is to be found in the Earth, or in the other planets, because they are not at all transported from the vicinity of the parts of the Sky that touch them . . .[5]

This definition of motion plays a central role in Descartes' entire formulation of physics, particularly in his discussion of the motion of bodies at the surface of the earth. Thus one cannot doubt his sincerity when he wrote:

> As to the censure of Rome, concerning the motion of the earth, I do not in this [Descartes' theory] see any resemblance. For I emphatically deny this motion . . . And all the passages from Scripture which do not recognize the motion of the earth are not considering the system of the world but only the mode of expression. So that, when I prove, as I do, that, properly speaking, the earth does not move, if motion is explained in the way I explain it, I completely satisfy these passages.[6]

There is much more to Descartes' description of the universe. For example, every star was supposed to have a vortex associated with it, and Descartes went to great pains to explain how the vortices of adjacent stars would have to be oriented so that they could coexist alongside one another. The origin of planets and comets was explained as an encrusting of stars by the third element and a subsequent breakdown of their vortices. Minor vortices surrounded the planets to account for the motion of their moons. The motion of the three elements near the surface of the earth explained qualitatively why objects fall toward the earth.

We shall not concern ourselves with the details of Descartes' physics. It is sufficient to note that he developed a theory of universal scope, explaining at once the motions of bodies in the heavens and on the earth.

To be sure, the Cartesian explanation of planetary motion was largely qualitative. Yet it was not clear that a more quantitative explanation was possible, and Descartes' theory did provide a plausible mechanism for the motion of the planets, a feature lacking in Kepler's perfect-solids model and in Newton's theory, which we shall soon discuss.

Suggested Reading

Koestler, A. *The Sleepwalkers*. New York: Macmillan, 1959. Contains extensive biographical material on Copernicus and Kepler and their part in the planetary problem. Beautifully done; reads like a novel.

Cohen, I. *The Birth of a New Physics*. New York: Doubleday, 1960. An excellent brief discussion of the work of Copernicus and Kepler.

Rosen, E. *Three Copernican Treatises*. 2nd ed. New York: Dover, 1959. A widely available sample of the writings of Copernicus.

Butterfield, H. *The Origins of Modern Science*. London: Bell, 1949. Chapter 2 discusses "The Conservatism of Copernicus."

Questions

1. In Figures 4–2 and 4–3, draw in the arrows for the remainder of the year. Use your results to discuss how the basic heliocentric model explains the observation that the sun moves all the way around the ecliptic in a year.
2. Continue the construction in Figure 4–5 for four more equally spaced time intervals, t_5, t_6, t_7, and t_8. Does Mercury appear to retrogress during this period?
3. Suppose that Mercury and the earth take the same time to orbit the sun and at time t_1 they are in the positions shown in the figure. Construct the analogues of Figure 4–5(a) and (b) for one full year and thus determine how Mercury would appear to move relative to the earth in this case.

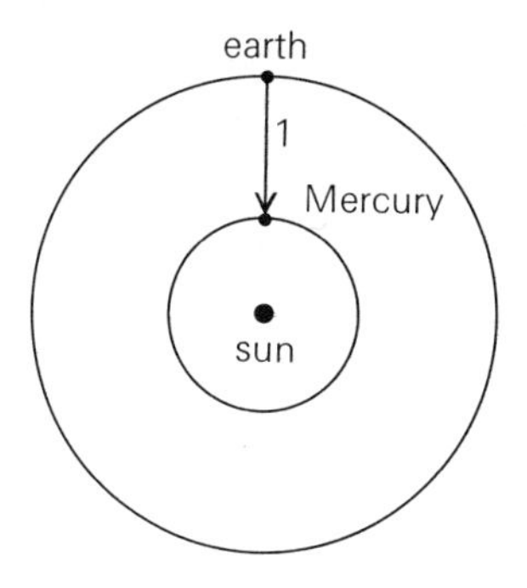

4. Same as Question 3, but suppose that the earth takes exactly four times as long to orbit the sun as does Mercury.
5. Same as Question 3, but suppose that the earth takes exactly half as long to orbit the sun as does Mercury.
6. By drawing a picture representing the motion of the earth and Venus in the basic heliocentric model, show how this model explains the observation that Venus is always a morning or an evening star, i.e., that Venus is always observed in the sky near the sun. Show also that the model does not predict the same behavior for an outer planet such as Mars.
7. Summarize the advantages of the Copernican model over that of Ptolemy and vice versa.
8. Find the ratio of the radii of the spheres that will fit inside and outside a cube. In Kepler's perfect sphere model this should correspond to the ratio of the radii of two planetary orbits. Compare your result with the planetary radii given on page 92. Are there any likely candidates?
9. Could Ptolemy's data, with its larger uncertainties, be fitted with Kepler's ellipses? If so, what was the effect of Tycho's increased precision of measurement?
10. Can experimental data prove the validity of a theory? For example, did the fact that Tycho's data for Mars coincided with Kepler's ellipse (within the 4 minutes uncertainty in the measurements) prove that the orbit was indeed an ellipse? Suppose the data had been one thousand times more precise (each measurement was characterized by an uncertainty of only .004 minutes) and it still coincided with the ellipse. Would that have *proved* that the orbit was an ellipse?
11. The maximum distance of a particular planet from its sun as it moves around its orbit is 10 million miles. The minimum distance is 5 million miles (see sketch). On the day when the

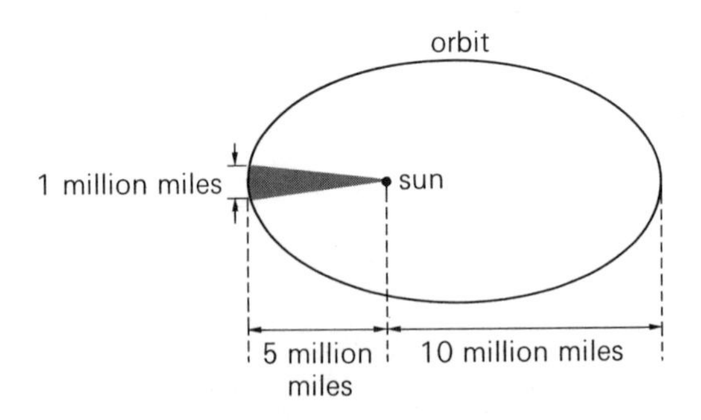

planet is closest to the sun, it moves 1 million miles, sweeping out the area indicated on the sketch. Approximately how far will the planet move on the day when it is farthest from the sun? (Notice that the area indicated is very nearly triangular, and recall that the area of a triangle is $\frac{1}{2} \cdot \text{base} \cdot \text{altitude}$.) What can you say about the speeds of the planet on these two days?

12. How would you respond to the statement, "Kepler's third law is obvious. Of course the planets farther away from the sun have longer periods. They have farther to travel." (What assumption about how a planet's speed depends on its distance from the sun is implicit in this statement? If one makes this assumption and further assumes the planet's orbit to be a circle—a special case of an ellipse—how would the period of a planet's motion be related to its distance from the sun?)

13. In another universe, there exists a star around which four planets *A*, *B*, *C*, and *D* move in circular orbits. Three different observatories on the star make measurements of the periods and radii of the four planetary orbits; their data and some calculated numbers are listed below. (Note that it is apparently easy to measure the periods with precision, but the radius measurements are more difficult.) What could you conclude about the relation between T and r from this data?

	T (years)	r (million miles)	T^2/r^2	T^2/r^3	T^2/r^4
Planet *A*	1.3	2.0	.42	.21	.105
	1.3	2.3	.32	.14	.060
	1.3	2.4	.29	.12	.051
Planet *B*	5.2	8.5	.37	.044	.0051
	5.2	8.3	.39	.047	.0056
	5.2	8.1	.41	.051	.0062
Planet *C*	8.7	12.3	.50	.081	.0066
	8.7	15.0	.34	.022	.0015
	8.7	13.0	.45	.034	.0027
Planet *D*	15.5	22.1	.49	.022	.0010
	15.5	25.0	.38	.015	.0006
	15.5	18.5	.70	.038	.0021

14. Descartes' whirlpool model and Kepler's perfect sphere model were both put forth as explanations of planetary observations. Compare and contrast these explanations. Which do you find more satisfying? Why?
15. The two figures are two different representations of modern planetary and solar data.[7]
    (a) Which figure presents the data *correctly*? Why?
    (b) Imagine yourself to be a physicist investigating the motion of the earth, sun, and planets. Consider these figures alone and your understanding of what "investigating the motion" means to a physicist. Which figure would interest you the most? Why?

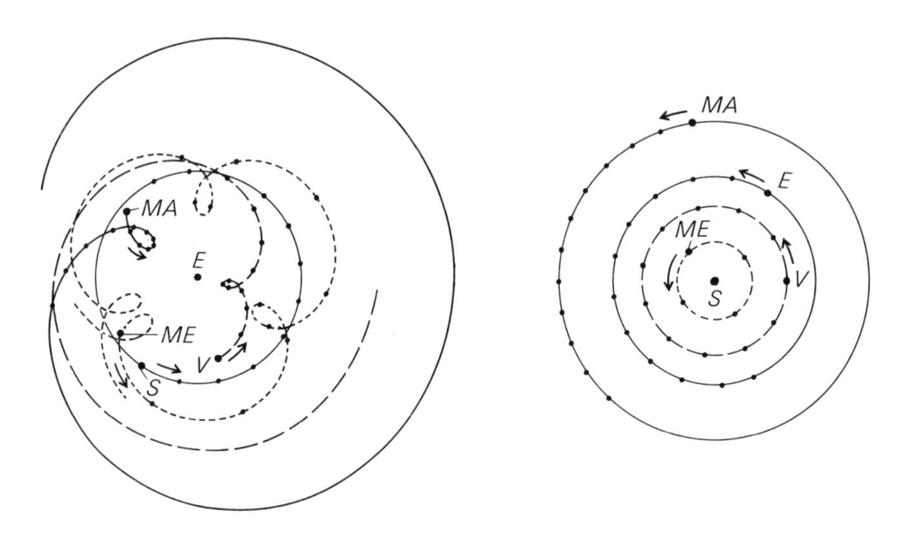

THE PROBLEM OF TERRESTRIAL MOTION

How comes it that some things are moved continuously, though that which has caused them to move is no longer in contact with them, as, for instance, things thrown?[1]

—Aristotle
Physics

In the stone or other projectile there is impressed something which is the motive force of that projectile . . . The projector impresses in a moving body a certain impetus . . . It is by that impetus that the stone is moved after the projector ceases to move.[2]

—Jean Buridan
Questions on the Eight Books of the Physics of Aristotle (1509)

When you throw it with your arm, what is it that stays with the ball when it has left your hand, except the motion received from your arm which is conserved in it?[3]

—Galileo Galilei
Dialogues Concerning the Two Chief World Systems (1632)

What sustains the horizontal motion of a projectile? What keeps it from falling straight down after it is released?

CHAPTER FIVE

A Language of Motion

The principal obstacle to general acceptance of the heliocentric universe remained men's reluctance to accept the idea of a moving earth. Many observations of the movement of objects at the surface of the earth seemed clearly incompatible with the motion of that earth. For example, a stone dropped from the top of a tower falls directly downward along the side of the tower. It was argued that if the earth were moving as the stone fell, the point of impact would be a great distance from the foot of the tower. Another observation frequently cited was that a cannonball shot eastward has about the same range as one shot westward. It was argued that if the earth were rotating from west to east, the ball shot eastward would be followed by the earth and travel only a short distance, while the ball shot westward would go much farther, since the earth would be moving eastward under it. These considerations and others like them seemed to preclude any motion of the earth. Even Kepler's seductively simple heliocentric ellipses could not overcome these down-to-earth difficulties.

To follow the changes in men's ideas that removed this obstacle, we must develop an appropriate language. In the study of celestial phenomena, the simple principle of uniform circular motion sufficed up to the time of Kepler. This principle is clearly not applicable to the varied motions of objects on the earth. Objects speed up and slow down, drop, collide, and roll. Such complex behavior would be difficult to visualize as a combination of uniform circular motions. To describe terrestrial motion, we must begin anew.

The language we need is quantitative. Although discussions of terrestrial motion began with qualitative, verbal concepts, from about the time of Kepler onward the quantitative, numerical aspects provided the greatest insights. Let us begin with an examination of the basic vocabulary and syntax of this quantitative language.

Motion in a Straight Line

Consider a typical terrestrial motion—an apple falling toward the surface of the earth (Figure 5–1). To describe the motion of

Figure 5–1. A falling apple.

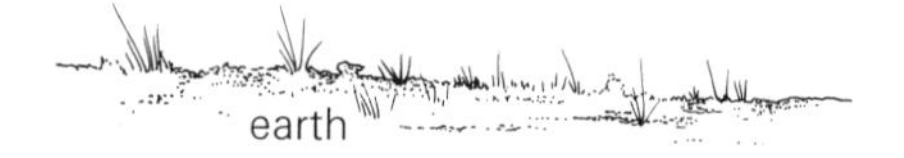

the apple, we indicate its *position* at each instant of *time*. The first step in a quantitative description of the motion is therefore to establish numerical specifications of position and time.

To specify position, it is convenient to establish a *frame of reference*. We simply lay along the direction of motion a rigid rod inscribed with equally spaced marks. The distance between the marks is arbitrary. Suppose that ours are 1 centimeter apart.* Let us define a *position coordinate* y as follows: We choose some arbitrary mark and call it $y = 0$; the mark immediately below is $y = 1$ centimeter; the next mark is $y = 2$ centimeters, etc., as shown in Figure 5–2. Similarly, the mark immediately above $y = 0$ is labeled $y = -1$ centimeter, the next mark $y = -2$ centimeters, etc. To specify the position of the center of the apple we simply specify the corresponding mark on the reference frame (i.e., on the rod). In Figure 5–2, the center of the apple is at $y = +5$ centimeters.

Of course, the choice of which mark to call $y = 0$ is arbitrary. We could have an equally satisfactory reference frame with any other mark chosen as $y = 0$.

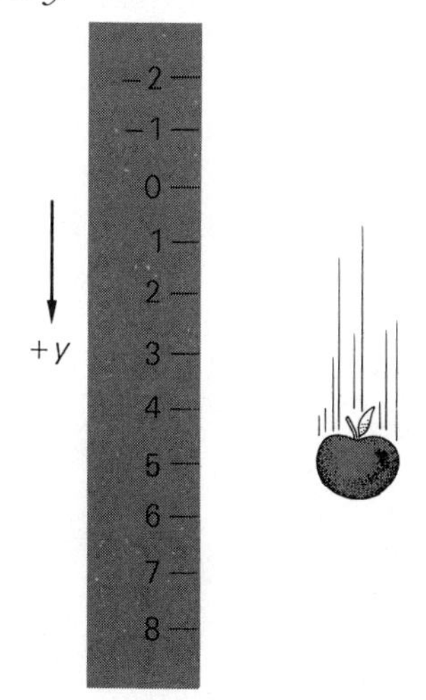

Figure 5–2. A frame of reference, consisting of a rigid rod with equally spaced marks. The center of the apple is at $y = +5$ centimeters.

* A centimeter is one hundredth of a meter; there are 2.54 centimeters in an inch. Centimeter is often abbreviated "cm" and meter "m."

We should note one other convention. An object moving along the line specified by the y-coordinate can either be moving from smaller values of y toward larger values (down in Figure 5–2) or from larger values of y toward smaller values (up in the figure). The former direction is conventionally called the $+y$ direction and the latter the $-y$ direction. Note that our choice of downward as the $+y$ direction is arbitrary; we could just as well have turned our measuring rod around, making the upward direction positive. With the convention we have chosen, we say that the apple in Figure 5–2 is moving in the $+y$ direction.

To specify time, we employ a periodic phenomenon such as a clock that ticks at equally spaced intervals. Suppose that our ticks are one second apart. We choose some tick and call it $t = 0$. The next tick will be $t = 1$ second, the next $t = 2$ seconds, etc. Similarly, the tick immediately preceding $t = 0$ is $t = -1$ second, the one before that is $t = -2$ seconds, etc. Using such a clock, we can associate a time with any event such as "center of the apple passes position $y = 5$ centimeters."

Of course, the choice of which tick to call $t = 0$ is arbitrary. We could have an equally satisfactory time scale with any other tick chosen as $t = 0$.

To these basic ideas of position and time, we can add a further concept that is closely related to them—*speed*. In common usage, the speed of an object expresses how far it goes in how much time. If a car moving steadily down the road covers 150 miles in 3 hours, we say its speed is 50 miles/hour. We shall say more about the concept of speed later. The primitive notion of distance divided by time will suffice for now.

Descriptions of Motion

Let us return to the falling apple and note that there are a number of alternative ways to describe its motion. We begin by assuming that it is falling with a constant speed, e.g., 10 cm/sec. (Of course, apples generally do not fall with a constant speed. We imagine one that does—perhaps because it has a parachute attached to it or is falling in a tank of oil—to begin with a case that is easy to discuss mathematically.) The following means could all be employed to describe the motion of the apple:

1. *Words.* We might say, "The apple is moving with a constant speed of 10 cm/sec in the $+y$ direction. At time $t = 0$ its center was at position $y = 5$ cm." In this simple case of motion with constant speed, the statement is quite succinct. But if the apple had been erratically changing its speed or its direction of motion, a great many more words might have been required to describe how its position changed over even a very short time interval. In that case, a verbal description would obviously be a very clumsy way to describe the motion.
2. *Pictures.* We could take pictures of the apple at regular intervals of time. Figure 5–3 represents a superposition of snapshots of the apple taken at one-second intervals. Once again, if the apple were moving more erratically, the snapshots would have to be taken more frequently to capture the full complexity of the motion.

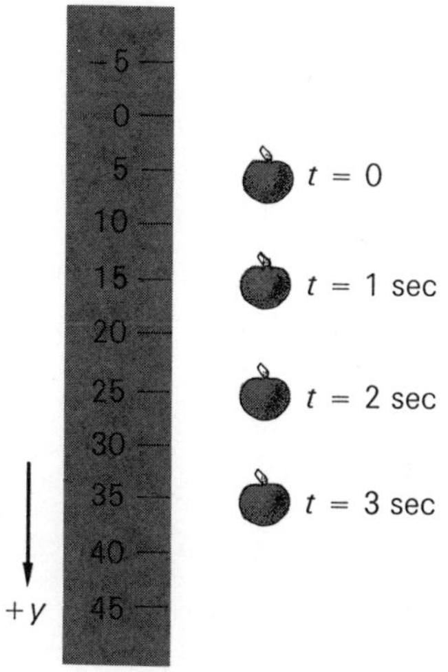

Figure 5–3. Multiple exposure of an apple falling with a constant speed of 10 cm/sec.

t (sec)	y (cm)
0	5
1	15
2	25
3	35

Table 5–1. Table of t and y for an apple falling with a constant speed of 10 cm/sec.

3. *Table of* y *vs.* t. A complete description of the motion would entail specifying the position y at every time t.

One way to display this information is a table like Table 5–1, which lists corresponding values of t and y. Again, if the motion is erratic, the table will have to be very long to show all the details.

4. *Graph of* y *vs.* t. A more convenient way to display the relation between y and t is a graph. Figure 5–4 shows one for our simple constant-speed example. It is a straight-line graph, corresponding to a steady motion of 10 centimeters each second (illustrated for two one-second intervals). Note that the values of y at times $t = 0, 1, 2,$ and 3 sec are 5, 15, 25, and 35 cm respectively, just the values listed in Table 5–1. In addition, the graph shows the position of the apple at all intermediate times.

   From either triangle in Figure 5–4, one can see that the *slope* of the straight line is 10 cm/1 sec = 10 cm/sec—just the speed of the apple. The *y-intercept* of the straight line is 5 cm, the position of the apple at $t = 0$. (Appendix 1 reviews the elementary properties of straight-line graphs.)

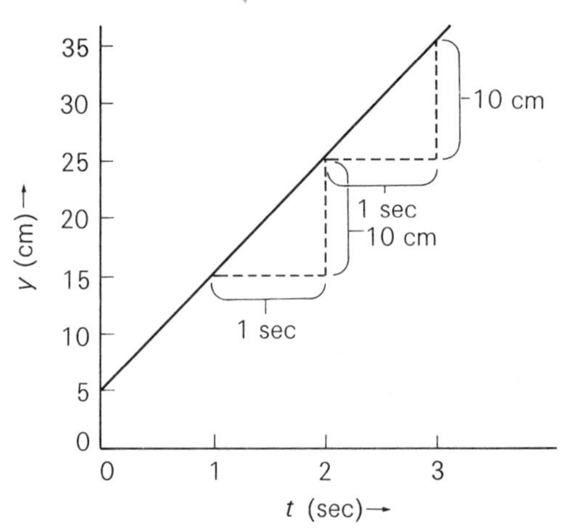

Figure 5–4. Graph of y vs. t for an apple falling with a constant speed of 10 cm/sec. It is a straight line with slope 10 cm/sec and y-intercept 5 cm.

5. *Equation.* In the simple case under consideration, there is another, very succinct way to express the relation between y and t—an algebraic equation:

$$y = 10t + 5 \qquad (y \text{ in cm, } t \text{ in sec})$$

(This is just the usual equation for a straight line, where 5 cm is the y-intercept and 10 cm/sec is the slope. If this is not clear, consult Appendix 1.) For complex motion, the algebraic relation between y and t may be extremely complicated. For all the examples we shall encounter, however, there will be a simple equation relating y and t.

Consider a car moving along a straight road. Let x designate the position of the center of the car. Describe in words and equations the motion of the car corresponding to each graph below.

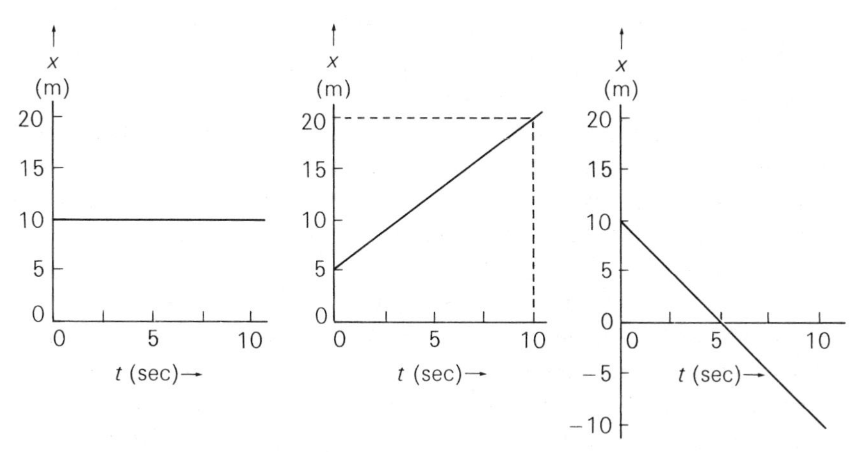

In each graph below, compare the motions of a car indicated by the three lines 1, 2, and 3.

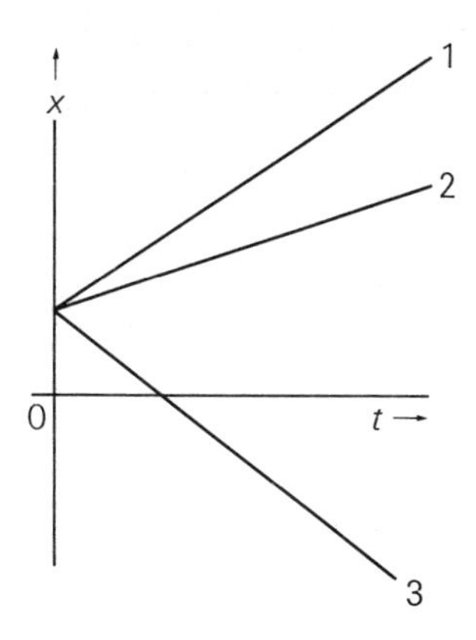

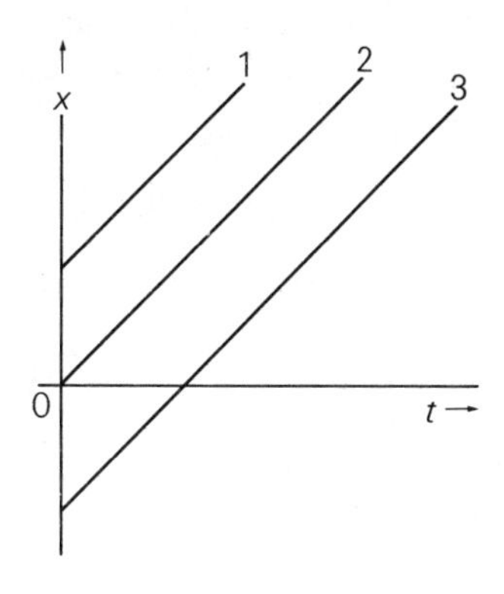

Describe in words the motions depicted in the graphs below.

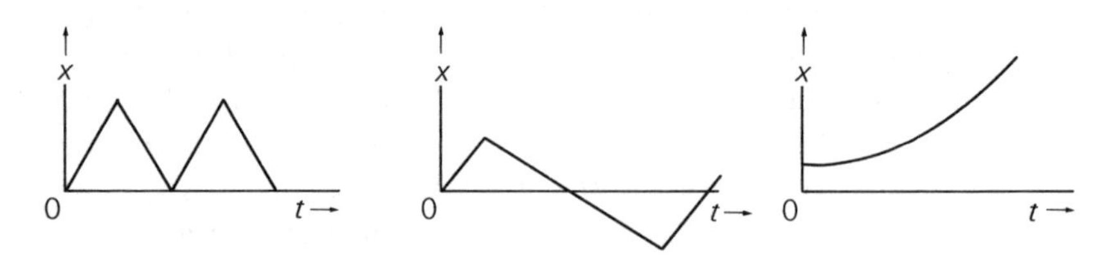

Velocity

In the last section, the notion of speed as distance traveled divided by time elapsed was employed to indicate how fast an object was moving. We considered a particularly simple example, an apple falling with constant speed. We now introduce a more general concept, *velocity*, which indicates not only how fast an object is moving, but the direction of its motion as well. This concept will prove useful in describing more complex motions than that in the apple example.

Let us review the description of motion we have developed. To specify position, one chooses a reference frame. The motion of an object is then described by indicating its position at each instant of time. A graph is a particularly succinct way to communicate this information. For example, Figure 5–4 describes the motion of our falling apple.

Suppose we are given such a description of the position y of an object at each time t. How can we determine how fast the object is moving? Consider a time interval Δt from an initial time t_i to a final time t_f:*

$$\Delta t = \text{final time} - \text{initial time} = t_f - t_i$$

For example, the time interval from $t = 2$ sec to $t = 3$ sec is

$$\Delta t = 3 \text{ sec} - 2 \text{ sec} = 1 \text{ sec}$$

* We use the Greek symbol Δ (delta) as a shorthand notation to specify a change in a quantity. Δx will always be defined to mean "final x minus initial x," whatever x symbolizes.

Suppose that the object is at position y_i at time t_i and has moved to position y_f at time t_f. Let Δy be the change in position of the object during this time interval Δt:

$$\Delta y = \text{final position} - \text{initial position} = y_f - y_i$$

For example, in Figure 5–4, the position at $t = 2$ sec is $y = 25$ cm; at $t = 3$ sec it is $y = 35$ cm. Therefore,

$$\Delta y = 35 \text{ cm} - 25 \text{ cm} = +10 \text{ cm}$$

Then we may define the *average velocity* $\bar{v}$ of the object during this time interval as the change in position Δy divided by the time interval Δt:

$$\text{average velocity } \bar{v} = \frac{\text{change in position}}{\text{time interval}} = \frac{\Delta y}{\Delta t}$$

As an example, let us calculate the average velocity of the apple in Figure 5–3 during the time interval from $t = 2$ sec to $t = 3$ sec. We saw above that for this interval, $\Delta t = 1$ sec and $\Delta y = 10$ cm. Thus

$$\bar{v} = \frac{\Delta y}{\Delta t} = \frac{+10 \text{ cm}}{1 \text{ sec}} = +10 \text{ cm/sec}$$

Hence the average velocity of the apple during this interval is $+10$ cm/sec.

> Consulting Figure 5–4, calculate the average velocity of the apple for the time interval from $t = 0$ to $t = 1$ sec; from $t = 0$ to $t = 2$ sec. Compare with the average velocity calculated above.

The result of this exercise illustrates an important property of constant-speed motion, the general validity of which follows from the fact that the corresponding position vs. time graph is a straight line: For the case of constant-speed motion, the average velocity is the same for any time interval. Or, to put it slightly differently, we find the same average velocity no matter how long or how short a time interval we use in the calculation.

Why have we gone to so much trouble to define such an elaborate concept as average velocity, when our intuitive notion of the speed of the apple furnishes the same value, 10 cm/sec? In fact, there is one small but very important difference between the two ideas. The concept of velocity automatically includes an additional piece of information: the direction of the motion. A positive $\bar{v}$ implies that the change in position during the time interval was in the direction chosen as positive; a negative $\bar{v}$ implies that the change in position was in the negative direction. Thus the positive value of $\bar{v}$ in the example above carries important information—that the motion is in the $+y$ direction.

On the other hand, consider an object moving upward (in the $-y$ direction) with a constant speed of 10 cm/sec. It begins at position $y = 5$ cm at time $t = 0$. Figures 5–5(a) and (b) illustrate its motion. Let us calculate the average velocity of the object during the one-second time interval from $t = 2$ sec to $t = 3$ sec. Here the change in position of the object is from -15 cm to -25 cm. Hence,

$$\Delta y = -25 \text{ cm} - (-15 \text{ cm}) = -10 \text{ cm}$$

—a negative change. That is, the object's net motion during the time interval has been in the $-y$ direction. Applying the definition of average velocity, we find:

$$\bar{v} = \frac{\Delta y}{\Delta t} = \frac{-10 \text{ cm}}{1 \text{ sec}} = -10 \text{ cm/sec}$$

Figure 5–5. Two descriptions of the motion of an object moving upward with a constant speed of 10 cm/sec.

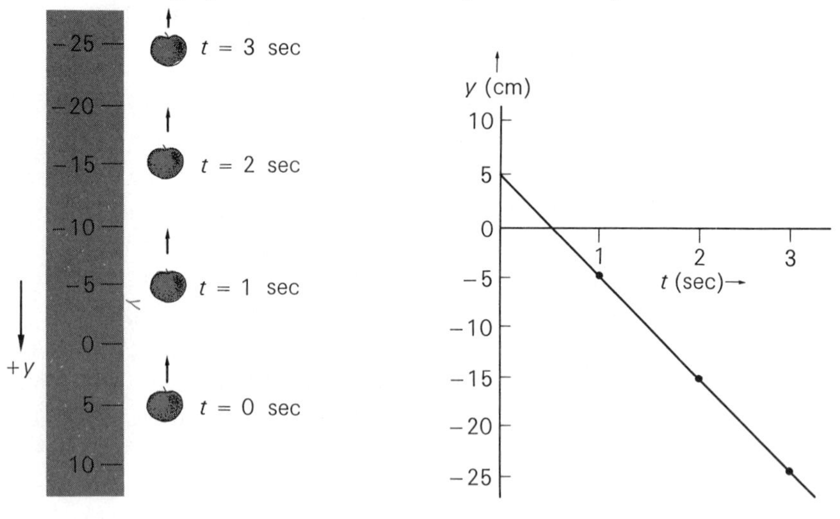

In this case the average velocity (change in position divided by time elapsed) is negative, -10 cm/sec. This indicates that the direction of motion is the $-y$ direction. The speed (distance traveled divided by time elapsed) is still $+10$ cm/sec. The difference arises because the "distance traveled" in a time interval is always a positive number, whereas the "change in position" may be positive or negative. Thus the average velocity of an object carries information about the direction of its motion as well as how fast it is moving, whereas its speed indicates only how fast it is moving.

Paradoxes of Motion

We have been discussing the average velocity of an object for an interval of time. Suppose, however, we ask how fast an object is moving at a given time. To the driver of an automobile this would probably seem a reasonable question, which a glance at the speedometer could answer. But is it really so reasonable? The very idea of motion involves a change in position over an interval of time; a snapshot, taken at a single instant of time, shows no motion. How can one even comprehend, much less quantify, motion at an instant of time?

This is one of the oldest and subtlest problems in the history of science, one that was instrumental in preventing the quantification of motion from Greek times until the Renaissance. To appreciate the conceptual difficulties, let us consider some related examples suggested by Zeno, a Greek philosopher of the fifth century B.C. Zeno's paradoxes, as they are called, were a source of puzzlement for centuries, as much in the time of Kepler as when they were first proposed.

Consider an object at rest that is about to be set in motion. Before it can cover a particular distance, it must first cover half that distance; before it can cover that half-distance, it must first cover half of that; and so on *ad infinitum*. Zeno concluded that there is no first distance through which the object can move, and thus it can never begin to move. And yet bodies at rest *can* be made to move.

Or there is the famous tale of the race between Achilles and the tortoise. Zeno supposed the tortoise to begin ahead of Achilles, who tries to overtake it. By the time Achilles reaches the tortoise's starting point, the tortoise has moved ahead a certain distance. By the time Achilles covers that distance, the tortoise has moved ahead again. And so it goes—Achilles can never overtake the tortoise. And yet we know Achilles could if the race were actually run.

These paradoxes remained unresolved for more than two thousand years. They were understood only after the invention of the branch of mathematics we call the calculus. The notion of the velocity of an object at a given time involves the same kinds of conceptual difficulties. To resolve them, we must introduce the mathematical notion of a *limit*.

Instantaneous Velocity

Perhaps most of us are not disturbed by the idea of associating a velocity with an instant of time because we often deal with objects that move with constant speed in a straight line. In that case, the same average velocity is obtained no matter what time interval is considered. Consequently it seems reasonable to associate this velocity with every instant of time. For example, in a car moving with a constant speed, the reading of the speedometer, which measures average speed over very short time intervals, does not change.

Most motions, however, are more complex than this. Usually the average velocity measured depends on the interval of time considered. For example, Figure 5–6 represents the motion of an apple held at rest at position $y = 10$ m until $t = 1$ sec, and then dropped.* It is clear from the figure that the apple is speeding up as it falls. In the one-second interval after $t = 2$ sec, the apple falls 15 m; in the one-second interval after $t = 4$ sec it falls 35 m; and in the one-second interval after $t = 6$ sec it falls 55 m.

* Again, we shall choose the $+y$ direction as downward.

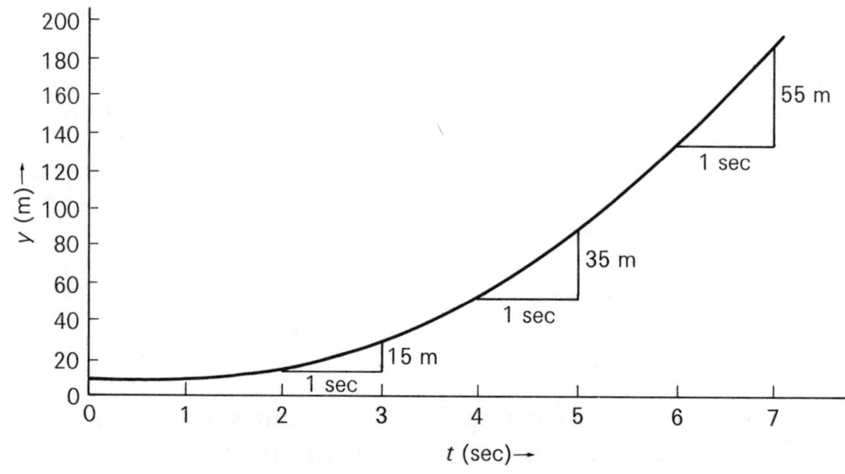

Figure 5–6. Graph of y vs. t for an apple that speeds up as it falls. During the one-second intervals following t = 2, 4, and 6 sec, it falls 15, 35, and 55 m, respectively.

Suppose we ask how fast the apple is moving at t = 2 sec. How can this question be answered? We might begin by calculating the average velocity for the 4-sec interval following t = 2 sec. Figure 5–7 and Table 5–2 illustrate such a calculation.

Figure 5–7. Distances fallen in different time intervals immediately following t = 2 sec. These distances are recorded in Table 5–2.

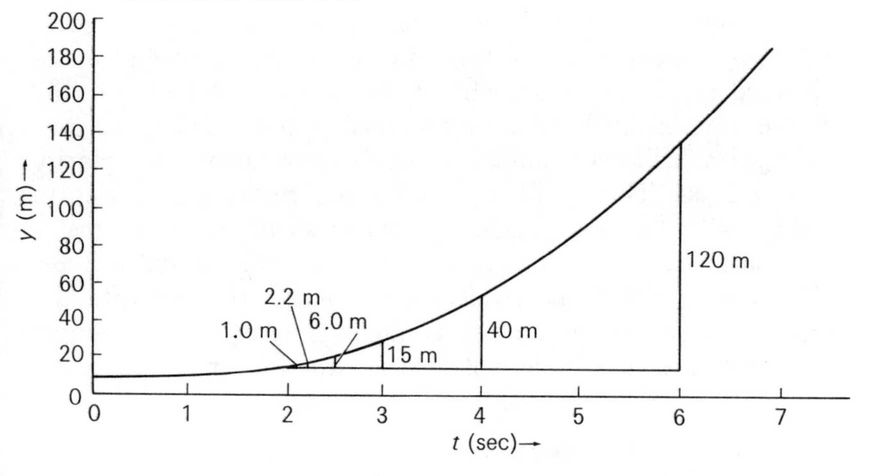

Table 5–2. Calculations from the data plotted in Figures 5–6 and 5–7.

t_i (sec)	t_f (sec)	Δt (sec)	Δy (m)	$\bar{v} = \Delta y/\Delta t$ (m/sec)
2	6	4	120	30
2	4	2	40	20
2	3	1	15	15
2	2.5	.5	6.0	12
2	2.2	.2	2.2	11
2	2.1	.1	1.0	10

In this 4-sec interval the apple moves from $y = 15$ m to $y = 135$ m, a change in position $\Delta y = 120$ m in a time $\Delta t = 4$ sec, or $\bar{v} = 30$ m/sec. But this calculation is clearly not adequate to determine the velocity of the apple at $t = 2$ sec. During the 4-sec time interval the apple was speeding up. The 120-m change in position the apple experiences in this time interval is consequently larger than the change in position that would have resulted in this interval had the apple maintained the velocity it had at $t = 2$ sec. Hence 30 m/sec is an overestimate.

We can improve our estimate by calculating the average velocity $\Delta y/\Delta t$ for a shorter time interval, like the 2-sec interval following $t = 2$ sec. This gives $\bar{v} = 20$ m/sec, as shown in Figure 5–7 and Table 5–2. However, since the speed of the apple is increasing throughout this interval as well, this estimate is also too large. Clearly we need to measure the change in position Δy for such a short time interval Δt that the speed has not changed appreciably from what it was at $t = 2$ sec. Presumably as the time intervals Δt are made smaller and smaller, the estimates $\Delta y/\Delta t$ will more and more closely approximate the velocity at $t = 2$ sec. This is illustrated by the remaining entries in Table 5–2. As Δt is made arbitrarily small, these estimates approach a limiting value we shall call the *velocity*, v, at $t = 2$ sec. (Often it is called the *instantaneous velocity*, to emphasize the distinction between it and the average velocity.) In the example we are discussing, Table 5–2 indicates the limiting value to be about 10 m/sec. In general, then, we write:

instantaneous velocity v

= limiting value of $\Delta y/\Delta t$ as Δt is made very small

A more succinct notation is:

$$v = \lim_{\Delta t \to 0} \frac{\Delta y}{\Delta t}$$

(This is read "limit as Δt approaches zero of $\Delta y/\Delta t$.")

The above definition is equivalent to the following operation: To determine the velocity of a moving object at time t, measure the average velocity $\Delta y/\Delta t$ using a time interval following t so short that the object does not speed up appreciably during the time interval.

The rigorous justification of this procedure is not easy. Questions arise: Will a limiting value always be found? How would the result differ if we had considered an interval Δt before, instead of after, the time t? The resolution of these questions and others like them was one of the triumphs of nineteenth-century mathematics.*

Suggested Reading

The references below are pertinent to this and the following chapter.

Arons, A. *Development of Concepts of Physics*. Reading, Mass.: Addison-Wesley, 1965. Chapter 1 discusses the concepts of velocity and acceleration with considerable care.

Holton, G. *Introduction to Concepts and Theories in Physical Science*. Reading, Mass.: Addison-Wesley, 1952. Chapter 1 is similar to Arons in scope, though somewhat briefer.

Cooper, Leon. *An Introduction to the Meaning and Structure of Physics*. New York: Harper & Row, 1968. The first two chapters present a less mathematical discussion than Arons or Holton.

Questions

1. A stone dropped from the top of a tower 20 m high takes about 2 sec to reach the ground. Geocentrists argued that if the earth were moving around the sun, the stone should strike the ground far from the base of the tower. Assuming

* It was shown that a limiting value always exists in situations of interest in physics; in addition, the same limiting value is obtained whether Δt is taken before or after the time t. In the language of the calculus, such a limiting value of $\Delta y/\Delta t$ at some time t is called the *derivative* of y at t.

that the earth moves in a circular path of radius 93 million miles, taking one year to complete an orbit, how far from the base of the tower would the stone hit according to this argument? (In this calculation you have neglected the rotation of the earth. Estimate whether this motion would contribute significantly to the effect.)

2. What kinds of properties must a device have to enable one to keep time with it? Can you give examples of such devices other than the usual present-day clocks? How was time measured before the invention of the pendulum and balance-wheel clocks?
3. Speculate on how the notion of a limiting value may be used to resolve Zeno's paradoxes. (You may wish to consult the literature on this subject.)
4. A car takes one hour to travel 60 miles due north on a straight road; it then immediately turns around and travels south, returning to its starting point in one more hour. What is the car's average speed? Show that, on the other hand, its average velocity for the two-hour interval is zero.
5. Use the result of the exercise on page 113 to show that for constant-speed motion the instantaneous velocity at a particular time t is the same as the average velocity for any time interval Δt immediately following t.
6. For the motion illustrated in Figure 5–6, estimate the instantaneous velocity at $t = 4$ sec.
7. Plot graphs of position vs. time and velocity vs. time for the following:

   (a) a drag racer
   (b) a swimmer (who swims two lengths of the pool)

   Estimate some realistic numbers for the positions, velocities, and times on your graphs.
8. Suppose the motion of an object is characterized by one of the accompanying position vs. time graphs. In each case sketch the corresponding velocity vs. time graph and give examples of such motion.

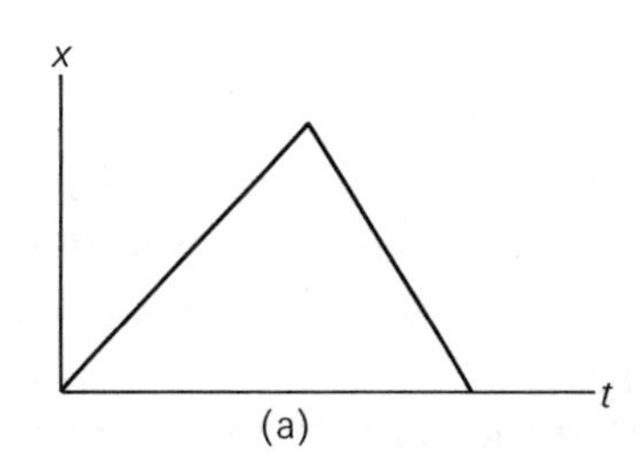

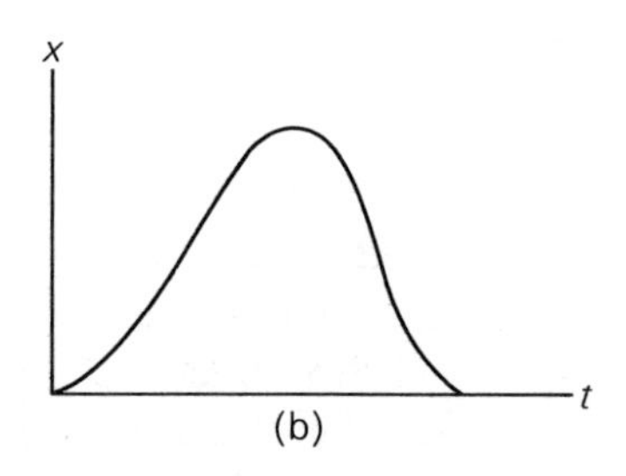

9. (a) Graph (a) shows velocity vs. time for two racing cars *A* and *B*. Assuming they begin the race from the same point at the same time, and the race ends at time *T*, which car is the winner? Explain your reasoning. Describe the race in words.
   (b) Graph (b) indicates the position vs. time for two racing cars *A* and *B*. If the race ends at time *T*, which car is the winner? Explain your reasoning. Describe the race in words.

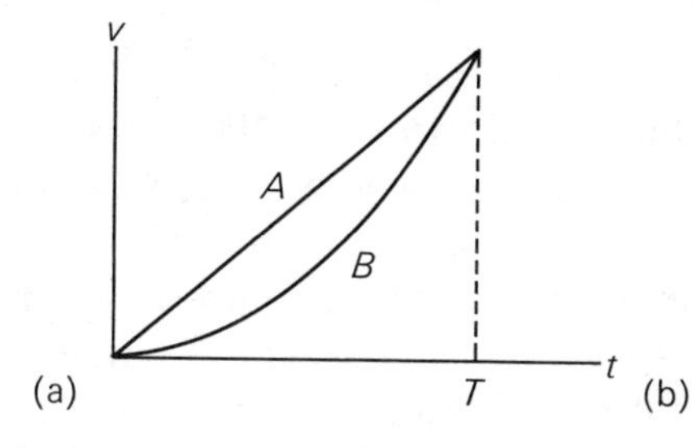

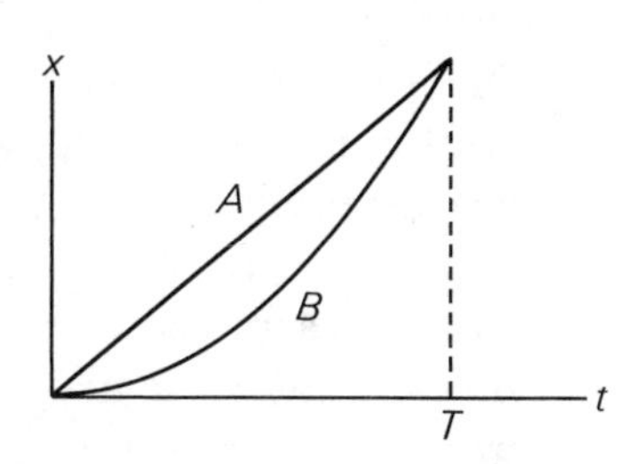

10. Suppose the motion of an object is characterized by this velocity vs. time graph. Sketch the corresponding position vs. time graph. (Is there more than one such graph?) Give examples of such motion.

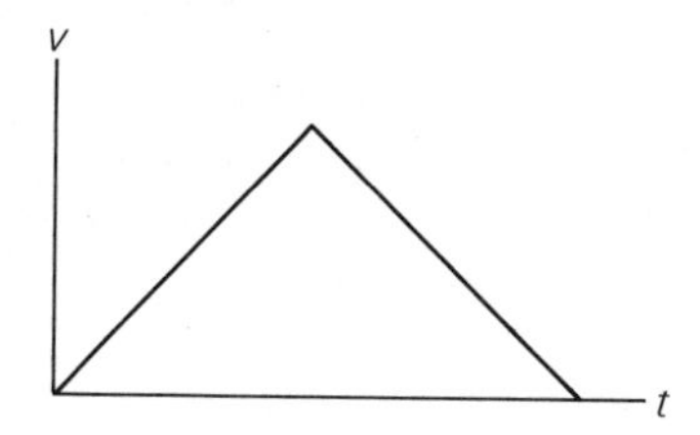

11. A ball is released from rest a distance h above the floor. It bounces off the floor, rebounding to height h.
    (a) Sketch a rough graph depicting the position vs. time for the ball. (Be sure to indicate clearly what position you have chosen to be $y = 0$ and what direction you have chosen as the $+y$ direction.)
    (b) Sketch a rough graph depicting the velocity vs. time for the ball. (Hint: When is the velocity zero? When is it positive and when negative?)
12. A Concorde SST (supersonic transport) flies from Los Angeles to New York, a distance of 3000 miles, at a speed of 1500 miles/hour. A Boeing 747 flies from New York to Los Angeles at a speed of 500 miles/hour. Suppose they both take off at the same time. Where will they meet? (Suggestion: First plot a position vs. time graph for each plane. Then write an algebraic equation for each graph. Use these equations to solve for the time when the positions are equal.)
13. A Minuteman missile, which has a speed of 5400 miles/hour, is accidentally fired. A Sprint antiballistic missile, which has a speed of 10,800 miles/hour, is sent after the Minuteman from the same base. If the Sprint has a maximum range of 30 miles, what is the maximum time that can elapse between the firing of the two missiles if the Sprint is to have a chance of catching the Minuteman? (Suggestion: First plot a position vs. time graph for each missile. Then write an algebraic equation for each graph. Use these equations to solve for the time when the Sprint would have to be fired to overtake the Minuteman 30 miles from the base.)

CHAPTER SIX

The Fall of Aristotle

Language is said to grow in response to a need, and the mathematical language of Chapter 5 is no exception. The basic ideas of length, time, and speed are simple and intuitive and were known even to the most primitive cultures. They adequately expressed the views of terrestrial motion prevalent as recently as Greek times. The sharpening of these ideas—for example, of speed into instantaneous velocity—came slowly, mainly as a concomitant of post-Renaissance attempts to understand the motion of objects at the surface of the earth.

To appreciate how the need for new concepts arose, we must examine older ideas of motion, chiefly those of the Greek philosopher Aristotle. We have already made brief reference to Aristotle's views on the nature of the heavens. His role in our story is most significant, however, in the realm of terrestrial motion.

The Aristotelian Approach

Aristotle's physics exemplifies an approach to the physical world quite different from the one followed by physicists today. We can separate into two general categories the various ways men have analyzed the world through the centuries. One is often called *reductionist*. It first reduces nature to its simplest and most easily understood units and then builds up an understanding of more complex phenomena in terms of interactions between these basic units. Most scientists today use this approach when they try to understand atoms as collections of electrons, neutrons, and protons or when they analyze chemical processes in terms of interactions between atoms, biological processes in terms of chemical reactions, and even human social behavior in terms of biological mechanisms.

The other category, into which Aristotle's science falls, might be termed *integrationist*. It can be typified by the claim that the whole is greater than the sum of its parts—that is, that there are aspects of a phenomenon that cannot be explained simply by interactions among its components. The Pythagorean attempt to understand music in terms of relations between the integers exemplifies the reductionist approach. Aristotle might have claimed, however, that there is far more to music and its power to affect us than these simple integers. Another example, which Aristotle often cited to illustrate this point, is the acorn. To see an acorn as merely a collection of atoms, even a particular and carefully specified one, is to miss the most significant thing about it—its potential of growing into an oak tree.

Aristotle sought to understand nature in terms of potentialities, and around this idea he constructed his philosophical system. Change occurs, he argued, when potentiality is transformed into actuality. Thus the acorn grows into an oak tree, and we understand this change if we view the acorn as a potential oak. He noted four basic types of change:

1. qualitative change (e.g., a stone being carved into a statue);

2. quantitative change (a piece of iron expanding when heated);
3. creation and destruction (the boiling of water, wherein the element water disappears and the element air, as vapor, appears); and
4. change in position (a stone falling).

Aristotle called all these types of change motion, though we now reserve the term for the last. To him, change in position was to be understood in the same sense as the other types of change. They are all fulfillments of potentiality. A stone has the potential of falling when released from our hand, while air does not. To our modern ears this may seem a strange way of speaking, but it is no less valid for that.

Natural Place

The key to the Aristotelian picture of the universe is the notion of *natural place*. Each of the four elements, Earth, Water, Air, and Fire, has a natural place. When in that place, it will remain at rest; removed to a distance, it will tend to move back (Figure 6–1). The natural place of the element Earth is the center of the universe (which happens to coincide with the center of the earth—that is how the earth got there). Thus Earth falls when released from the hand.

Water also tends to fall because its natural place is a spherical shell about the center of the earth. On the other hand, Fire and Air tend to rise because their natural places are spherical shells at the top of the sublunary world. (See Figure 6–1.) However, because the four elements are continually interfering with each other, they are prevented from separating into their natural spheres. Thus the Water in a crucible does not fall because it is constrained by the Earth forming the crucible.

Aristotle explained the fall of a stone in the following way: When we pick up a stone (whose major constituent is the element Earth), we move it farther away from its natural place, the center of the universe. When we release it, the stone strives to reach its natural place, moving in a straight line toward the

Figure 6–1. The natural places of the four elements, according to Aristotle. The natural place of Earth (La terre) is at the center, surrounded by that of Water (L'eau), Air (L'air), and Fire (Le feu). (From a publication of 1528.)

center of the universe. The potentiality of falling inherent in the stone becomes actuality.

One might question whether this is any explanation at all. It seems tautological to say that the stone falls because Earth has a tendency to fall. Wouldn't a satisfactory explanation necessarily involve a *mechanism*, some sort of invisible strings or magnets or vortices that push or pull the stone downward?

Aristotelian Explanation

Aristotle did not propose any mechanism responsible for the tendencies of the four elements to move to their natural places; what is more, he did not feel that any mechanism was necessary. To understand this, we need to discuss more carefully the Aristotelian notion of explanation. To Aristotle, there were four kinds of explanation:

> An explanatory factor, then, means (1) from one point of view, the material constituent from which a thing comes; for example, the bronze of a statue, the silver of a cup, and their kinds. From another point of view, (2) the form or pattern of a thing, that is, the reason which explains what it was to be that thing; for example, the factors in an octave are based on the ratio of two to one and, in general, on number. This kind of factor is found in the parts of a definition. Again, (3) the agent whereby a change or a state of rest is first produced; for example, an advisor is "responsible" for a plan, a father "causes" his child, and, in general, any maker "causes" what he makes and any agent causes what it changes. Again, (4) the end or the where-for; so, when we take a walk for the sake of our health, and someone asks us why we are walking, we answer, "in order to be healthy," and thus we think we have explained our action.[1]

Thus, for example, the motion of a cart pulled by a horse can be discussed in terms of these explanatory factors, or *causes*. There is (1) a *material cause*, the wood of which the cart is made; (2) a *formal cause*, the form of the cart, which enables it to roll; (3) an *efficient cause*, the agent or mover of the cart, in this case the horse; and (4) a *final cause*, the purpose for which the cart is being moved—to transport grain to market, for example.

To Aristotle, all four of these factors were important in understanding a phenomenon, but the last, the final cause, was the most fundamental, particularly with regard to natural processes. He implied that we cannot really understand a natural process unless we know its purpose or end. This emphasis on purpose, or *teleology* (from the Greek *teleos*, end) is striking in Aristotle, but should not be misunderstood. In his view, everything in nature has a purpose, but this does not mean a purpose in a conscious plan of some transcendent God—nothing as grand as that. Rather, Aristotle saw nature as fitting together, each phenomenon with its own part in the order. The final cause explains how the one fits in with the many—the "ecology" of the one in the whole.

Let us now examine the Aristotelian explanation of the falling stone to see how it differs from that of the moving cart. The cart's motion is an example of what Aristotle called *con-*

strained motion, motion that is not natural. The cart's natural motion would be straight downward, but the road beneath the cart prevents it from moving that way. All constrained motion, Aristotle said, requires a mover—something external to the object must continually be pushing or pulling on it. In constrained motion, the formal cause—the form of the moving object—allows the motion, but the efficient cause—the mover—does the moving.

In natural motion, however, Aristotle saw the formal, efficient, and final causes as one and the same. The form of stone includes its tendency to move downward when released from a point above the surface of the earth; the form not only allows but also brings about the falling motion. No external agent is needed. Thus any further inquiry into a mechanism drawing the stone downward was unnecessary.

The Leaning Tower

The first serious challenge to the Aristotelian description of terrestrial motion came in the seventeenth century. There were two particularly weak points in Aristotle's description, one concerning the fall of objects when released above the surface of the earth, the other concerning the motion of projectiles. We will examine the former in this chapter, the latter in Chapter 7.

Aristotle wrote that "the greater the mass of fire or earth the quicker always is its movement toward its own place."[2] This seems to assert, for example, that if two stones are released from the same height above the ground, the one with the greater weight will reach the ground first. He further wrote that "the downward movement of a mass of gold or lead, or of any other body endowed with weight, is quicker in proportion to its size."[3] This seems to imply that if one stone weighs twice as much as another, it will reach the ground twice as fast.

Simon Stevin of Bruges and the Italian Galileo Galilei were apparently the first to challenge these assertions effectively by citing actual observations of falling objects. In 1605 Stevin, alluding to the above passages from Aristotle's *Physics*, wrote:

> First of all, Aristotle, with his adherents, thinks that when two similar bodies of the same density fall in air, their rate of fall is in proportion to their relative weights. . . . But the experiment against Aristotle is like this: Take two balls of lead (as the eminent man Jean Grotius, a diligent investigator of Nature, and I formerly did in experiment) one ball ten times the other in weight; and let them go together from a height of 30 feet down to a plank below—or some other solid body from which the sound will come back distinctly; you will clearly perceive that the lighter will fall on the plank, not ten times more slowly, but so equally with the other that the sound of the two in striking will seem to come back as one single report. And the same thing happens with bodies of equal magnitude, but differing in weight as ten to one. Wherefore the alleged proportion of Aristotle is foreign to the truth.[4]

Galileo (1564–1642) performed similar free-fall experiments, and first published comments on them in *De Motu*, written in 1590. There is an oft-repeated though probably apocryphal story that Galileo made his observations by dropping objects from the Leaning Tower of Pisa. His most famous comment on the result of his experiments appeared in his *Dialogues Concerning Two New Sciences* (1638) in the following dialogue:

> Salviati: I greatly doubt that Aristotle ever tested by experiment whether it be true that two stones, one weighing ten times as much as the other, if allowed to fall, at the same instant from a height of, say, 100 cubits, would so differ in speed that when the heavier had reached the ground, the other would not have fallen more than 10 cubits . . .
>
> Sagredo: . . . I, who have made the test, can assure you that a cannon ball weighing one or two hundred pounds, or even more, will not reach the ground by as much as a span ahead of a musket ball weighing only half a pound . . .
>
> Simplicio: Your discussion is really admirable; yet I do not find it easy to believe that a bird shot falls as swiftly as a cannon ball.
>
> Salviati: Why not say a grain of sand as rapidly as a grindstone? But, Simplicio, I trust you will not follow the example of many others who divert the discussion from its main intent and fasten upon some statement of mine which lacks a hair-

> breadth of the truth and, under this hair, hide the fault of another which is as big as a ship's cable. Aristotle says* that an iron ball of one hundred pounds falling from a height of one hundred cubits reaches the ground before a one-pound ball has fallen a single cubit. I say that they arrive at the same time. You find, on making the experiment, that the larger outstrips the smaller by two finger breadths . . . now you would not hide behind these two fingers the ninety-nine cubits of Aristotle, nor would you mention my small error and at the same time pass over in silence his very large one.[5]

Galileo's strong position against the Aristotelian predictions about the fall of objects led to an intense and sometimes acrimonious controversy between Galileo and his followers and those of Aristotle. Interestingly, it is now clear that, in a sense, both sides were right since they were really not talking about the same thing. Aristotle and his followers tended to focus on resisted motion, in which a falling object passes through some resistive medium such as air or water. This motion is what we commonly observe; the resistance of the medium explains why, for example, a leaf and a stone fall in different times.

Aristotle would have considered unresisted motion an unrealistic abstraction. In fact, he explicitly argued that in a void, with no resistance, all bodies *would* move with the same speed. This he took to be an argument against the existence of a void:

> We see that, other things being equal, heavy and light bodies move with unequal velocities over an equal space in the ratio which their magnitudes have to each other. Hence they must do so even when moving through a void. But this is impossible. Why should one of them move faster than another? To be sure, one of them does necessarily move faster in a [space] full [of matter] because the greater body divides the medium faster by reason of its shape or of its [upward or downward] tendency. But [in a void] everything would, accordingly, move with equal velocity. But this is impossible.[6]

Galileo, on the other hand, focused on the opposite extreme. For the free fall of most objects, feathers and leaves aside, his

* In fact, nowhere in the writings of Aristotle do we find this stated explicitly. As mentioned earlier, Aristotle did say, "the downward movement of a mass . . . is quicker in proportion to its size."

experiments showed that the time of fall is virtually the same. The "two finger breadths" by which the hundred-pound ball outstrips the one-pound ball—so important to the Aristotelians because it resulted from the resistance of the air—was minimized by Galileo as merely indicating how close to the ideal of unresisted motion he had come. Apparently, unresisted motion has a particular simplicity: all objects would fall in the same time in a void.

Of course the void was, even as late as the time of Galileo, an unattainable idealization. Galileo showed, however, that an analysis stemming from this idealization gave a new understanding of the real world.

Acceleration

Once he had determined that different objects fall through the same distance in virtually the same time, Galileo investigated the details of this motion. Clearly, objects speed up as they fall. Is there any regularity in the way their speed increases?

Galileo recognized two particularly simple ways in which the speed of a falling object might change: First, its speed might be proportional to the distance it has fallen. That is, after falling 100 centimeters the object would have acquired twice the speed it had after falling only 50 centimeters. Second, the speed might be proportional to the time it has fallen. That is, after falling for ten seconds the object would have acquired twice the speed it had after only five seconds.

It is perhaps not obvious that these two imagined sorts of behavior are mutually exclusive, and it is certainly not clear *a priori* that either applies to the fall of real objects. Galileo demonstrated experimentally that the second kind of motion—speed proportional to time—was a good approximation to the observed fall in air of objects near the surface of the earth. This result, striking in its simplicity, was of great importance to the later work of Newton. To understand how Galileo arrived at it, we must extend our mathematical language to include the concept of *acceleration*.

The notion of acceleration is familiar to most of us. It is commonly used to describe how fast the speed of something changes in time. To say that an automobile accelerates at a rate of 10 miles/hour per second means that each second its speed increases by 10 miles/hour. Note that, like our original intuitive notion of speed as distance divided by time, this idea of acceleration pertains to an interval of time. If a car accelerates from rest to 50 miles/hour in 5 seconds, and we calculate an acceleration of 10 miles/hour per second, we are really talking about an average acceleration over the entire 5-second time interval.

We can define average acceleration in a manner quite parallel to the way we defined average velocity in Chapter 5. Consider a time interval Δt between some initial time t_i and some final time t_f. Suppose that an object has velocity v_i at time t_i and velocity v_f at time t_f. Let Δv be the change in velocity of the object during this time interval Δt:

$$\Delta v = \text{final velocity} - \text{initial velocity} = v_f - v_i$$

Then we can define the *average acceleration* $\bar{a}$ of the object during this time interval as the change in velocity Δv divided by the time interval Δt:

$$\text{average acceleration } \bar{a} = \frac{\text{change in velocity}}{\text{time interval}} = \frac{\Delta v}{\Delta t}$$

Notice that this definition involves the velocity rather than the speed of the object. One reason for this choice will be discussed in Chapter 7.

As an example of the use of this definition, consider the motion shown in Figure 6–2. A stone is released from rest at some time before $t = 0$. Its speed at $t = 0$ is 1 m/sec, and at $t = 1$ sec it is 2 m/sec. What is the average acceleration of the stone during this one-second time interval? Let us choose the $+y$ direction to be downward. With this choice, the velocities of the stone at the beginning and the end of the time interval are $v = 1$ m/sec and $v = 2$ m/sec respectively. Hence the average acceleration of the stone during this interval is:

$$\bar{a} = \frac{\Delta v}{\Delta t} = \frac{2\text{ m/sec} - 1\text{ m/sec}}{1\text{ sec}} = \frac{+1\text{ m/sec}}{1\text{ sec}} = +1\text{ m/sec}^2{}^{*}$$

* By m/sec^2 we mean m/sec per second. Velocity has units of m/sec; thus acceleration, telling how much velocity changes per second, has units of m/sec per second.

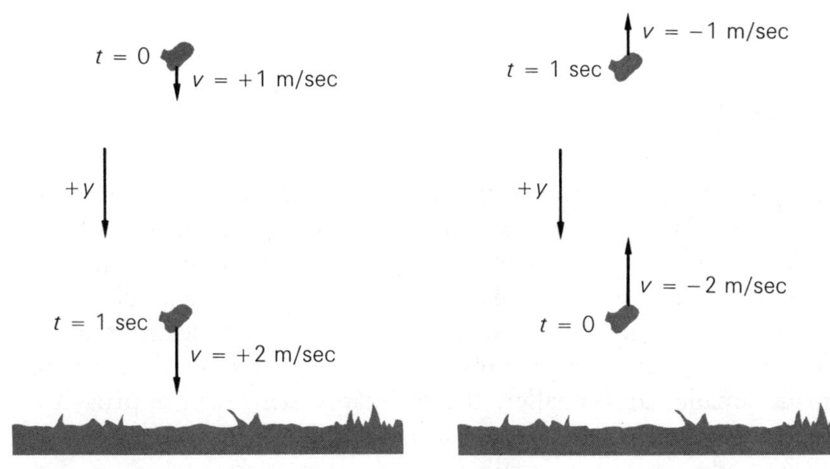

Figure 6–2. A falling stone. Its velocity increases from $+1$ m/sec at $t = 0$ to $+2$ m/sec at $t = 1$ sec.

Figure 6–3. A stone thrown upward. Its velocity changes from -2 m/sec at $t = 0$ to -1 m/sec at $t = 1$ sec.

As a second example, consider the motion illustrated in Figure 6–3. At time $t = 0$ a stone is thrown upward (in the $-y$ direction) with a speed of 2 m/sec. One second later, it has slowed to 1 m/sec. What is the average acceleration of the stone during this one-second time interval? In this case, the velocity of the stone changes from $v = -2$ m/sec at the beginning of the time interval to $v = -1$ m/sec at the end of the time interval. The change in velocity, Δv, is positive. Hence the average acceleration of the stone during the time interval is positive:

$$\bar{a} = \frac{\Delta v}{\Delta t} = \frac{-1 \text{ m/sec} - (-2 \text{ m/sec})}{1 \text{ sec}} = \frac{+1 \text{ m/sec}}{1 \text{ sec}} = +1 \text{ m/sec}^2$$

An important lesson can be drawn from these examples. Notice that each time the acceleration was positive. It follows from our definition that the acceleration is positive if an object is speeding up while traveling in the $+y$ direction *or* if an object is slowing down while traveling in the $-y$ direction.

> Make a similar statement concerning the conditions under which the acceleration is negative.

Instantaneous Acceleration

Suppose next that the acceleration of our object is changing with time. We can define the instantaneous acceleration a of the object at a given time t much as we defined the instantaneous velocity of an object in Chapter 5. We estimate the acceleration at time t by calculating $\Delta v/\Delta t$, using a short time interval Δt immediately following t. Presumably as the time intervals Δt are made smaller and smaller, the estimates $\Delta v/\Delta t$ will approach a limiting value. We call the limiting value a of these estimates, as Δt is made arbitrarily small, the *instantaneous acceleration* (or just the *acceleration*) at time t:

instantaneous acceleration a
= limiting value of $\Delta v/\Delta t$ as Δt is made very small

Or, more briefly,

$$a = \lim_{\Delta t \to 0} \frac{\Delta v}{\Delta t}$$

Compare this definition with that for instantaneous velocity in Chapter 5. Notice that just as velocity measures the change of position with time, so acceleration measures the change of velocity with time. That is, the relation between a and v is exactly the same as the relation between v and y.

In concluding this discussion of acceleration, let us mention one interesting side issue. We have defined velocity in terms of the change in position with time, and acceleration in terms of the change in velocity with time. One might reasonably expect that we would next discuss the change of acceleration with time, and then the change of that with time, and so on *ad infinitum*. These quantities can be easily defined mathematically, but we shall not consider them. Why not? The reason is pragmatic. As we shall see, the concepts of velocity and acceleration have proved useful in organizing observations of motion in nature in a way that leads to a simple and appealing theoretical explanation. The other concepts have not been used in physics simply because we have done very well without them. There is no *a priori* justification for their exclusion and no guarantee that there will never be a theory that uses them.

Uniformly Accelerated Motion

The concept of acceleration can be used to make a more concise statement of Galileo's result, that in free fall the velocity of an object increases in proportion to the time it has fallen. Figure 6–4(a) represents this kind of motion. At $t = 0$ the velocity has some value v_0; this might result from the object's being thrown downward at $t = 0$ or from its being released from rest before $t = 0$. The velocity then increases in proportion to the time t. Hence the graph is a straight line.

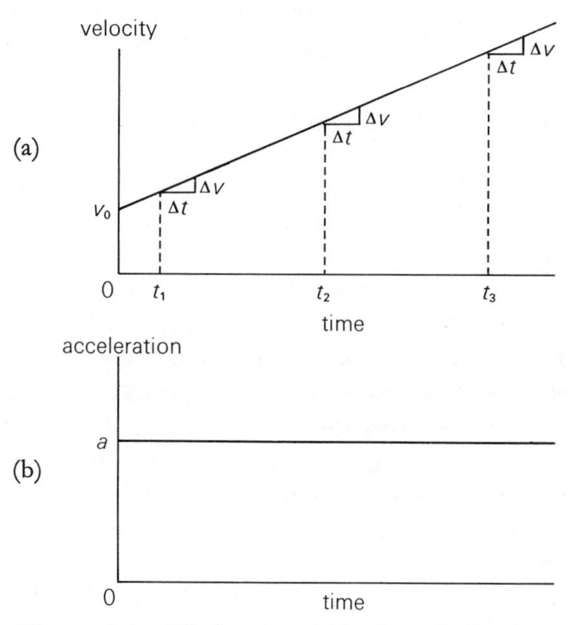

Figure 6–4. Motion in which the velocity increases in proportion to the time. (a) The velocity vs. time graph is a straight line. (b) The corresponding acceleration is constant.

What does this imply about the acceleration? To find the acceleration at some instant of time t_1, we calculate $\Delta v/\Delta t$ for some very short time interval Δt immediately following t_1. An interval Δt and the corresponding change in velocity Δv during this interval are shown in Figure 6–4(a). The figure also shows identical time intervals Δt and the corresponding changes in velocity immediately following two other times t_2 and t_3. Since the graph is a straight line, the three triangles in the figure are identical. Thus $\Delta v/\Delta t$ is the same at each of the three times. Since these times were arbitrarily chosen, it is clear that the same value of $\Delta v/\Delta t$ would be obtained at any instant of time. We therefore conclude that the acceleration at each instant of time is the same. This is shown in Figure 6–4(b). We have thus established that *a velocity that increases in proportion to time corresponds to a constant acceleration.*

In principle, one could determine whether a freely falling object has a constant acceleration simply by measuring its velocity at a number of successive times and determining whether the velocity increases in proportion to time. In practice, such a procedure would be difficult even with modern apparatus, and in the seventeenth century it would have been nearly impossible. Recall what is involved in a velocity measurement: One must measure the change in position Δy corresponding to an interval of time Δt; furthermore, Δt must be taken small enough so that in it the velocity does not change appreciably. An accurate measurement of such a small Δt and the corresponding Δy would be exceedingly difficult for a rapidly moving object.

Fortunately, another method of testing whether a given motion is uniformly accelerated can be devised and was, in fact, used by Galileo. For an object moving with constant acceleration one can derive a simple relation between its position and the time. This relation, involving only the easily measured quantities y and t, can then be used to test whether an observed motion is in fact uniformly accelerated; it acts as a "fingerprint" of constant acceleration.

Such a relation between the distance traveled by an object and the time elapsed would be quite simple if the velocity of the object were constant. Then the distance would be just the velocity times the time. But in uniformly accelerated motion the velocity is not constant. It increases in proportion to the time, as shown in Figure 6–4.

For the motion shown in Figure 6–4, it should be clear that during the time interval from 0 to t, the object will move a distance greater than $v_0 t$, the distance it would have moved if it had maintained its initial velocity v_0 throughout the interval, and less than vt, the distance it would have moved had it traveled with its final velocity v throughout the interval. But where between these extremes is the correct distance?

This problem had been solved by William Heytesbury of Merton College, Oxford. In 1335 he wrote, "the moving body . . . will traverse a distance exactly equal to what it would traverse in an equal period of time if it moved uniformly as its mean degree [of velocity]."[7] By "mean degree" of velocity he meant what we would write in modern notation as $(v_0 + v)/2$, the velocity the object actually achieves halfway through the interval. (See Figure 6–5.) In other words, Heytesbury claimed that during the time interval from 0 to t, the object will move a distance $[(v_0 + v)/2]\, t$. This is exactly halfway between the two extremes noted above.

While Heytesbury apparently considered his statement intuitively obvious, several of his colleagues at Merton College

Figure 6–5. The mean degree of velocity. If the velocity increases in proportion to time, the velocity at the middle of a time interval is midway between the initial and final velocities.

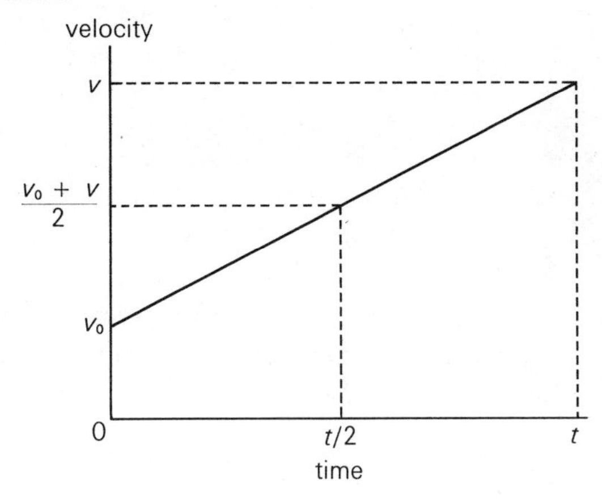

attempted to prove it formally. None of their proofs were entirely convincing, however, and three hundred years later Galileo himself could do no better. Only after the development of the calculus, so important to the mathematics of motion, was a satisfactory demonstration of the statement possible. In the meantime, the science of physics continued to use Heytesbury's relation, which seemed reasonable but had not been proved rigorously.

We may put this relation into an algebraic form as follows: Suppose an object moves with uniform acceleration from a position y_0 at $t = 0$ to another position y at a later time t. Heytesbury's relation, which has been called the "Merton College rule," can then be written

$$y - y_0 = \left(\frac{v_0 + v}{2}\right) t \tag{1}$$

This is very nearly the relation between position and time that we are seeking. It lacks only an explicit relation between the instantaneous velocity v and the time t. We can find such a relation from the graph in Figure 6–4(a), which specifies how the velocity varies with time for uniformly accelerated motion. Since the graph is a straight line with slope a and intercept v_0, it is described by the equation

$$v = at + v_0 \tag{2}$$

We may substitute this expression for v into equation (1):

$$y - y_0 = \frac{v_0 + (at + v_0)}{2} t = \frac{2v_0 + at}{2} t$$

$$= v_0 t + \tfrac{1}{2}at^2$$

or, finally,

$$\boxed{y = y_0 + v_0 t + \tfrac{1}{2}at^2} \tag{3}$$

This is the fingerprint relation of uniformly accelerated motion. The three terms can be interpreted rather simply: y_0 is of course just the position of the object at $t = 0$. $v_0 t$ is the change in position that can be attributed to the object's initial velocity v_0—that is, the change in position that would have occurred had there been no acceleration. $\frac{1}{2}at^2$ is the additional change in position due to the acceleration of the object.

Equations (2) and (3) together provide the basic mathematical description of uniformly accelerated motion. Given the initial position y_0 and initial velocity v_0 along with the acceleration a, equations (2) and (3) tell us the velocity and position at any later time.

Free Fall as Uniformly Accelerated Motion

Equation (3) becomes particularly simple for an object dropped from rest above the surface of the earth. If we call the point of release $y = 0$, the time of release $t = 0$, and choose the $+y$ direction to be downward, then $y_0 = 0$, $v_0 = 0$, and we have simply

$$\boxed{y = \tfrac{1}{2}at^2} \qquad (4)$$

Thus, in uniformly accelerated motion beginning from rest, the distance traveled is proportional to the square of the time. This result seems to have been derived first by Galileo in one of his two principal books, *Dialogues Concerning Two New Sciences*, which appeared in 1638.

We can best appreciate Galileo's demonstration that free fall does correspond to uniformly accelerated motion by reading his own description. He begins by describing experiments done with an inclined plane:

> A piece of wooden moulding or scantling, about 12 cubits long, half a cubit wide, and three finger-breadths thick, was taken; on its edge was cut a channel a little more than one finger in breadth; having made this groove very straight, smooth, and polished, and having lined it with parchment, also as smooth and polished as possible, we rolled along it a hard, smooth, and very round bronze ball. Having placed this board in a sloping position, by lifting one end some one or two cubits above the other, we rolled the ball, as I was just saying, along the channel, noting . . . the time required to

> make the descent. . . . Having performed this operation and having assured ourselves of its reliability, we now rolled the ball only one-quarter the length of the channel; and having measured the time of its descent, we found it precisely one-half of the former. Next we tried other distances, comparing the time for the whole length with that for the half, or with that for two-thirds, or three-fourths, or indeed for any fraction; in such experiments, repeated a full hundred times, we always found that *the spaces traversed were to each other as the squares of the times*, and this was true for all inclinations of the plane. . .[8]

Thus Galileo's measurements corresponded to equation (4), allowing him to conclude that such motion was uniformly accelerated. Since this result—distance traveled is proportional to t^2—seemed independent of the angle of tilt of the plane, he inferred that motion on a vertical plane (i.e., free fall) would also correspond to uniform acceleration. Thus Galileo concluded from his inclined plane experiments, to a precision unobtainable with the much more rapid motion of true free fall, that free fall corresponds to uniform acceleration. What is more, in the free-fall experiments described previously, Galileo found that there was only a negligible difference in the times of fall of objects of different weight. These two results enabled him to conclude that *all objects fall at the surface of the earth with the same uniform acceleration.*

In modern times, the value of the constant acceleration of objects in free fall has been given the symbol g. Measurements show its value to be approximately

$$g = 980 \text{ cm/sec}^2$$

Such a simple result, that all objects released near the surface of the earth fall with an acceleration of 980 cm/sec^2, leads us immediately to ask *why*. This question also occurred to Galileo, who dealt with it in the following statement:

> The present does not seem to be the proper time to investigate the cause of the acceleration of natural motion concerning which various opinions have been expressed by various philosophers, some explaining it by attraction to the center, others to repulsion between the very small parts of the body,

while still others attribute it to a certain stress in the surrounding medium which closes in behind the falling body and drives it from one of its positions to another. Now, all these fantasies, and others too, ought to be examined; but it is not really worth while. At present it is the purpose . . . merely to investigate and to demonstrate some of the properties of accelerated motion . . .[9]

Suggested Reading

In addition to the references listed at the end of Chapter 5, the following are specifically relevant to this chapter:

Arons, A. *Development of Concepts of Physics.* Reading, Mass.: Addison-Wesley, 1965. Chapter 2 discusses Greek physics and Galileo's free-fall experiments.

Cooper, Lane. *Aristotle, Galileo, and the Tower of Pisa.* Ithaca: Cornell University Press, 1935. A short but interesting treatise on the free-fall experiment and its background.

Clagett, M. *The Science of Mechanics in the Middle Ages.* Madison: University of Wisconsin Press, 1959. A study of physics, mainly in the area of terrestrial motion, between the time of Aristotle and that of Galileo. It includes numerous translated excerpts from original manuscripts.

Questions

1. Discuss what an Aristotelian might hold to be the material, formal, efficient, and final causes of one of the following:
   (a) The atomic bomb
   (b) The manned lunar landing
   (c) The Safeguard ABM system
   (d) The Vietnam war
2. Aristotle used the terms "explanation" and "cause" interchangeably (cf. page 127). Today we generally associate different meanings with the two words. What is the distinction between explanation and cause in contemporary usage? Do any of Aristotle's causes fit the modern meaning of the word? Do any seem better labeled explanations?

3. Speculate on how an Aristotelian would explain the following observations:
   (a) A helium-filled balloon rises while an air-filled balloon falls.
   (b) A balloon that rises when filled with warm air falls when it cools off.
   (c) An upward-thrown ball continues to rise after it leaves the thrower's hand.
4. In Aristotelian physics the fall of objects toward the center of the earth was natural motion, requiring no further explanation. Is there an analogous idea in the world view we hold today? Is there any kind of motion that we hold to be natural in the sense that it needs no explanation (and any deviation from this kind of motion must be explained)?
5. Aristotle's statement, "The downward movement of . . . any . . . body endowed with weight is quicker in proportion to its size . . ." is somewhat ambiguous. (E.g., what exactly is meant by "downward movement" and "size"?) Suggest some interpretations of the statement using precise concepts such as the time of fall, the velocity of the body, the weight of the body, the volume of the body, etc. Discuss experiments one might perform to test these hypotheses.
6. An acorn drops from a branch of a tree 20 meters above the ground.
   (a) Neglecting air resistance, how long will it take the acorn to reach the ground?
   (b) What will the velocity of the acorn be just before it strikes the ground?
7. One second after the acorn in Question 6 has fallen, a squirrel jumps after it from the same branch.
   (a) When the squirrel jumps, how far below him is the acorn?
   (b) When the acorn hits the ground, how far above it is the squirrel? (Assume that the initial velocity of the squirrel as he leaves the branch is zero.)
8. Galileo, in his *Dialogues Concerning Two New Sciences*, states that an object falling with constant acceleration has the

following property: "If we take any equal intervals of time whatever, counting from the beginning of the motion, . . . the spaces . . . traversed in these intervals will bear to one another the same ratio as the series of odd numbers, 1, 3, 5, 7, . . ." Imagine an object starting from rest with a constant acceleration of 2 m/sec^2. Calculate its position at the end of 1, 2, 3, and 4 sec, and show that Galileo's claim is borne out.

9. Suppose a ball is thrown vertically upward. It rises and then falls back to earth. What is the velocity of the ball when it reaches the highest point in its trajectory? What is its acceleration at that time?
10. Suppose an ostrich egg is thrown vertically upward. It rises and then falls back to earth. Draw a rough sketch of the position vs. time for this motion. Recalling Galileo's result for free fall, sketch graphs of velocity vs. time and acceleration vs. time for this motion.
11. In his experiments, Galileo noted that objects of different weight do not fall to the ground in precisely the same time (cf. page 129: "a span," "two finger breadths"). He attributed these differences to the effects of air resistance, but without the vacuum pump could not directly substantiate this rationalization. A skeptic might have proposed instead that the differences were due to gravity itself. What justification did Galileo have for not searching for some formula relating the fall time to the weight of the object?
12. The accompanying graph represents an object moving with a constant velocity v_0. The area under the graph from $t = 0$ to the later time t is simply $v_0 t$ as shown. Note that this is equal to the object's change in position during the time interval from $t = 0$ to time t. Now consider the graph in Figure 6–5, which represents an object moving with

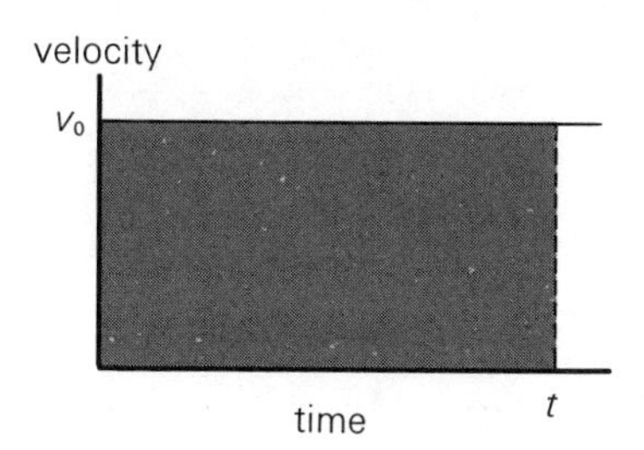

constant acceleration. Show that the area under this velocity vs. time graph from $t = 0$ to time t is also equal to the change in position of the object during the time interval from $t = 0$ to time t [cf. equation (1)]. (Using the methods of calculus, it can be demonstrated that the area under the velocity vs. time graph is always equal to the change in position.)

13. Using modern apparatus, we can measure the time of fall for objects near the earth's surface with a far greater precision than Galileo could. In addition, we can more closely approach Galileo's ideal of unresisted motion by allowing objects to fall in a long glass cylinder from which air has been removed with a vacuum pump.

    (a) Suppose the following times (in seconds) were measured for the free fall of three objects, a lead sphere, a steel sphere, and a cork sphere, in a vacuum through a distance of 1.00 m.

lead (200 gm)	steel (100 gm)	cork (20 gm)
.415	.464	.478

    What could you conclude from these data alone about the validity of the hypothesis: all objects fall at the same rate in a vacuum independent of their weight? What could you conclude from these data alone about the validity of the hypothesis: the time of fall of an object varies inversely as its weight? (E.g., the time of fall of an object of twice the weight is one-half as great.)

    (b) Suppose instead that you obtained the following data:

trial	lead	steel	cork
1	.415	.464	.478
2	.483	.491	.442
3	.455	.413	.490
average	.451	.456	.470

What could you now conclude concerning the hypotheses? Explain.

(c) Suppose you had obtained the following data:

trial	lead	steel	cork
1	.452	.452	.452
2	.452	.452	.452
3	.452	.452	.452
4	.452	.452	.452
5	.452	.452	.452
6	.452	.452	.452
7	.452	.452	.452
8	.452	.452	.452
9	.452	.452	.452
10	.452	.452	.452

Now what could you conclude concerning the hypotheses? Explain.

14. Galileo cited his experimental measurements of the motion of a bronze ball down a smooth channel (p. 139) as evidence in support of his claim that such motion corresponds to a uniform acceleration.

(a) The following are data typical of those Galileo might have obtained:

trial	time to roll 10 cm	time to roll 40 cm	time to roll 90 cm	time to roll 160 cm
1	1.0 sec	2.1 sec	3.0 sec	4.1 sec
2	.9	2.0	2.9	4.0
3	1.0	2.0	3.1	4.0

Are these data consistent with Galileo's claim?

(b) Suppose that Galileo had available to him more precise modern instruments for the measurement of position and time, and obtained the following data:

trial	time to roll 10 cm	time to roll 40 cm	time to roll 90 cm	time to roll 160 cm
1	1.00 sec	2.02 sec	3.05 sec	4.10 sec
2	1.01	2.02	3.06	4.08
3	0.99	2.01	3.05	4.09

Could Galileo use these more accurate data to support his claim? Would these data disprove his claim? Would the increased accuracy of these data be of any use to him?

CHAPTER SEVEN

Free Fall on a Moving Earth

In Chapter 6 we focused on a principal finding of Galileo and his contemporaries—that in the absence of air resistance all objects dropped near the surface of the earth fall vertically with the same constant acceleration g. This result would seem to bear on the question of whether the earth is moving or at rest. The fact that objects fall directly downward appears to speak against any movement of the earth. Let us rephrase the standard argument:

Imagine a stone dropped from a bridge, as shown in Figure 7–1. According to a stationary observer on the bridge, it falls vertically as indicated in Figure 7–1(a). How does the path of the stone appear to a moving observer—watching from a passing boat, for instance? Figure 7–1(b) shows the answer: the path looks curved.

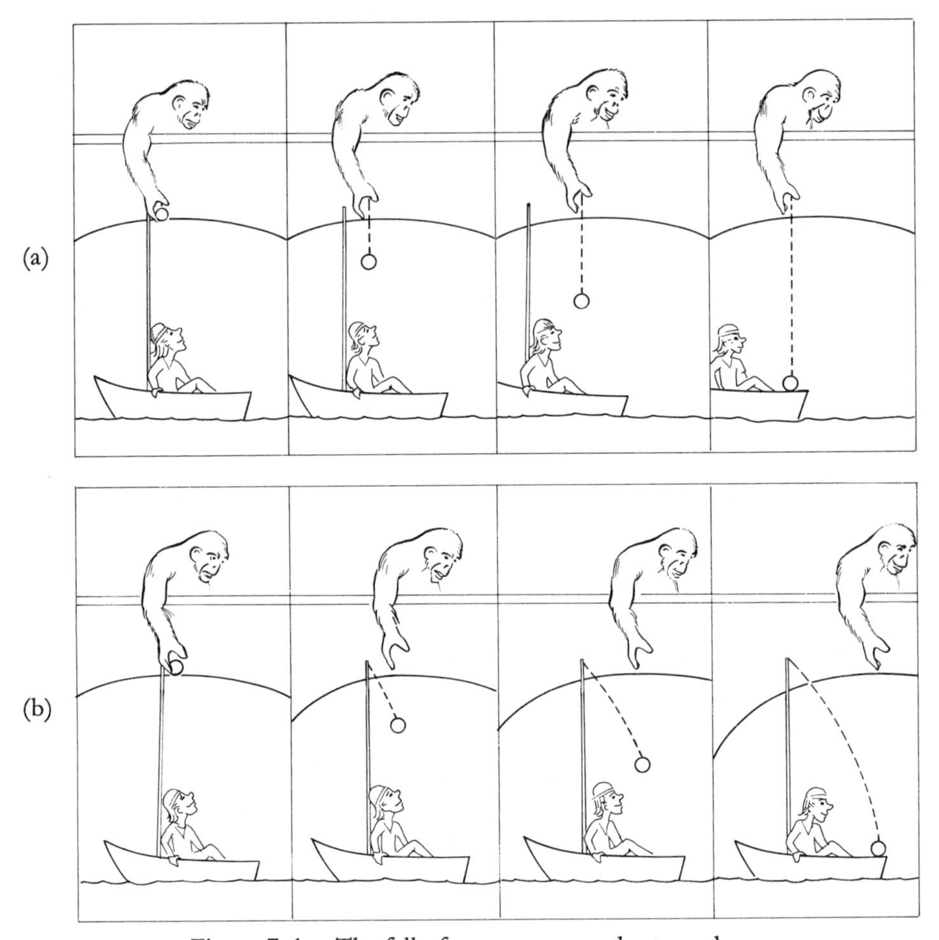

Figure 7–1. The fall of a stone as seen by two observers, one moving and the other at rest. (a) The stone falls vertically according to an observer at rest on the bridge. (b) The stone falls in a curved path according to an observer on a passing boat.

Suppose the earth were moving. Wouldn't an observer on the moving earth watching the falling stone see the same thing as an observer on the moving boat, namely a curved path? But objects are in fact observed to fall directly downward. Consequently the earth must be at rest. Or so went the argument.

Consider again the curved path in Figure 7–1(b). It resembles the path that an object projected horizontally follows as it drops earthward. In fact, a study of the motion of such projectiles led to a reconciliation between observed terrestrial motion and the idea of a moving earth. For this reason, we now turn our attention to projectile motion.

Motion: The Aristotelian View

We have seen that the ideas of terrestrial motion prevalent at the time of Galileo were basically Aristotle's. For Aristotle, these motions could be divided into two categories. Natural motion, in a straight line toward or away from the center of the universe (and thus of the earth), was attributed to the tendency of objects to seek their natural place. Constrained motion was attributed to the action of some outside agent or force; once that force was removed the constrained motion stopped.

The motion of projectiles, however, was difficult to reconcile with this point of view. "If everything that is in [constrained] motion . . . is moved by an external agent, how is it that some things (such as projectiles) can be in continuous [non-natural] motion after they have ceased to have contact with their mover?"[1] Aristotle tried to answer his question by suggesting that the air through which the projectile moves is the mover. The air in front of a moving cannonball is pushed out of the way; behind the cannonball a void is left into which the displaced air rushes, thrusting the ball forward. This was never a very satisfactory solution. It implies, for example, that the shape of the back end of the projectile should make a great difference in how rapidly it is pushed forward by the inrushing air. Of course there is no such effect.

Progress on this question did not come until the fourteenth century, with the introduction of the idea of *impetus*. The cannonball was imagined to receive from the cannon a quantity of impetus, which somehow kept it going during its flight until the impetus ran out and the projectile fell to earth. The details of the various theories of impetus are not important; it suffices to note that all were attempts to answer what was a basic question in the context of the Aristotelian world view: For constrained motion like that of a projectile, what is the mover?

Motion: The Galilean View

The first effective challenge to this approach resulted from Galileo's study of free fall, which called into question Aristotle's distinction between constrained and natural motion. Consider a particularly simple projectile: a stone thrown vertically upward. The stone rises, slowing down until it reaches some maximum height; then it falls, speeding up as it drops. To the followers of Aristotle, the downward fall of the stone was natural motion, caused by the stone's seeking its natural place at the center of the universe. The upward motion, however, was constrained motion, with its cause at least indirectly related to the initial throw of the stone. The entire flight of the stone, then, consisted of two distinct parts: first upward constrained motion, then downward natural motion.

Galileo's work, however, suggested a different way to look at the stone's motion. In his study of free fall, Galileo had found that all objects fall with the same constant acceleration g. Equally important for these considerations, he found that the upward motion of an object thrown vertically also corresponded to the same constant downward acceleration g. His argument was essentially that discussed on page 133: Let us choose the downward direction as positive, and note that during both the upward and downward parts of the stone's motion, its velocity is becoming more positive (or, to put it another way, less negative). This means that the acceleration is positive throughout the motion. Thus, to Galileo, the stone moves with a single motion from the instant it leaves the hand until it returns to the earth. The entire motion can be viewed as uniformly accelerated motion with acceleration g in the downward direction.

This result shows why we defined the concept of acceleration as a change with time of velocity, not speed. This definition means that a positive acceleration corresponds to either a speeding up in the positive direction or a slowing down in the negative direction. The stone example gives a rationale for the definition. It reveals a unity in the entire motion of a vertically projected stone.

The contrast between the Aristotelian and Galilean descriptions of the flight of the stone provides a counter-example to the usual, perhaps oversimplified, view of scientific thought. Galileo saw a simplicity not apparent to Aristotle. But one might charge that it is only a simplicity imposed by a highly abstract and artificially defined concept of acceleration. It seems almost *created* by Galileo. What does this imply about the notion of scientists discovering facts about nature? Did Galileo prove Aristotle wrong?

Motion in Two and Three Dimensions

Our discussion of a vertical projectile used the mathematical vocabulary for one-dimensional motion that we developed in Chapters 5 and 6. To understand Galileo's reconciliation of observed terrestrial motions with a moving earth, we must turn to his study of more general projectile motion, and extend our conceptual vocabulary for the description of motion to two and three dimensions.

Let us focus on a particularly simple example of two-dimensional motion. Consider a car moving along a road with a bend in it, as shown in Figure 7–2. To specify the position of

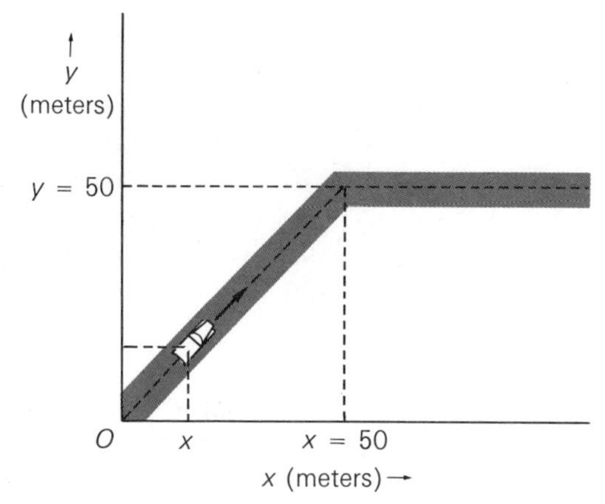

Figure 7–2. A car moving along a road. Its position can be specified by its x- and y-coordinates.

the car we need a reference frame, this time consisting of two perpendicular rigid rods with equally spaced marks. The two rods are known as the x-axis and y-axis respectively, while the point of intersection of the rods O is called the *origin* of the reference frame. To specify the position of a point, one draws perpendicular lines from the point to each of the rods. The distances from the origin O along the x-axis and y-axis to the perpendiculars are called, respectively, the x-coordinate and y-coordinate of the point. Thus in Figure 7–2 the origin O is specified by the coordinates $x = 0$ and $y = 0$ and the bend in the road is specified by the coordinates $x = 50$ meters and $y = 50$ meters.

If an object is moving, either or both of its coordinates change with time. We might describe the motion with words or snapshots. A more succinct quantitative description could be given in a table of values of x and y at various times t. Or we could construct graphs of x vs. t and y vs. t.

Table 7–1. The position of the car in Figure 7–2 at various times.

t (seconds)	x (meters)	y (meters)
0	0	0
1	10	10
2	20	20
3	30	30
4	40	40
5	50	50
6	60	50
7	70	50
8	80	50
9	90	50
10	100	50

A table corresponding to a typical motion of the car along the road shown in Figure 7–2 is given in Table 7–1. Describe this motion in words. On the following graphs, sketch x and y vs. t.

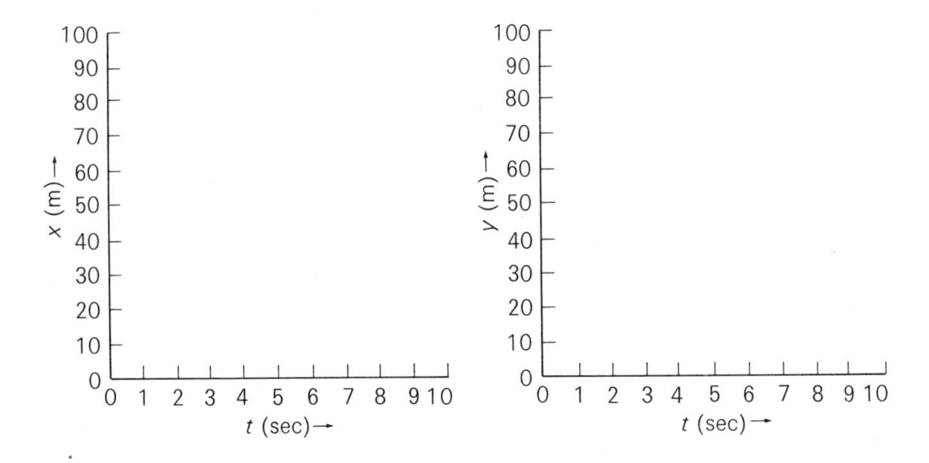

Such a specification of the way x and y depend on t gives a complete description of motion in two dimensions. We may describe motion in three dimensions by the obvious generalization of this idea: To specify position we use a reference frame with three mutually perpendicular axes labeled x, y, and z. The motion of an object is then described by the dependence of x, y, and z on t.

Vectors

In describing the motion of objects, another mathematical concept, the *vector*, is useful. An elementary example is a *displacement vector*, which we may illustrate with the motion of the car described in Table 7–1.

During the first 5 sec, the car moves from the position $x = 0$, $y = 0$ to the position $x = 50$ m, $y = 50$ m. In Figure 7–3, we represent this change in position by the arrow $\vec{D}_1$ from its position at the beginning of the time interval to its position at the end of the time interval. $\vec{D}_1$ is the displacement vector for the time interval from $t = 0$ to $t = 5$ sec.

Magnitude and Direction: The vector $\vec{D}_1$ is characterized by a magnitude and a direction. Its magnitude, which we label D_1, is

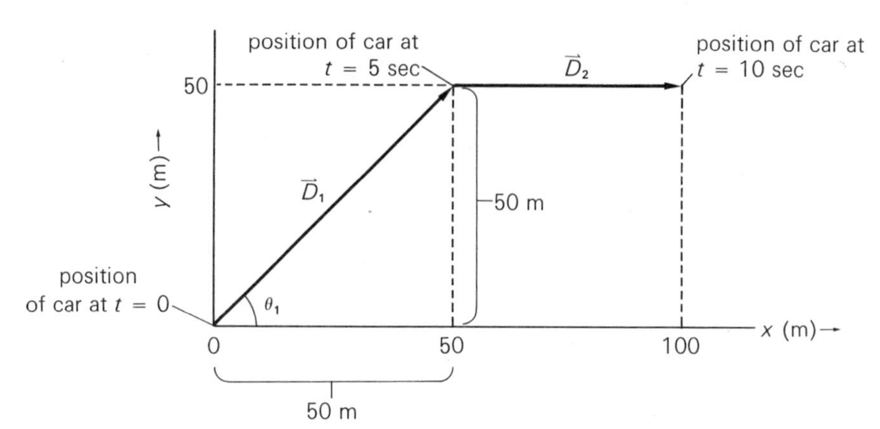

Figure 7–3. Displacement vectors $\vec{D}_1$ and $\vec{D}_2$ corresponding to time intervals from $t = 0$ to $t = 5$ sec and $t = 5$ to $t = 10$ sec, respectively.

the length of the displacement. Using the Pythagorean theorem, we find:

$$D_1 = \sqrt{(50\text{ m})^2 + (50\text{ m})^2} = 70.7\text{ m}$$

The direction is specified by the angle θ_1 between the vector and some fixed reference direction, which may be chosen arbitrarily. If we choose that reference direction to be the $+x$ direction, then:

$$\theta_1 = 45°$$

Figure 7–3 also shows the displacement vector $\vec{D}_2$ for the interval from $t = 5$ sec to $t = 10$ sec. The magnitude and direction of $\vec{D}_2$ are:

$$D_2 = 50\text{ m}$$

$$\theta_2 = 0°$$

Equality: We may specify a vector by giving its magnitude and direction. This means that it can be represented by an arrow of a given length and direction. We can say that two vectors are equal if they have the same magnitude and the same direction. For example, Figure 7–4 shows four arrows of the same length and direction as the displacement vector $\vec{D}_1$ of Figure 7–3. Thus each of these arrows is an equally good representation of the vector $\vec{D}_1$.

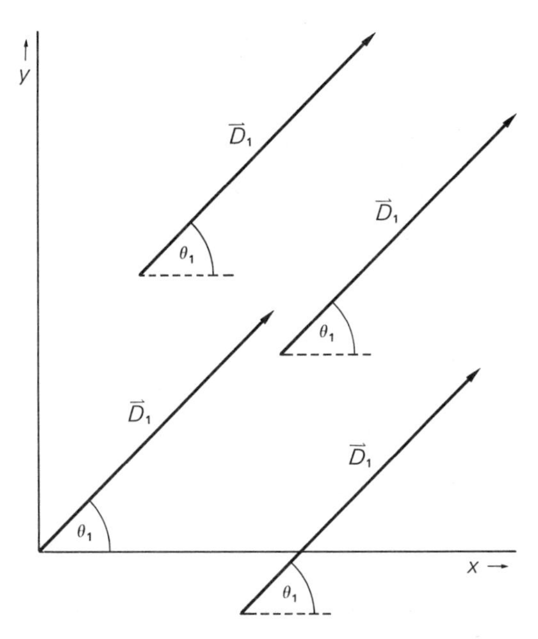

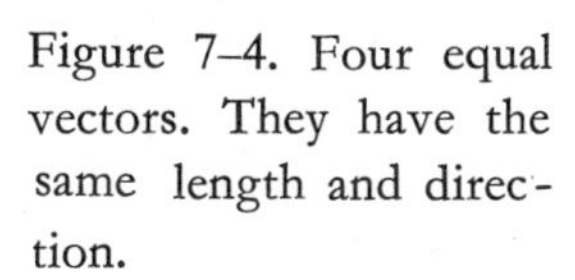
Figure 7–4. Four equal vectors. They have the same length and direction.

Addition: How might we define the addition of two vectors? A useful way is suggested by the displacement vectors we have been discussing.

Consider the ten-second interval between $t = 0$ and $t = 10$ sec. The displacement vector $\vec{D}$ during that interval is shown in Figure 7–5. The net effect of a displacement $\vec{D}_1$ followed by a displacement $\vec{D}_2$ is the displacement $\vec{D}$. We might say that $\vec{D}_2$ added to $\vec{D}_1$ equals $\vec{D}$. In fact, this observation provides a motivation for the definition of vector addition. We write:

$$\vec{D} = \vec{D}_1 + \vec{D}_2$$

with the understanding that this means that $\vec{D}$ is the vector from the tail of $\vec{D}_1$ to the head of $\vec{D}_2$ as depicted in Figure 7–5. We shall refer to this as the *vector law of addition.*

We shall see that displacements are not the only things that can be represented as vectors. The essential requirements for identifying an entity as a vector are that it must be specifiable by a magnitude and a direction (that is, by an arrow of a particular length) and two such entities must add according to the vector law of addition (Figure 7–6). These properties define a vector.

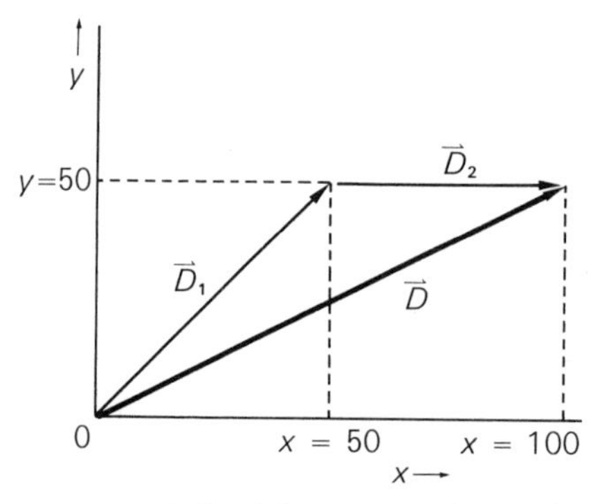

Figure 7–5. The vector law of addition. The tail of $\vec{D}_2$ is placed at the head of $\vec{D}_1$. $\vec{D}_1 + \vec{D}_2$ is the vector drawn from the tail of $\vec{D}_1$ to the head of $\vec{D}_2$.

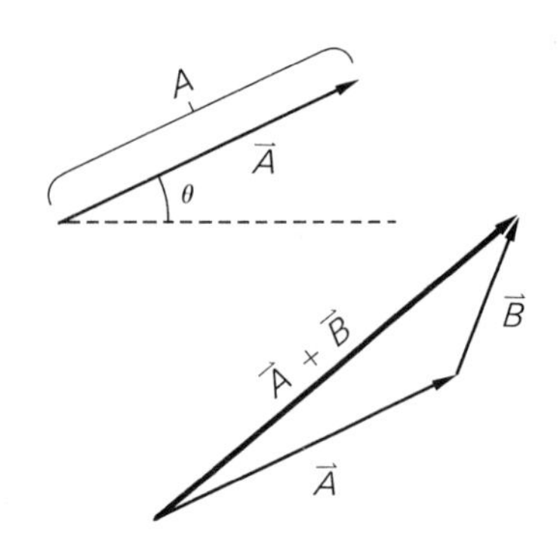

Figure 7–6. The defining characteristics of a vector: magnitude, direction, and vector law of addition.

Multiplication: Some other properties of vectors are implicit in this definition but deserve explicit mention. Suppose we add the vector $\vec{A}$ to the same vector $\vec{A}$ as shown in Figure 7–7(a). Just as when we add a number to itself we get twice the number, it is natural to refer to the vector $\vec{A} + \vec{A}$ as $2\vec{A}$. $2\vec{A}$ is a vector in the same direction as $\vec{A}$ but with twice the magnitude. Similarly Figure 7–7(b) indicates that the vector $3\vec{A}$ is a vector of three times the magnitude of $\vec{A}$ and in the same direction. Figure 7–7(c) shows that the vector $\frac{1}{2}\vec{A}$ (the vector that, when added to itself, gives $\vec{A}$) has one-half the magnitude of $\vec{A}$ and the same direction. These examples can be generalized to the statement that the vector $c\vec{A}$, where c is some number, is a vector having magnitude $c \cdot A$ and the same direction as $\vec{A}$.

Division: Division of a vector by a number presents no new problem, since division by a number c is the same as multiplication by $1/c$. For example, the vector $\vec{A}/2$ is the one shown in Figure 7–7(c).

Subtraction: With ordinary numbers, subtraction is a special case of addition. For example, $10 - 3 = 10 + (-3) = 7$. Subtraction is merely the addition of a negative number. A negative number, in turn, is whatever must be added to the corresponding positive number to give 0. For example, -3 is that number which, when added to $+3$, gives 0.

The negative of a vector can be defined in a similar way. We take the vector $-\vec{A}$ to be that vector which, if added to $\vec{A}$, gives a zero vector—that is, a vector with zero length. Figure 7–8 shows the vector $-\vec{A}$ corresponding to a given vector $\vec{A}$. Note that $-\vec{A}$ has the same magnitude as $\vec{A}$ but the opposite direction.

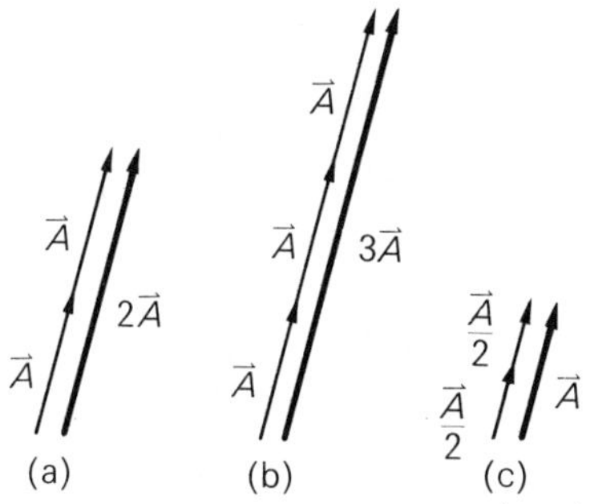

Figure 7–7. The vectors $2\vec{A}$, $3\vec{A}$, and $\frac{1}{2}\vec{A}$ have the same direction as $\vec{A}$ but twice, three times, and half its magnitude, respectively.

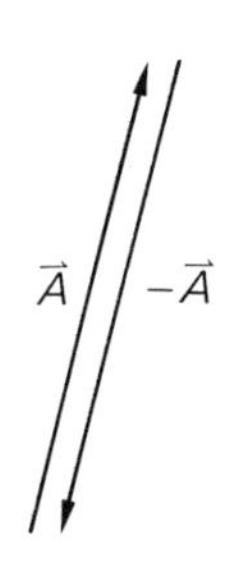

Figure 7–8. The negative of a vector. $-\vec{A}$ has the same magnitude as $\vec{A}$ but the opposite direction. Note that $\vec{A} + (-\vec{A})$ is the vector with zero length.

The vector $\vec{A} - \vec{B}$ is then obtained by adding the vector $-\vec{B}$ to $\vec{A}$:

$$\vec{A} - \vec{B} = \vec{A} + (-\vec{B})$$

This is illustrated in Figure 7–9.

Figure 7–9. The subtraction of vectors. (a) shows the vectors $\vec{A}$ and $\vec{B}$, and (b) indicates the construction of the vector $\vec{A} - \vec{B} = \vec{A} + (-\vec{B})$.

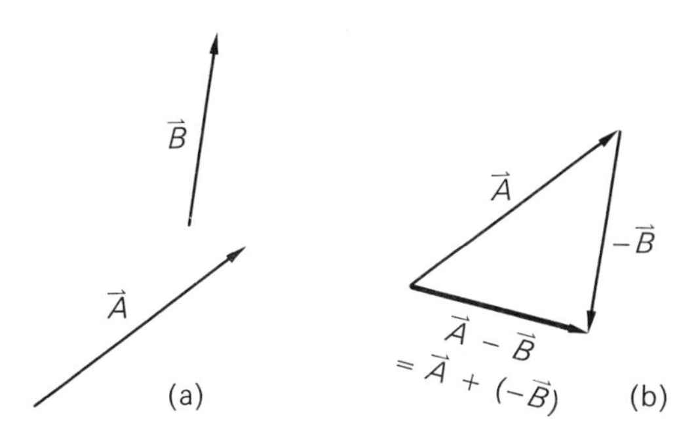

Vector Description of Two-Dimensional Motion

We have already discussed a simple way to describe the motion of an object in two dimensions: We select a frame of reference consisting of perpendicular x- and y-axes and specify the position of the object at each instant of time by its x- and y-coordinates. The motion is then completely described by specifying how x and y change with time.

An alternative method of description, equivalent but often simpler, uses the vector ideas we have just developed. Figure 7–10 shows a typical trajectory of an object, indicating its position at a time t. The vector $\vec{r}$ has been drawn from the origin O of the reference frame to the position of the object at time t. $\vec{r}$ is the *position vector*. Specifying the vector $\vec{r}$ at all times provides a complete description of the object's motion.

The Velocity Vector

We can now define a *velocity vector* to describe how rapidly the position is changing with time. Consider the position of the object at some initial time t_i and at a later final time t_f. The position vectors $\vec{r}_i$ and $\vec{r}_f$ at the two times are shown in Figure 7–11. The displacement vector $\overrightarrow{\Delta r}$ for the time interval $\Delta t = t_f - t_i$ is also shown in Figure 7–11. Recalling the law of vector addition, we see that

$$\vec{r}_f = \vec{r}_i + \overrightarrow{\Delta r}$$

or

$$\overrightarrow{\Delta r} = \vec{r}_f - \vec{r}_i$$

Dividing the displacement $\overrightarrow{\Delta r}$ during the time interval by the length of the interval Δt we obtain a vector

$$\vec{v}_{\text{ave}} = \frac{\overrightarrow{\Delta r}}{\Delta t}$$

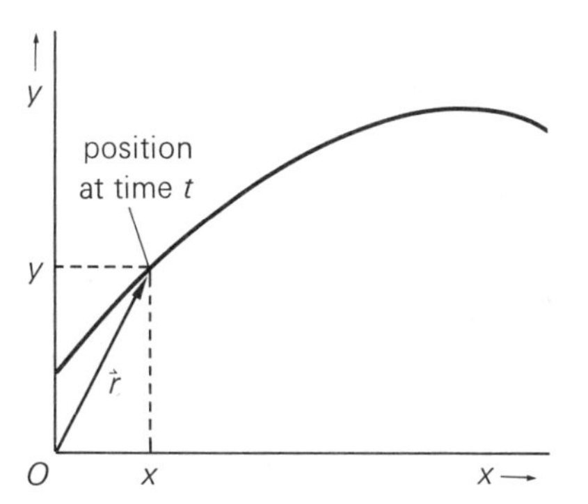

Figure 7–10. The position vector $\vec{r}$ drawn from the origin to the position of the object at time t.

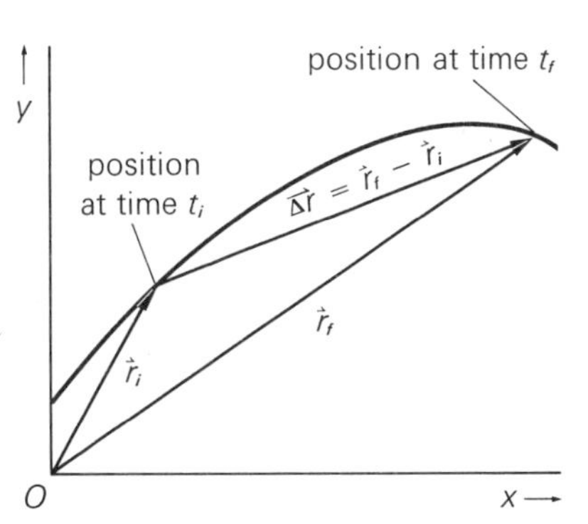

Figure 7–11. The displacement vector $\overrightarrow{\Delta r}$, corresponding to the time interval from t_i to t_f.

which has magnitude $\Delta r/\Delta t$ and is in the direction of $\overrightarrow{\Delta r}$.* We shall call this the average velocity vector during the interval Δt. It is clear from Figure 7–11 that Δr, the magnitude of $\overrightarrow{\Delta r}$, is a rough measure of the distance traveled by the object during the interval Δt. Thus $\Delta r/\Delta t$, the magnitude of $\vec{v}_{\text{ave}}$, is a rough measure of the speed of the object during this time interval. The direction of $\vec{v}_{\text{ave}}$, along $\overrightarrow{\Delta r}$, indicates roughly the direction of motion of the object during the interval.

Consider the average velocity vectors that result from making the time interval following t_i smaller and smaller. For example, Figure 7–12 illustrates displacement vectors $\overrightarrow{\Delta r_1}$, $\overrightarrow{\Delta r_2}$, and $\overrightarrow{\Delta r_3}$, corresponding to shorter and shorter time intervals Δt after time t_i. As the time interval grows shorter the magnitude of the displacement vector (its length) more closely approximates the distance traveled by the object during Δt and its direction more closely approximates the direction of motion at time t_i. Consequently, it is reasonable to define the *instantaneous velocity vector* $\vec{v}$ (commonly referred to as simply the *velocity vector*) of an object at a time t_i as the limiting value of $\overrightarrow{\Delta r}/\Delta t$ as Δt becomes very small:

instantaneous velocity vector $\vec{v}$
= limiting value of $\overrightarrow{\Delta r}/\Delta t$ as Δt is made very small

* Remember that the result of dividing a vector $\vec{A}$ by a number c is a vector in the same direction as $\vec{A}$ but with a magnitude $1/c$ times as large.

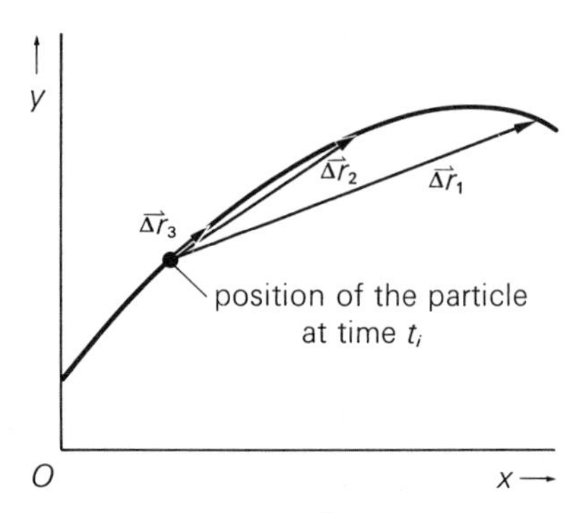

Figure 7–12. The limit of $\overrightarrow{\Delta r}$ as Δt becomes very small. $\overrightarrow{\Delta r_1}$, $\overrightarrow{\Delta r_2}$, and $\overrightarrow{\Delta r_3}$ are the displacement vectors for successively shorter time intervals after t_i. Note that the smaller the Δt, the more closely the magnitude of $\overrightarrow{\Delta r}$ approximates the distance traveled during the time interval and the more closely its direction approximates the direction of the object's motion at time t_i.

More succinctly, we write:

$$\vec{v} = \lim_{\Delta t \to 0} \frac{\overrightarrow{\Delta r}}{\Delta t}$$

The interpretation of $\vec{v}$ should be clear from our discussion. *Its magnitude is the speed of the object*, the distance traveled per unit time. *Its direction is the direction of motion at time t.* In fact, it is more convenient to think of $\vec{v}$ in terms of these properties than the abstract definition above. In the exercises below, simply think of $\vec{v}$ as a vector whose magnitude is the speed of the car and whose direction is the direction in which the car is moving.

Consider the car whose motion is described by Table 7–1. What is the magnitude and direction of the velocity vector of the car at $t = 2$ sec? At $t = 8$ sec?

Consider a car moving with constant speed counterclockwise around a circular track of radius 100 m, as shown in the figure. Suppose that at $t = 0$, the car is at the position indicated, and suppose it takes 100 sec for one revolution.

(a) What are the magnitude and direction of the velocity vector at $t = 0$? at $t = 12.5$ sec? at $t = 25$ sec? at $t = 50$ sec?

(b) Represent these four vectors by drawing arrows in the figure.

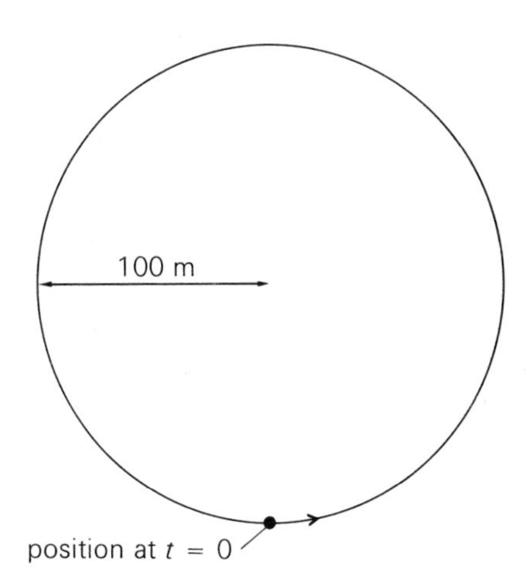

The Acceleration Vector

Similarly, we can recast the concept of acceleration in vector form. Suppose that during a time interval Δt the average velocity vector changes by $\overrightarrow{\Delta v} = \vec{v}_f - \vec{v}_i$. Then we can define the average acceleration vector $\vec{a}_{\text{ave}}$ as the ratio of $\overrightarrow{\Delta v}$ to Δt:

$$\vec{a}_{\text{ave}} = \frac{\overrightarrow{\Delta v}}{\Delta t}$$

As the time interval Δt is taken smaller and smaller, $\overrightarrow{\Delta v}/\Delta t$ approaches a limiting value we define as the *instantaneous acceleration vector* at time t:

instantaneous acceleration vector $\vec{a}$
= limiting value of $\overrightarrow{\Delta v}/\Delta t$ as Δt is made very small

$$\boxed{\vec{a} = \underset{\Delta t \to 0}{\text{limit}} \frac{\overrightarrow{\Delta v}}{\Delta t}}$$

It takes some practice to get a feeling for the direction of the acceleration vector. Consider a stone thrown vertically upward. Let us determine the direction of the acceleration vector while the stone is rising. Figure 7–13(a) shows the velocity $\vec{v}_i$ of the stone at a time t_i and its velocity $\vec{v}_f$ at a slightly later time t_f.

The magnitude of $\vec{v}_f$ is less than that of $\vec{v}_i$ since the stone is slowing down. In Figure 7–13(b) we construct $\overrightarrow{\Delta v} = \vec{v}_f - \vec{v}_i$ and find it is directed downward. It follows from the definition of the acceleration vector, $\vec{a} = \overrightarrow{\Delta v}/\Delta t$, for Δt small, that $\vec{a}$ is in the same direction as $\overrightarrow{\Delta v}$, and hence we conclude that $\vec{a}$ is directed downward while the stone is moving upward.

Now consider the downward motion of the stone. Again we examine the velocity $\vec{v}_i$ at a time t_i and the velocity $\vec{v}_f$ at a later time t_f, as shown in Figure 7–14(a).* Now the ball is speeding up, so the magnitude of $\vec{v}_f$ is greater than the magnitude of $\vec{v}_i$. As we can see from Figure 7–14(b), $\overrightarrow{\Delta v} = \vec{v}_f - \vec{v}_i$ is once again directed downward. Hence $\vec{a}$ is also directed downward as the stone is descending.

In fact, throughout its entire motion the stone is undergoing the *same* acceleration downward. Even at the instant it reaches the top of its trajectory, when its velocity is momentarily zero, it has this downward acceleration. These results correspond exactly to our earlier results for the same problem, obtained without vector concepts.

Inertia

The final break with Aristotelian physics came when Galileo considered horizontal motion, i.e., motion parallel to the surface of the earth. As we have seen, Aristotle felt that such motion required an external influence to sustain it, as in the example of a horse pulling a cart on a horizontal road. If the horse stops pulling, the cart comes to a stop.

* Note that $\vec{v}_i$ and $\vec{v}_f$ now stand for the initial and final velocity vectors during the downward part of the motion. They are not the same as the vectors identified in Figure 7–13 by these symbols.

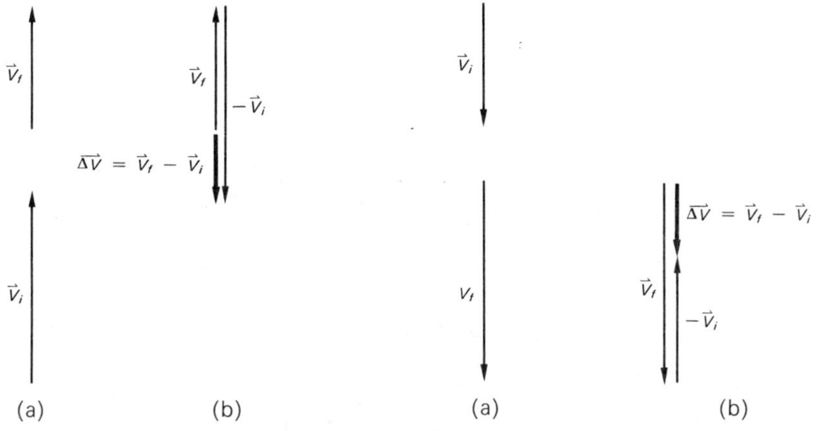

Figure 7–13. The upward part of the motion of an upward thrown stone. (a) indicates that the stone is slowing down, since the magnitude of the velocity vector $\vec{v}_f$ at the end of a short time interval is less than that of the velocity vector $\vec{v}_i$ at the beginning of the interval. (b) illustrates the construction of the vector $\overrightarrow{\Delta v} = \vec{v}_f - \vec{v}_i$. The fact that $\overrightarrow{\Delta v}$ is downward implies that the acceleration vector $\vec{a}$ is downward.

Figure 7–14. The downward part of the motion of an upward thrown stone. (a) indicates that the stone is speeding up, since the magnitude of the velocity vector $\vec{v}_f$ at the end of a short time interval is greater than that of the velocity vector $\vec{v}_i$ at the beginning of the interval. (b) illustrates the construction of the vector $\overrightarrow{\Delta v} = \vec{v}_f - \vec{v}_i$. The fact that $\overrightarrow{\Delta v}$ is downward implies that the acceleration vector $\vec{a}$ is downward.

Galileo suggested an almost diametrically opposed interpretation. Consider an object initially set in motion in the horizontal direction. In the absence of any external influence such as friction, the object will not slow to a stop, Galileo asserted, but will continue to move with constant speed. Let us examine his argument from the *Two World Systems*:

> Suppose you have a plane surface as smooth as a mirror and made of some hard material like steel. This is not parallel to the horizon, but somewhat inclined, and upon it you have placed

> a ball which is perfectly spherical and of some hard and heavy material like bronze. . . . On the downward inclined plane, the heavy moving body spontaneously descends and continually accelerates, and to keep it at rest requires the use of force.

Next he imagined the plane tilted upward:

> On the upward slope, force is needed to thrust it [the ball] along or even to hold it still, and motion which is impressed upon it continually diminishes until it is entirely annihilated. . . Now . . . what would happen to the same movable body placed upon a surface with no slope upward or downward[?] . . . There being no downward slope, there can be no natural tendency toward motion; and there being no upward slope, there can be no resistance to being moved, so there would be an indifference between the propensity and the resistance to motion.

Thus, he concluded, the ball on the level plane will either remain at rest if placed initially at rest, or

> if given an impetus in any direction . . . it would move in that direction. But with what sort of movement? One continually accelerated, as on the downward plane, or increasingly retarded as on the upward one?

Neither, he answers; on the level plane "I cannot see any cause for acceleration or deceleration, there being no slope upward or downward." Consequently, if the ball were placed initially in motion, it would continue to move with constant speed "as far as the extension of the surface continued without rising or falling."[2]

Of course, if the surface is extended without rising or falling, it will continue to parallel the surface of the earth, forming in the extreme a circle around the earth. Thus in the absence of friction a body set in motion on a level surface will move forever with constant speed on a circle about the center of the earth. It is therefore uniform circular motion that Galileo saw as natural motion—motion continuing of itself, needing no explanation. A ship on a frictionless ocean, given an initial push, will coast on around the world in a great circle. We may view this as the last dying gasp of the Greek ideal of uniform circular motion. But at the same time, as we shall discuss, this idea of circular *inertia* lies at the heart of Galileo's reconciliation of terrestrial motion with the heliocentric view of the universe.

The notion of inertia—the tendency of objects to sustain constant-velocity motion in the absence of external agents—was a revolutionary change from the Aristotelian point of view. As in the case of the falling stone, where he ignored air resistance, here again Galileo argued in terms of an idealized limit never achieved in practice—no friction. Aristotle had avoided such abstraction; the world is not this way. In his view, a prototype of realistic motion would be a horse pulling a wagon. There it is clear that motion requires a continued pull. Which view better represents reality? For most everyday experiences, Aristotle's clearly does. And yet in the end Galileo's was the more fruitful, for it led to a more detailed understanding of the physical world.

Galileo's Conjecture

We can now examine Galileo's analysis of the general motion of projectiles. Consider the situation shown in Figure 7–15. A ball is shot horizontally from a cannon at the edge of a cliff. Galileo saw the motion of the ball as a combination of two motions—

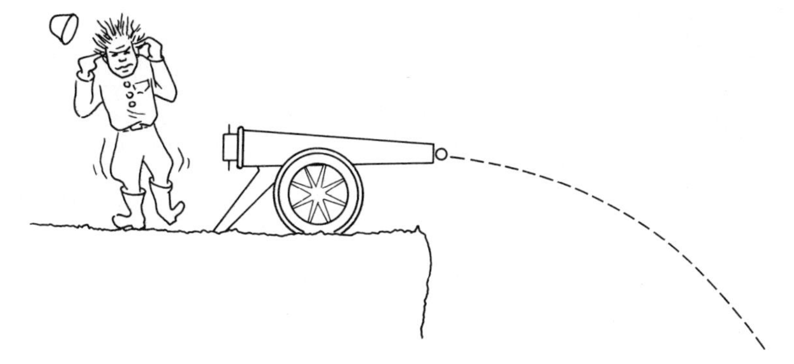

Figure 7–15. A cannon fired from the top of a cliff.

one horizontal and the other vertical. Moreover, he speculated that it might be possible to consider these motions as completely independent, each behaving as if the other were not present. That is, the vertical position coordinate would change with time just as in the case of a freely falling body—with a downward acceleration of g. At the same time, the horizontal position coordinate would change with time just as in the case of an object moving horizontally without resistance—with a constant velocity.

This would be a very simple kind of motion; could it be that nature really follows this scheme? Galileo examined in detail the motion of real projectiles, comparing it with that predicted by his conjecture. An elementary consequence of his conjecture is that the path of the projectile should be a particular kind of curve—a parabola.

To see that this result follows from the independence of the horizontal and vertical motions, we choose a suitable frame of reference. A convenient one is shown in Figure 7–16; its origin is the position of the cannonball as it leaves the barrel of the cannon and its $+y$-axis points down. Each point on the trajectory of the ball then has a positive x-coordinate and a positive y-coordinate.

Consider the horizontal motion first. The ball is shot horizontally with an initial velocity we shall call v_0. According to Galileo's conjecture, the horizontal motion is independent of

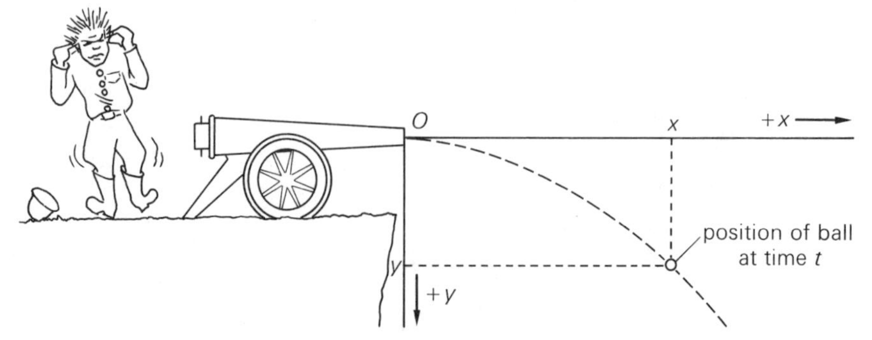

Figure 7–16. The motion of a cannonball fired horizontally. The path will be a parabola if Galileo's conjecture is correct.

the fact that the ball is falling. Thus the velocity in the x direction remains constant, namely v_0, throughout the motion. We can express this result algebraically. Let us call $t = 0$ the time the ball leaves the cannon barrel. At time $t = 0$ the x-coordinate of the ball is $x = 0$. Since its x motion corresponds to constant velocity v_0, its x-coordinate at any later time t will be:

$$x = v_0 t \qquad (5)$$

Similarly, according to Galileo's conjecture, the vertical motion is unaffected by the horizontal motion. Consequently the ball moves in the y direction as a freely falling body—with constant acceleration g. Since its y-coordinate at $t = 0$ is $y = 0$ and it has no initial velocity in the y direction, its y-coordinate at a later time t will be [cf. equation (4) on page 139]:

$$y = \tfrac{1}{2}gt^2 \qquad (6)$$

Equations (5) and (6) specify how the position coordinates of the ball vary with time. They therefore provide a complete description of its motion. Whether the motion of a real cannon-ball projected horizontally is actually that described by these equations is a matter for experimental investigation. Galileo himself suggested such an experiment:

> Barring the accidental impediment from the air, I consider it certain that if, when one ball left [horizontally from] the cannon, another one were allowed to fall straight down from the same height, they would both arrive on the ground at the same instant, even though the former would have traveled ten thousand yards and the latter a mere hundred.[3]

Figure 7–17 depicts the results of a modern experiment designed to test Galileo's conjecture. In this experiment a ball is projected horizontally while at the same time another ball is allowed to fall from rest. A multiple-exposure photograph displays the positions of the two balls at equally spaced times.

Note that at each time the vertical coordinates of the two balls are the same. Thus the vertical position of the projected ball does change with constant acceleration g as described by equation (6). Notice also that the distances between successive horizontal positions of the projected ball all appear to be equal.

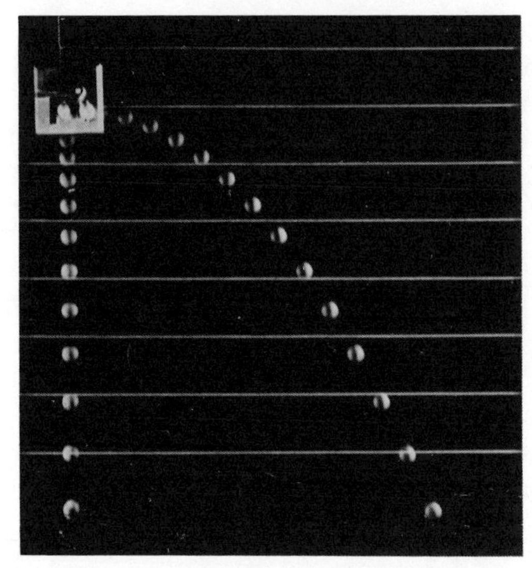

Figure 7–17. A multiple-exposure photograph of the motion of two balls, one projected horizontally at the same instant the other is dropped from rest. (Photograph from PSSC *Physics*, D. C. Heath and Company, Lexington, 1965.)

Hence the horizontal position of the projected ball does display the constant-velocity motion described by equation (5). Thus, Galileo's conjecture is supported by this experiment.

By eliminating the variable t from equations (5) and (6), we can find an expression for the trajectory of a projectile like that shown in Figure 7–17. Let us solve equation (5) for t:

$$t = \frac{x}{v_0}$$

This can be substituted into equation (6):

$$y = \frac{g}{2}\left(\frac{x}{v_0}\right)^2$$

or

$$y = \left(\frac{g}{2v_0^2}\right)x^2$$

This equation specifies the path of the projectile.

Note that y is proportional to x^2, which means that the path is a parabola. This is a curious and surprising result. Not only is the path exceedingly simple, but it corresponds to one of the most fundamental curves known in abstract mathematics—the intersection between a cone and a plane parallel to the edge of the cone (Figure 7–18). Who but a Pythagorean would have guessed that there would be a connection between such a mathematical abstraction and the physical world?

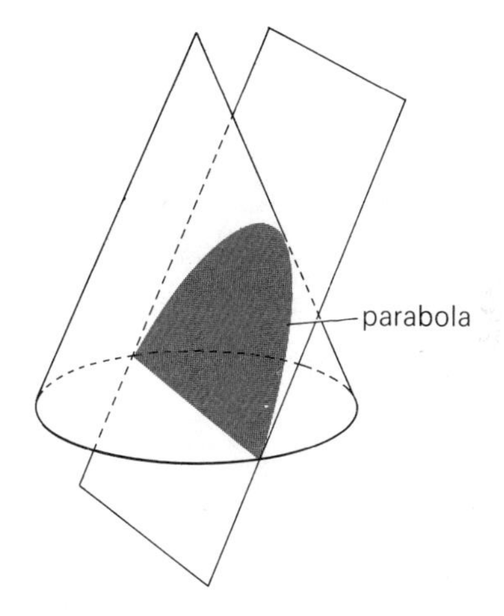

Figure 7–18. A parabola as the intersection of a plane and a cone.

The experiment of Figure 7–17 has been cited as evidence in support of Galileo's analysis of projectile motion. It is important to realize, however, that neither Galileo nor modern observers would claim that real projectiles move in precisely parabolic paths. There are significant discrepancies, particularly apparent in the motion of long-range artillery shells. Galileo, noting these discrepancies, emphasized their small magnitude and attributed them to the effects of air resistance. Aristotelians of that day, however, could cite these discrepancies as evidence against not only the Galilean conjecture, but the whole Galilean approach to the study of terrestrial motion.

Projectiles: A Vector View

We can cast Galileo's conjecture in an especially succinct and useful form in terms of the vector ideas we have developed: *The acceleration* $\vec{a}$ *of a projectile is a vector of magnitude* g *and direction vertically downward toward the center of the earth*, whatever the projectile's initial horizontal motion. This is equivalent to our previous statement of the conjecture, as we shall now see.

We begin with an elementary mathematical theorem: Any vector can be represented as the sum of a vector in the horizontal direction and a vector in the vertical direction. Figure 7–19 illustrates this representation of a vector $\vec{A}$. The horizontal vector is known as the *horizontal component* of $\vec{A}$ and the vertical vector as the *vertical component* of $\vec{A}$.

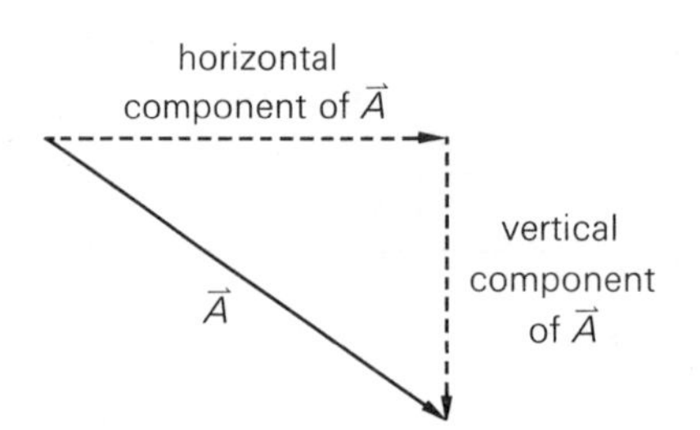

Figure 7–19. The horizontal and vertical components of a vector.

The horizontal and vertical components of the position vector $\vec{r}$ of an object have a simple interpretation. As can be seen in Figure 7–20, the length of the horizontal component is simply the object's x-coordinate, while the length of the vertical component is its y-coordinate.

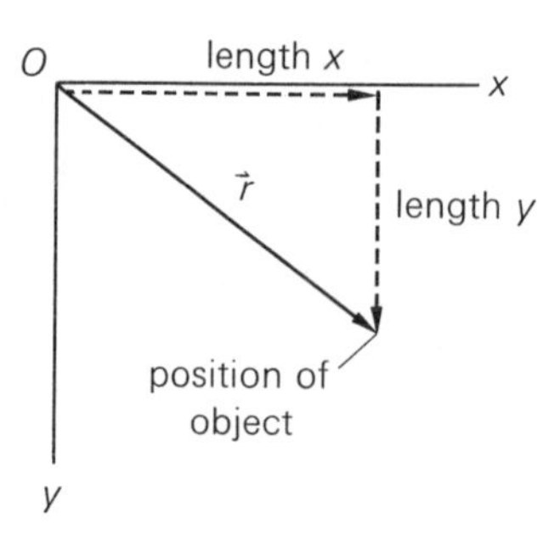

Figure 7–20. The horizontal and vertical components of the position vector. They are the x- and y-coordinates, respectively.

The horizontal and vertical components of the velocity vector $\vec{v}$ also have a simple interpretation. Recall that $\vec{v}$ was defined to be the ratio $\overrightarrow{\Delta r}/\Delta t$, for very short Δt. Figure 7–21 shows the change $\overrightarrow{\Delta r}$ in the position vector corresponding to a

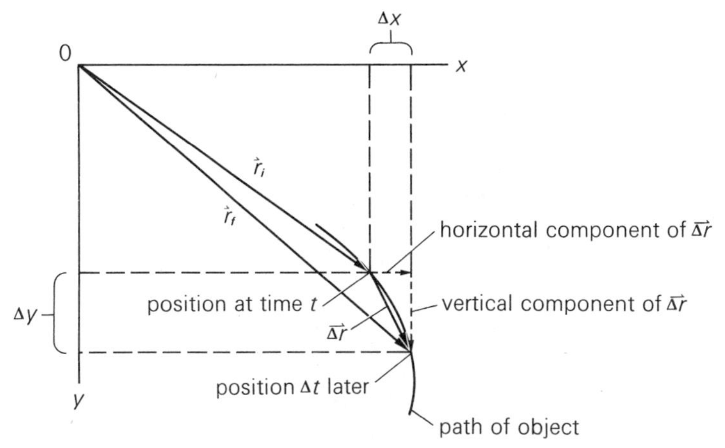

Figure 7–21. The horizontal and vertical components of the vector $\overrightarrow{\Delta r}$. They are Δx and Δy, respectively.

very short Δt. Note that the horizontal component of $\overrightarrow{\Delta r}$ corresponds to a small change Δx in the x-coordinate of the object's position, while the vertical component of $\overrightarrow{\Delta r}$ corresponds to a small change Δy in the y-coordinate. Thus the velocity vector $\vec{v}$ has a horizontal component whose magnitude is $\Delta x/\Delta t$ and a vertical component whose magnitude is $\Delta y/\Delta t$. Thus the horizontal and vertical components of $\vec{v}$ specify how fast the x- and y-coordinates change with time.

Let us apply these ideas to the description of a horizontally projected object. Equations (5) and (6) specify how x and y change with time in this case. Equation (5) indicates that the horizontal component of the velocity vector has a constant magnitude v_0. Equation (6) indicates that the y-motion corresponds to a constant acceleration g. This is equivalent to saying that the vertical component of the velocity vector has a magnitude that increases in proportion to time. In other words, in each time interval Δt, the magnitude of the vertical component of the velocity increases by $\Delta v = g\Delta t$. In Figure 7–22(a) we have used this prescription to construct the velocity vectors at a sequence of equally spaced times. The vectors have the same horizontal component, but their vertical components increase in proportion to time.

What is the acceleration vector $\vec{a}$ corresponding to the velocity vectors of Figure 7–22(a)? From Figure 7–22(b) it is clear that in

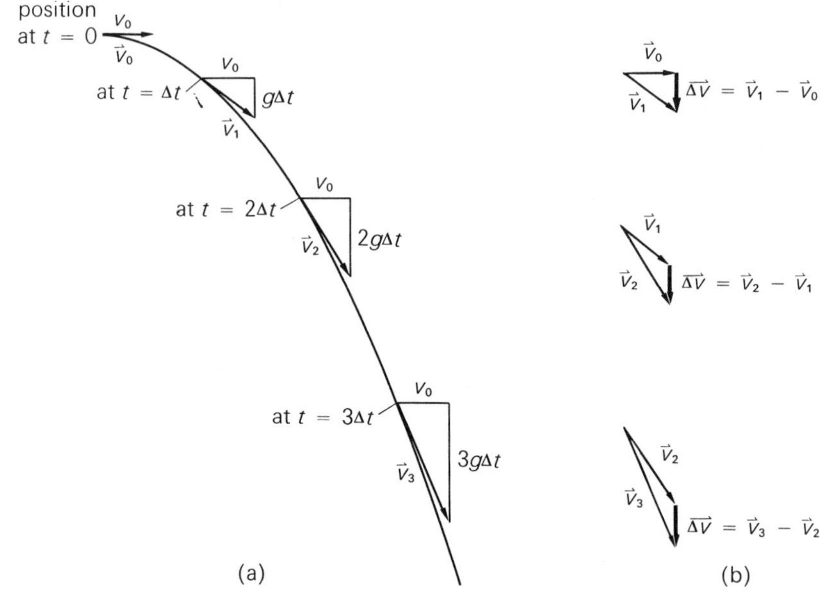

Figure 7–22. The velocity vector for a horizontally projected object and its change with time. (a) Construction of the velocity vectors $\vec{v}_0$, $\vec{v}_1$, $\vec{v}_2$, and $\vec{v}_3$ of a projectile at $t = 0$, Δt, $2\Delta t$, and $3\Delta t$, respectively. (b) The changes in velocity in each of the time intervals Δt is the same, $\overrightarrow{\Delta v}$. Thus the acceleration $\overrightarrow{\Delta v}/\Delta t$ remains constant during this motion.

each time interval Δt the change $\overrightarrow{\Delta v}$ in the velocity vector of the projectile is the same—in the downward direction with magnitude $\Delta v = g\Delta t$. This means that the acceleration vector $\vec{a}$, defined as $\overrightarrow{\Delta v}/\Delta t$ for very small Δt, is the same at each instant of time. It has a constant magnitude $\Delta v/\Delta t = g$ and is directed downward toward the center of the earth throughout the motion. Thus we have established the vector form of Galileo's conjecture.

Falling Objects and the Motion of the Earth

We began this chapter with a discussion of the apparent incompatibility between the heliocentric notion of a moving earth and the observation that objects released from rest fall vertically.

Our argument was by analogy: we noted that a stone dropped vertically from a bridge would appear to move in a curved path if viewed from a passing ship, and argued that by the same token if the earth were moving, an observer on earth should see a falling body as moving on a curved path. Let us reconsider the issue in the light of the Galilean idea of inertia.

In the first place, have we really considered the proper analogy? We are interested in how the fall of a stone, released from above the surface of the earth, will appear to an observer on the earth. The important point to note is that prior to its release the stone is fixed to the earth. If the earth is moving, then prior to its release the stone is moving too. Wouldn't the correct analogy be an observation of a stone released from the ship by an observer on the ship?

Suppose a stone were released from the top of the mast of a ship, as illustrated in Figure 7–23. At the instant of release the stone has not yet acquired any velocity in the vertical direction. However, the stone does have an initial horizontal velocity—that of the ship. The ship's motion projects the stone. Once released, the stone follows the projectile path described by Galileo. Its vertical motion is downward with constant acceleration g. However, at the same time it continues to move in the horizontal direction with a constant velocity—that of the ship.

Figure 7–23. A stone is released from the mast of a moving boat.

Thus, with respect to the shore, the stone describes a parabolic path as shown in Figure 7–24(a). But since the horizontal velocity of the stone and the ship are the same, the stone falls vertically with respect to the ship. This is shown in Figure 7–24(b), which

Figure 7–24. The fall of a stone as seen by two observers, one at rest on the shore and the other moving with the boat. (a) The stone falls in a curved path according to an observer at rest on the shore. (b) The stone falls vertically according to an observer moving with the boat.

we can imagine to be taken by a camera moving along with the ship. Notice that the motion in Figure 7–24(b) is exactly the same as what we would observe if the ship were not moving and the stone were released from the top of the mast. One cannot tell from the motion of the stone relative to the ship whether or not the ship is moving.

It is clear that the same argument will hold for the motion of the earth. A stone dropped from a bridge will still drop vertically downward even if the earth is moving, just as the stone dropped from the mast of the ship did in the previous example. The stone is not left behind; its initial velocity in the horizontal direction is maintained, allowing it to keep up with the earth as it falls.

With this reasoning a major argument against the motion of the earth was discredited. Or so it might seem. Yet the furious and bitter controversy between Galileo and the latter-day followers of Aristotle was evidence that an objective reconciliation between the two world views was not possible. When Galileo was brought before the Inquisition, his opponents, men of learning and intelligence like himself, testified to the error of his views. In 1634 Galileo was forced by the Church to recant.

Suggested Reading

Arons, A. *Development of Concepts of Physics.* Reading, Mass.: Addison-Wesley, 1965. Chapter 4 treats vectors and projectile motion.

Holton, G. *Introduction to Concepts and Theories in Physical Science.* Reading, Mass.: Addison-Wesley, 1952. Chapter 3 presents a similar treatment.

de Santillana, G. *The Crime of Galileo.* Chicago: University of Chicago Press, 1955. Galileo's conflict with the Church is related in an interesting and thoughtful fashion.

Brecht, B. *Galileo.* In *Seven Plays.* New York: Grove Press, 1961. A well-known play on the same subject.

Galilei, G. *Dialogues Concerning Two New Sciences.* Translated by H. Crew and A. deSalvio. New York: Macmillan, 1914. Galileo's major work on terrestrial motion, surprisingly readable.

Galilei, G. *Dialogue Concerning the Two Chief World Systems.* Translated by S. Drake. Berkeley: University of California Press, 1953. Galileo's other major work, in which he implicitly supports the Copernican over the Ptolemaic system.

Butterfield, H. *The Origins of Modern Science.* London: Bell, 1949. Chapter 1 concerns the concept of impetus.

Questions

1. State clearly, from a Galilean point of view, what is wrong with the arguments against the motion of the earth presented on page 147.
2. Investigate the details of some of the impetus theories proposed by Aristotelians to account for projectile motion.
3. Scientists sometimes disagree with each other in their testimony before congressional committees on the feasibility and desirability of technological programs such as the supersonic transport, the antiballistic missile, etc. Compare

and contrast such disagreements with those between Galileo and his Aristotelian opponents.

4. Sketch a crude map of the United States. Indicate on your sketch the following displacement vectors:

$\vec{D}_1$: from Los Angeles to Dallas
$\vec{D}_2$: from Dallas to Boston
$\vec{D}_3$: from Boston to Los Angeles

How are these three displacement vectors related?

5. An object has velocity $\vec{v}_i$ at time t_i and velocity $\vec{v}_f$ at a later time t_f. The directions of these velocity vectors are different. Consider the change in velocity:

$$\overrightarrow{\Delta v} = \vec{v}_f - \vec{v}_i$$

Is the magnitude of $\overrightarrow{\Delta v}$ equal to the difference between the magnitudes of $\vec{v}_f$ and $\vec{v}_i$? Explain.

6. Is it necessarily true that an object moving in two dimensions with a non-zero acceleration must either be speeding up or slowing down? Explain.

7. In the exercise on page 160, you constructed the velocity vectors at different instants of time for a car moving in a circle with constant speed. If properly constructed, the velocity vectors corresponding to different instants of time should point in different directions.

   (a) Is the car accelerating? In what direction?
   (b) If the car moves with a greater constant speed around the same circle, will the magnitude of its acceleration be larger or smaller than in the previous case? Explain.
   (c) If the car moves with the same constant speed around a larger circle, will the magnitude of its acceleration be larger or smaller than in the previous case? Explain.

8. A car races with constant speed around an oval track, as shown in the figure. Its position vectors at times t_1 and t_2 are $\vec{r}_1$ and $\vec{r}_2$ respectively.

   (a) Sketch the instantaneous velocity vectors at times t_1 and t_2.
   (b) Construct the average velocity vector during the time interval from t_1 to t_2.

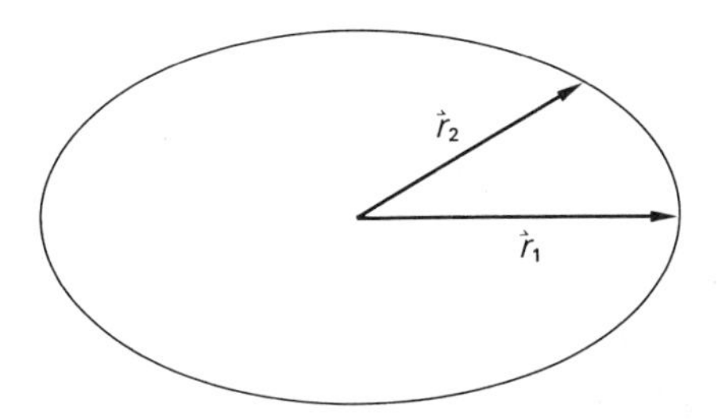

(c) Construct the average acceleration vector during this time interval.

9. An object moves along the path shown below in such a way that its speed doubles each second. Its positions at t = 1 sec, t = 2 sec, and t = 3 sec are shown in the figure. Draw and label each of the following:

(a) The position vector at t = 2 sec.
(b) The position vector at t = 3 sec.
(c) The average velocity vector during the time interval from t = 2 sec to t = 3 sec.
(d) The instantaneous velocity vector at t = 2 sec. (Assume an arbitrary magnitude.)
(e) The instantaneous velocity vector at t = 3 sec.
(f) The average acceleration vector during the time interval from t = 2 sec to t = 3 sec.

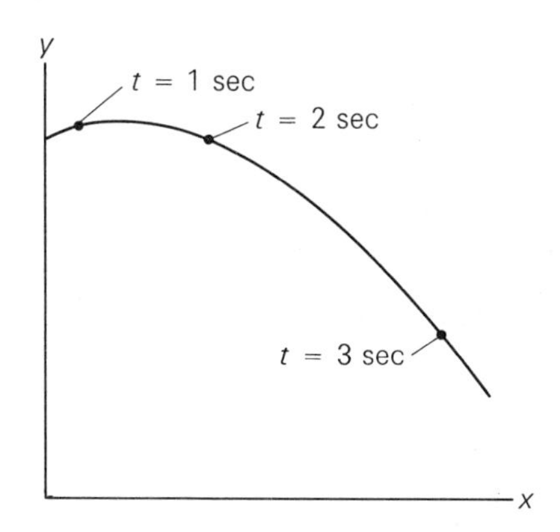

10. Suppose one of the engines falls off an SST (supersonic transport) flying at a speed of 1500 miles/hour at an altitude of 10 miles.

(a) How far is the point at which the engine hits the ground from the point at which it fell off the plane?
(b) At the time the engine hits the ground, how far will it be from the SST? Assume the SST continues to fly in a straight line with its original speed.

11. Suppose level ground is 100 m below the cannon barrel in Figure 7–16. The cannon fires a ball with an initial speed $v_0 = 100$ m/sec. Using the value $g = 9.8$ m/sec^2 for the gravitational acceleration, calculate how long it will take the cannonball to reach the ground. Also determine how far from the base of the cliff the ball will strike the ground. Does your result agree with that calculated from the equation of the ball's path on page 168?
12. A train moves along a straight track at a constant speed of 30 m/sec. A sinister figure atop one of the cars releases a canister of microfilm from rest (relative to the train) a distance of 5 m above the track. At the instant before the canister strikes the track, what is the velocity of the canister relative to the train? relative to the track?
13. A hunter sits in a tree, 100 feet above the ground. A monkey is hanging from a branch of another tree, also 100 feet above the ground. The hunter shoots the monkey (using a soft, slow, harmless bullet, of course). The monkey, startled by the flash of the gun, releases the branch and starts to fall at the same instant that the bullet leaves the gun barrel. Applying Galileo's analysis of the motion of falling objects and projectiles, explain why the bullet will strike the monkey as he falls to the ground.
14. The hunter and monkey experiment described in Question 13 is often performed in physics courses as a demonstration. A toy gun is mounted high above the floor and aimed at a tin-can "monkey" suspended from an electromagnet at the same height as the gun. The "bullet" breaks an electrical circuit as it leaves the gun barrel, turning off the electromagnet and releasing the "monkey." In many cases, when the instructor adjusts the apparatus before class, he finds that he must aim the gun slightly below horizontal to hit the "monkey." Explain what might be responsible for this apparent failure of Galileo's theory. Does such a failure constitute grounds for rejecting the theory?

THE NEWTONIAN SYNTHESIS

What makes planets go around the sun? At the time of Kepler some people answered this problem by saying that there were angels behind them beating their wings and pushing the planets around an orbit . . . The answer is not far from the truth. The only difference is that the angels sit in a different direction and their wings push inwards.[1]

—Richard P. Feynman
The Character of Physical Law

You probably believe it is gravity that causes the planets to orbit the sun. What is the nature of gravity? What possible connection could there be between a "scientific" concept like gravity and the metaphysical notion of angels?

CHAPTER EIGHT

The Creation of Force

Thus far we have examined two problems central to man's understanding of the physical world—terrestrial motion and celestial motion. Initially these two problems were considered unrelated, reflecting not only two different regions of space but also the Aristotelian philosophical position that the sublunary and superlunary worlds were quite distinct and governed by entirely different principles.

The synthesis of these two problems into one, the problem of motion in general, was the historic contribution of Isaac Newton, who was born in England in 1642, the year of Galileo's death. Destined to become recognized as the chief architect of modern science, Newton was brought up in a small country town in Lincolnshire and attended a nearby grammar school, studying the Latin classics that were the educational fare of the day. He was apparently a withdrawn youth, not popular among his classmates. His unusual curiosity about the physical world led him to mechanical experiments and inventions, many of some ingenuity. An eighteenth-century biographer of Newton reported that

> These fancies sometimes engrossed so much of his thoughts, that he was apt to neglect his book, and dull boys were now and then put over him in form. But this made him redouble his pains to overtake them, and such was his capacity, that he could soon do it, and out-strip them when he pleased; and it was taken notice of by his master.[1]

In 1661 Newton enrolled at Trinity College, Cambridge, as a "subsizar," a scholarship student who performed a number of menial tasks to pay his keep. His studies included natural philosophy, theology, and mathematics, and in them his unusual mind soon attracted his tutors' attention. During the following three years at Cambridge his thoughts began to turn seriously to the problems of celestial and terrestrial motion.

There were many unanswered questions about both problems. For terrestrial motion, Galileo had described the motions of falling bodies and projectiles with some success, but had evaded the questions of why all objects fall with the same uniform acceleration and why horizontally moving objects continue to move with constant speed. Should a theory of terrestrial motion answer such questions, or should one be content with the description Galileo had given? If a theory should answer these "why" questions, what sort of answers are adequate? Should one seek a mechanism to explain the uniform acceleration of free fall? the constant speed of horizontal motion? Or should one search for some modification of the Aristotelian scheme to account for the Galilean results?

The problem of celestial motion involved similar ambiguities. Perhaps the most striking feature of our study thus far is the failure of men to agree on just what the problem of celestial motion is. By all accounts, the planets are a key puzzle, but should a theory of the planets, as Ptolemy believed, describe their motion? Or should it, following Kepler, explain the number of planets and the sizes of their orbits by geometric principles like the five perfect solids? Or should the focus be that of Descartes, who sought a description of the mechanism behind the motion?

Newton found the whole state of affairs with regard to celestial and terrestrial motion unsatisfactory. He was not willing to accept the descriptions of Kepler and Galileo as sufficient. What was needed, he felt, was a different approach, which would deal with the causes of the motions. Descartes' theory was a step in the right direction, for he had attempted to describe a mechanism that moved the planets. On the other hand, Descartes had not established any connection between this mechanism and Kepler's laws of planetary motion.

The Principia

Newton was a reticent man, not given to discussing his feelings publicly. He guarded his scientific ideas jealously until he was ready to publish them. For this reason, we know very little of what went on in his mind as he worked. His uncertainties, his false tries, his failures remain hidden from us. Most of what we know about his theories of motion comes from his major work, the *Philosophiae Naturalis Principia Mathematica* (*Mathematical Principles of Natural Philosophy*), usually called simply the *Principia*, which was published in 1686, when Newton was 44. It contains in detail essentially all Newton's work on the motion of bodies and of the heavens—a creation so successful, original, and far-reaching that the *Principia* may well be called the single most important book in the history of science. Its style is cold and rigid; the format is that of Euclidean geometry, a deductive structure of axioms, theorems, and proofs. The proofs are almost entirely geometric, and frequently are challenging even for a mathematically trained person today. Now we can give equivalent but far simpler and more elegant proofs using calculus,* and we no longer use precisely the arguments Newton presented. Nevertheless, although the *Principia* was written three hundred years ago, its general plan and approach

* In fact, Newton invented the calculus (as did Leibniz, working independently) for just this purpose. However, he employed geometric proofs in the *Principia* to obtain maximum acceptance by his contemporaries, who were not yet conversant with the new mathematics.

to understanding nature are clearly recognizable as the basis of much of the physics taught today.

Interactions

At the heart of Newton's approach is his concern not with how an object moves *per se* but rather with what causes it to move the way it does. Presumably the answer has to do with its interactions with other objects. Someone pulls on it, a wind blows it, a magnet attracts it. In each case, the surroundings affect the object by some sort of push or pull. The central question of the Newtonian scheme is: How can we describe the interactions that produce the motion?

While asking the right questions is the first and probably the most important step in establishing a theory, it is by no means the whole story. We have seen that Descartes had also concentrated on describing the interactions responsible for planetary motion, but he proceeded in an entirely different direction from that of Newton. What comes next in the creation of a theory is a highly complex process we might loosely call intuitive. It involves various metaphysical preconceptions and a complicated conglomerate of clues from empirical data. For the Ptolemaic theory of epicycles, the Aristotelian notions of perfect circles and a fixed earth were essential guides. For Kepler's theory of perfect solids, an important factor was a feeling akin to the Pythagorean view that nature reflects geometric simplicity. Descartes felt that interactions must take place through contact and was struck by the similarity between the orbits of the planets and the vortices in an emptying basin of water. Although we have little direct knowledge of the intuitive ideas that lay behind Newton's formulation, we know enough about their culmination in the *Principia*, and about the general state of physics in Newton's time, to attempt a plausible account of these ideas.

What is the relation between a force* on an object and the object's resultant motion? Aristotle had concluded that a force

* The term force will presently be given a more careful definition. For the present we use it as a synonym for push or pull.

is necessary to sustain constrained motion. To keep an object moving with constant velocity one must continue to apply a force. When a horse pulls a wagon over rough ground, the harder the horse pulls, the faster the wagon moves; the smoother the ground, the faster the motion for a given pull. The speed of the wagon remains constant as the horse continues to pull (except for an initial transition period when the wagon is accelerated from rest). If the horse stops pulling, the wagon slows to a stop.

Having been brought up in a world pervaded by Newtonian ideas, we might object that the motion of the wagon is resisted by friction and that is why it is necessary to continue pulling to keep it moving. We might say that one really ought to consider how the wagon would move if the friction were reduced to zero. This objection makes good sense if one has tacitly adopted the Newtonian viewpoint that the basic goal is describing the forces that produce motion. Aristotle, as we have seen, did not view motion in these terms; to him, since one could never actually remove friction, it was pointless and even misleading to think it away. What is the idea of friction, an Aristotelian might ask, but another way to express the universal tendency of objects to come to a stop when they are no longer pushed or pulled?

As we have seen, Galileo's orientation was quite different from Aristotle's. He was struck by the observation that as resistance is reduced, a body in motion will continue to move even if there is no force acting on it. We have already discussed his example of a sailing ship. Though suddenly becalmed, it continues to move. If there were no resistance at all, Galileo speculated, the ship would move with constant speed along its natural path, a great circle about the earth. He generalized such observations to the principle that the natural motion of bodies at the surface of the earth is motion along a great-circle path at constant speed. To Galileo, frictional and other resistive forces, although unavoidable in the motion of any real object, were merely complications that obscured this simple natural motion.

Newton, unlike Galileo and Aristotle, was concerned with the nature of the forces that cause motion rather than with a description of the motion. Given this perspective, it was natural for him to examine situations in which a single force is isolated. From his viewpoint, the motion produced when a horse pulls a wagon over rough ground is a consequence of two forces: the pull of the horse and the frictional resistance resulting from the contact between the wagon wheels and the earth. Because two forces are acting, one cannot learn the detailed effect of either one by observing the motion. To investigate the effect of the pull of the horse, one would try to observe a situation in which the frictional force was negligibly small. To investigate the frictional force, one would observe the motion after the horse had stopped pulling.

Consider another situation—a steel ball falling in a viscous medium, for example in a tank of water. Again two forces are acting: the resistance of the medium and the force that draws the object toward the surface of the earth. Clearly if one wanted to investigate the interaction that pulls the object toward the earth, the data obtained from this experiment would not be very useful because they are complicated by the presence of the resistive force. While an Aristotelian might regard a ball falling in water as a prototype of realistic, resisted motion and therefore worthy of study, a Newtonian would focus instead on an experiment in which the resistive force was not present, e.g., the fall of an object in air, or, better yet, in a vacuum.

These examples illustrate an important facet of the interrelation between observation and theory in physics. According to a naïve description of the scientific process, physicists first observe nature, then dream up an hypothesis to account for their observations. In practice, however, theoretical considerations exert an important influence on what data are observed. The questions Aristotle asked make the viscous medium experiment pertinent, while in the Newtonian context, with its focus on interactions, this experiment is not of fundamental relevance.

Newton's First Law

Galileo had abstracted from observations of the motion of actual objects a principle of inertia—in the absence of a resistive force an object would continue to move with constant speed along a great-circle path on the surface of the earth. From

Newton's point of view, there was still a further abstraction to be made. The object whose motion Galileo described was not free from the influence of forces. It was attracted toward the earth. For Newton it was more relevant to inquire about the motion of an object subject to no forces at all.

Newton argued that without the resistance of the air and the attraction of the earth, which had come to be called the force of gravity, a projectile would continue to move in a straight line with constant speed. He hypothesized that this is true for all objects in the universe. This hypothesis appears in the *Principia* as Law I:

> *Every body continues in its state of rest, or of uniform motion in a straight line, unless it is compelled to change that state by forces impressed on it.*[2]

This law singles out a particular kind of motion—that with constant velocity—as "natural motion." Any deviation from constant-velocity motion indicates that one or more forces are acting on an object. By examining deviations from constant velocity when only one force is acting, we can learn about that force. We shall now consider the characteristic relation between forces and the motions they produce in such experiments.

At the very beginning of the *Principia*, Newton stated such a relation, calling it Law II. However, this law is only justified *a posteriori*—by the success of the theoretical edifice Newton built upon it. How he imagined such a law in the first place is not completely clear; Newton generally did not talk of such things. In what follows we shall examine critically some of the concepts involved in this law, trying to illuminate the kind of intuitive ideas that lie behind it.

Intuition About Force

The first problem is that *a priori* we have no quantitative measure of force. How can we know when one push is twice (or 3.746 times) as large as another? We experience pushes and pulls in a complex physiological and psychological manner. Trying to decide on the basis of feel when one push is twice another is like trying to decide whether one sound is twice as loud as another, or one room twice as warm as another. To proceed, we must *define* what we mean by a force of a certain strength.

We have, however, some intuitive feeling for this matter. For example, it seems reasonable to say that the pull exerted on an object by two equally stretched identical springs is twice that exerted by one of the springs alone. (See Figure 8–1.) For the time being, let us rely on intuitive notions to guide our inquiry, realizing that such intuition is usually an important element in developing a new concept.

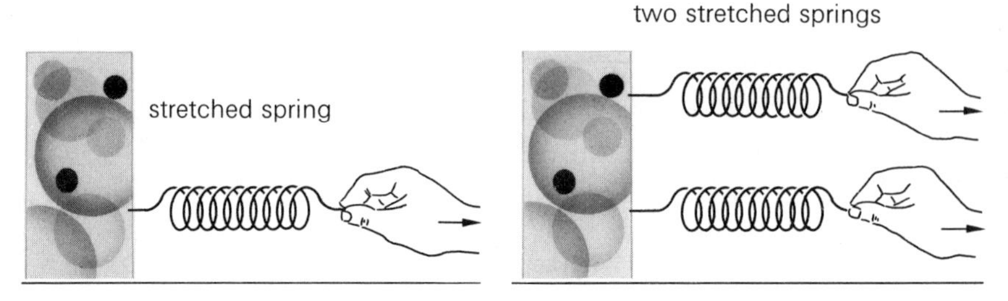

Figure 8–1. It is plausible to assume that two equally stretched, identical springs exert twice the force of one alone.

Consider the simplest case of motion produced by a force. Suppose there is a single constant force acting on an object. What motion results? First we must decide how to produce this situation in the laboratory. To Newton, the free-fall experiment investigated by Galileo seemed a case of an object moving under the influence of a single constant force, gravity. That this force is constant may be seen by holding an object at rest first near the ceiling and then near the floor. In each case the downward gravitational pull that must be counterbalanced by one's hand is at least roughly the same. This can be made more quantitatively convincing by hanging the object from the end of a spring. The extension of the spring is the same near the ceiling and at the floor. (See Figure 8–2.) It is therefore plausible that the force on a falling body is approximately constant throughout its motion.

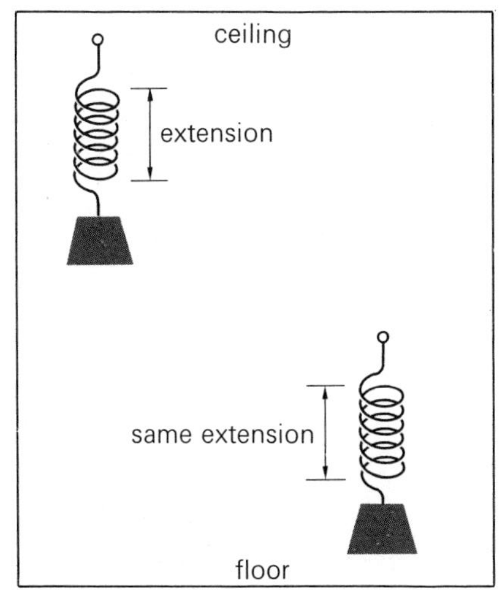

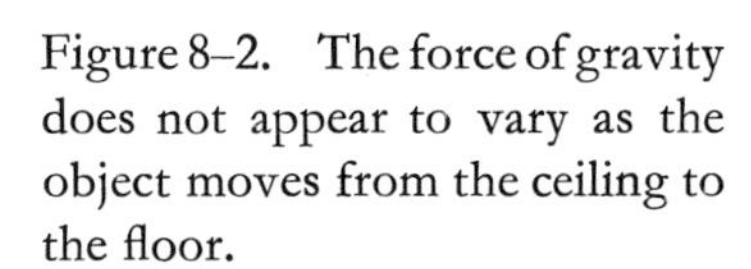
Figure 8–2. The force of gravity does not appear to vary as the object moves from the ceiling to the floor.

What motion results when a constant force is applied to an object? As Galileo found, the object falls with a constant acceleration. Perhaps this is a general relation. Perhaps in general a constant force applied to an object produces a constant acceleration of the object in the direction of the force.

How would the motion be affected if the size of the constant force were changed? Unfortunately, we cannot "turn up" gravity, so the free-fall experiment is not appropriate to answer this question. In seeking an alternative experiment, we find it is rather difficult technically to produce a constant force whose magnitude can be adjusted. Let us nevertheless consider an idealized arrangement to produce such a force—a stretched spring pulling on an object in such a manner that throughout the object's motion the extension of the spring does not vary. (See Figure 8–3.) Since we would agree that the pull of a spring depends on its extension, it is a reasonable presumption that such a spring would exert a constant force on the object. If we further arrange for the object to be placed on a horizontal, virtually frictionless surface, the only force affecting its motion will be that provided by the spring. This arrangement is easy to talk about, though rather difficult to achieve in practice.

Suppose we were to apply a constant force to a block initially at rest by pulling on it with a spring extended a distance L.*

* In fact, experiments essentially equivalent to this one can be performed with the aid of an air track, a device in which an object rides on a cushion of air, rendering its motion nearly frictionless.

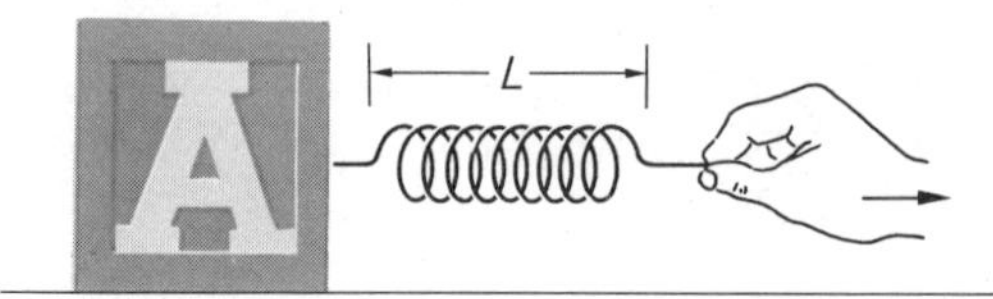

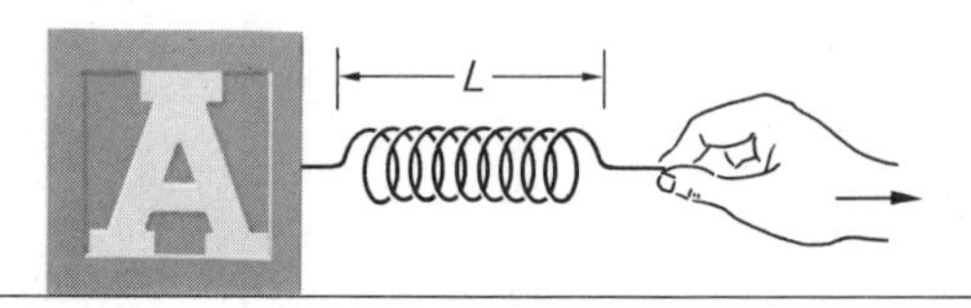

Figure 8–3. It is plausible to assume that the force exerted on the object remains constant if the extension of the spring is kept constant.

We would find that the block moves with a constant acceleration A, reinforcing our speculation that a constant force produces motion with constant acceleration. We are next interested in learning how the motion is changed if the size of the constant force is changed. Figure 8–4 depicts the appropriate experiment. Suppose we were to increase the force by pulling on the same block with two springs identical to the one employed in the first experiment, with each extended the same distance L. If we performed this experiment we would find that the block again underwent uniformly accelerated motion, but this time the acceleration would be twice as great, $2A$. Similarly if we pulled

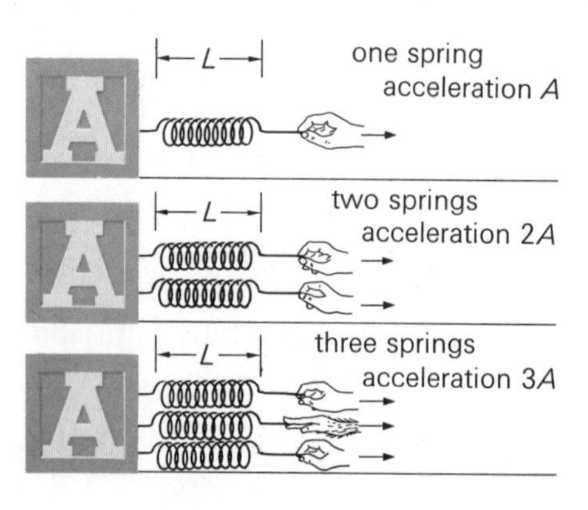

Figure 8–4. Results of pulling on the same block with one, two, and three identical springs, each extended the same distance L.

on the same block with three identical springs, each extended the same distance L, we would find a constant acceleration $3A$. Clearly we could continue our observations with 4, 5, 6, . . . springs. Also, we know that if no spring pulls on it, the block will remain at rest—i.e., its acceleration will be zero.

These results are summarized in the table and graph of Figure 8–5, where we have indicated the values of the acceleration a corresponding to the number of springs pulling on the block. What we really want to talk about, however, is not number of springs, but force. The useful thing about these spring experiments is that they suggest a possible way to assign numbers to forces that is simple and agrees with our intuition. If we say arbitrarily that our original spring, extended to length L, exerts a force f, would it not be reasonable to say that two such springs exert a force $2f$, three a force $3f$, etc.?* What is

Figure 8–5. The acceleration of a block vs. the number of springs pulling the block. The graph is a straight line.

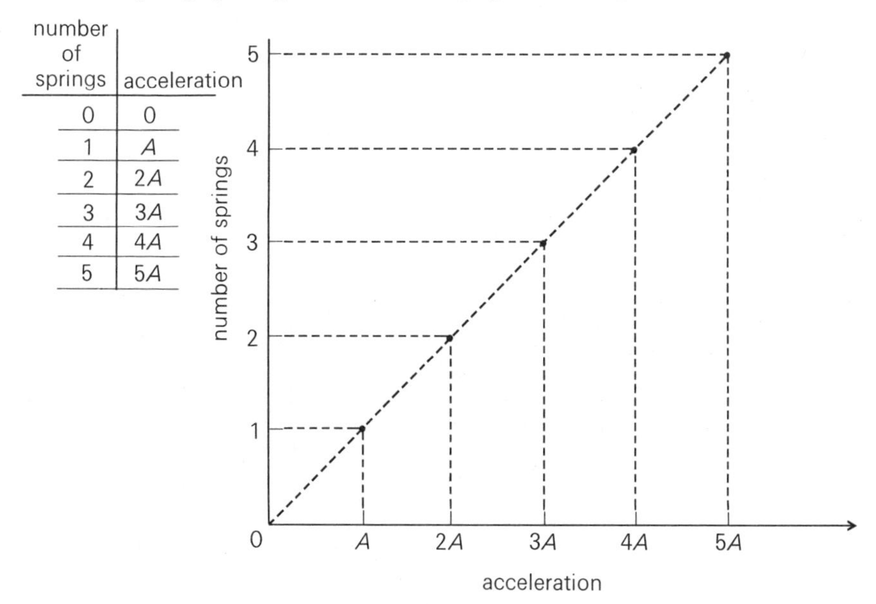

number of springs	acceleration
0	0
1	A
2	$2A$
3	$3A$
4	$4A$
5	$5A$

* Note, however, that this does not have to be true. The presence of a second spring might somehow weaken or get in the way of or in some way affect the first.

more, the graph suggests a way to determine the magnitude of forces due to in-between springs. If the force we call f brings about an acceleration A, and the force called $2f$ brings about $2A$, would it not be reasonable to call the force that produces an acceleration of $1.5A$ a force of $1.5f$? And to say that the force producing an acceleration $3.746A$ has a magnitude of $3.746f$?

These considerations make it seem reasonable to adopt this procedure as our definition of the magnitude of a force. Figure 8–6 illustrates the procedure. This figure is almost the same as Figure 8–5, with one principal change—it is no longer a summary of experiments on springs, but rather specifies magnitudes of forces. To measure the force of a different type of spring, or of our original spring extended to a different length, we simply pull on our block with it, measure the acceleration, and read the force off the graph.

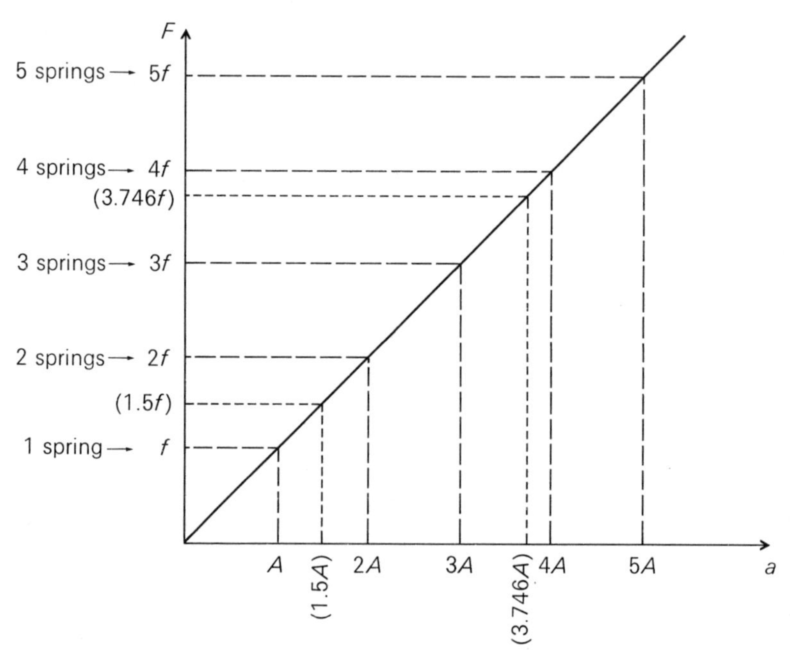

Figure 8–6. Definition of the magnitude of a force in terms of the acceleration it produces.

An equivalent, but more succinct, way to put this is to say that we have decided to assign a number F to a force according to the relation

$$F = ma$$

where m is a constant proportionality factor, the slope of the graph of F vs. a in Figure 8–6.

Mass

Let us examine the meaning of the constant m in the relation $F = ma$. In particular, let us ask what happens when we apply forces to objects other than the block we have been considering. Suppose we perform the same series of acceleration experiments as before (with the same springs extended the same distance) on a number of different objects—e.g., a Ping-Pong ball, a golf ball, and a billiard ball. The results we might obtain from such experiments are indicated schematically in Figure 8–7. Note that for each force the Ping-Pong ball undergoes the largest acceleration, the golf ball the next largest, and the billiard ball the least.

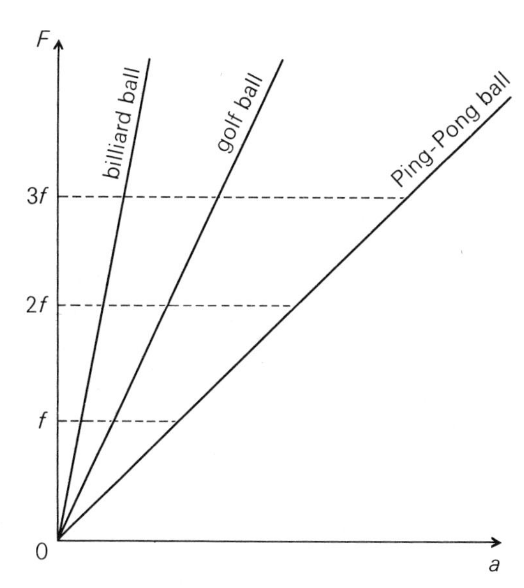

Figure 8–7. Acceleration vs. force for three different objects. The slope m is largest for the billiard ball, next largest for the golf ball, and least for the Ping-Pong ball.

For each object the F vs. a data lie on a straight line indicating that in each instance F and a are proportional, but the slopes of the lines (the proportionality constants m in $F = ma$) are different. The slope m is smallest for the Ping-Pong ball, larger for the golf ball and largest for the billiard ball. Thus m is a characteristic of the object to which the forces are applied. It is conventional to refer to m as the *mass* of an object.

To understand the significance of mass, it is helpful to write the equation describing each graph in the form

$$a = \frac{F}{m}$$

For a given force F, the acceleration is large if the denominator m is small, and small if m is large. Hence the mass m is a measure of the sluggishness of an object. An object that experiences a large acceleration for a given force (and hence is not very sluggish) has a small mass; one that experiences a smaller acceleration for the same force (and hence is more sluggish) has a larger mass. Thus it follows directly from its definition as the proportionality constant in the $F = ma$ relation that mass is fundamentally a measure of the resistance of an object to acceleration by a force.

Defined in this way, mass seems a rather esoteric property of an object. From our experience, however, we all have an intuition about mass. What characteristics of a body are correlated with its resistance to acceleration? Its color? Its feel? Its size? Having pushed and pulled numerous objects we have a feeling for the answer to this question. Generally speaking, the bigger the object (in the sense of containing more stuff) the larger the push or pull required to move it. This was borne out by the spring experiments with the Ping-Pong, golf, and billiard balls. Thus, loosely speaking, the bigger the object the larger its mass.

Units of Force and Mass

A few formal details of the program of attaching numbers to forces and masses remain to be discussed. The basic relation $F = ma$ has been used twice, first to assign magnitudes to forces, then to assign magnitudes to masses.

Our first series of experiments was done with a single object, which we might call our standard. We measured the acceleration of this standard object and used Figure 8–6 to assign magnitudes to the corresponding forces. Numerically, the procedure would be as follows: In the metric system (which we shall adopt), we would use as a standard object a cylinder of platinum kept in a vault in Paris. The mass of this object has been arbitrarily designated as one *kilogram*. We apply to it some force we wish to measure. Suppose the resulting acceleration is 1 m/sec². According to our convention for assigning numbers to forces, the strength of the force here is $F = ma = 1 \text{ kg} \times 1 \text{ m/sec}^2 = 1 \text{ kg m/sec}^2$. Thus, the unit of force is kg m/sec², which is called a *newton*. One newton is the force that produces a 1 m/sec² acceleration of the standard object. A force that gives the standard object an acceleration of 2 m/sec² is 2 newtons; a force that gives the standard object an acceleration of 3.746 m/sec² is 3.746 newtons.

Now that we have found a way to attach numbers to forces in terms of the one-kilogram standard object, we can use these known forces to determine the masses of other objects. We apply a known force (e.g., a spring extended to a length L, whose force has been measured with the one-kilogram object) to a new object, and measure the object's resulting acceleration. Again we use $F = ma$, but now we know both F and a, and so m can be determined. Suppose, for example, we apply a one-newton force to an object and measure its acceleration to be 2 m/sec². The mass of this object is then:

$$m = \frac{F}{a} = \frac{1 \text{ kg m/sec}^2}{2 \text{ m/sec}^2}$$

$$= 1/2 \text{ kg}$$

In general, if a one-newton force gives an object an acceleration of A m/sec², the mass of that object is $1/A$ kg. This procedure can in principle determine the mass of any object once the standard mass has been chosen. We simply measure the acceleration of the object when a one-newton force is applied to it.

Finally, to measure the strength of an unknown force F (in newtons), acting on any object of known mass m (in kilograms), we measure the acceleration of the object a (in m/sec²), and calculate:

$$F \text{ (newtons)} = m \text{ (kg)} \times a \text{ (m/sec}^2\text{)}$$

Acceleration as a Measure of Force

Let us survey what we have accomplished thus far. The initial problem was to establish a way to measure the strengths of forces—i.e., attaching numbers to pushes and pulls. This is clearly a prerequisite to the Newtonian program of describing the forces that influence the motion of objects.

Basically, it was a matter of definition. We had to create a new concept, quantifying our qualitative notion of push or pull. In principle, being a creation of man, the concept could be defined arbitrarily. In practice, however, we are severely limited by pragmatic considerations. A quantitative measure of force should coincide with our intuition about the nature of forces as gleaned from experience. And so we tried our spring experiments. We had a feeling that in those experiments we knew how to compare the strengths of forces. From them, we arrived at a suggestive and appealingly simple conclusion: intuitively, the strength of the force acting on an object seems proportional to its acceleration. Newton himself had to work with considerably less precise and somewhat more indirect observations, but his results also suggested this relation between applied force and resultant acceleration.

All this points to a way to measure force in general. Newton proposed that we regard the acceleration of a body as a measure of the strength of the force acting on it, calculating the force by the relation $F = ma$, *for all kinds of forces*, not just springs. Thus, to carry out the Newtonian program of describing the nature of a force, be it gravity or magnetism or friction or another interaction, one should observe the acceleration of an object under the influence of the force.

One caution is in order. While the prescription $F = ma$ gives us a quantitative measure of force that agrees with the limited experience where we feel we know how forces behave (e.g., springs), this does not guarantee that the same will be true with other kinds of forces. For example, how can we be sure that the properties of spring forces also characterize the attraction of a ferrous material to a magnet, the forces that bind matter together, the gravitational attraction that holds man to the earth, the resistive forces that oppose motion, and the other pushes and pulls one encounters? We cannot pull on an object with two earths to see whether the acceleration doubles. We can only tentatively apply $F = ma$ to the investigation of forces like gravity and see if it is indeed a useful element of a successful theory. The ultimate test is pragmatic—will it work, will it be fruitful? *A priori* we cannot be sure.

Force as a Vector

There is another dimension to the relation between force and acceleration that we have not yet mentioned, although it has been implicit in our discussion. If a force is applied to an object, the object accelerates, and we have agreed to measure the magnitude of the force in terms of the magnitude of the acceleration by $F = ma$. We should note further that the direction of the object's acceleration always seems to be the direction of the applied push or pull. It seems reasonable to extend our concept of force to include this specification of direction as well. This is easily accomplished by writing the relation between force and acceleration as the vector equation:

$$\vec{F} = m\vec{a}$$

Since m is a number, the vector $m\vec{a}$ is in the direction of the vector $\vec{a}$. Hence, this equation says that the magnitude of the force is given by the mass times the magnitude of the acceleration, $F = ma$, and that the direction of the force is the same as the direction of the acceleration.

Although we have discussed $\vec{F} = m\vec{a}$ only for a constant force, it may also be applied to a situation where the force is varying with time. If the force changes with time, the motion is more complicated than uniform acceleration. Nevertheless, our prescription for measuring force asserts that the force $\vec{F}$ acting on an object at any given instant of time is related to the instantaneous acceleration $\vec{a}$ of the object at that time by $\vec{F} = m\vec{a}$. In fact, most forces in nature are not constant.

Superposition of Forces

So far we have discussed examples in which only one force at a time acts on an object. What if more than one force is present? Consider a boy pulling a wagon, as illustrated in Figure 8–8(a). Part (b) shows four separate forces acting on the wagon: $\vec{F}_B$, the pull of the boy; $\vec{F}_f$, the frictional force between the wheels and the ground that impedes the motion of the wagon (and hence is directed opposite to its motion, as shown); $\vec{F}_G$, the

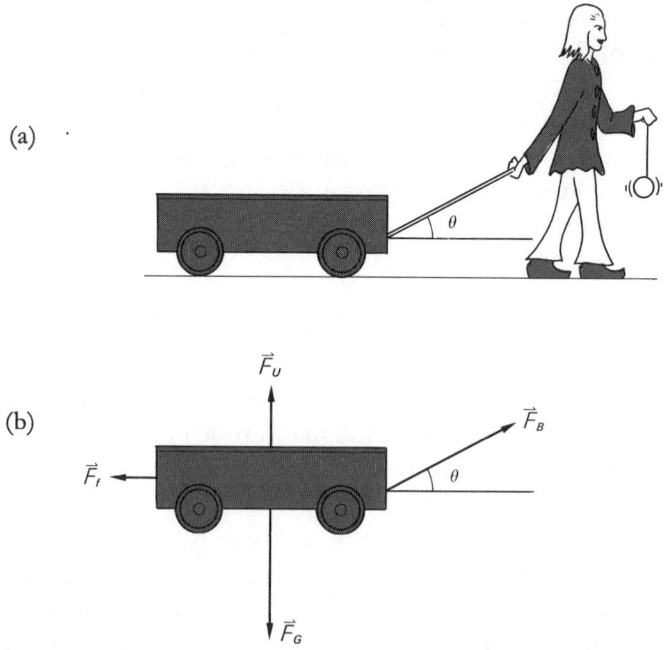

Figure 8–8. Several forces acting at once on an object. (a) A boy pulling a wagon. (b) The forces acting on the wagon.

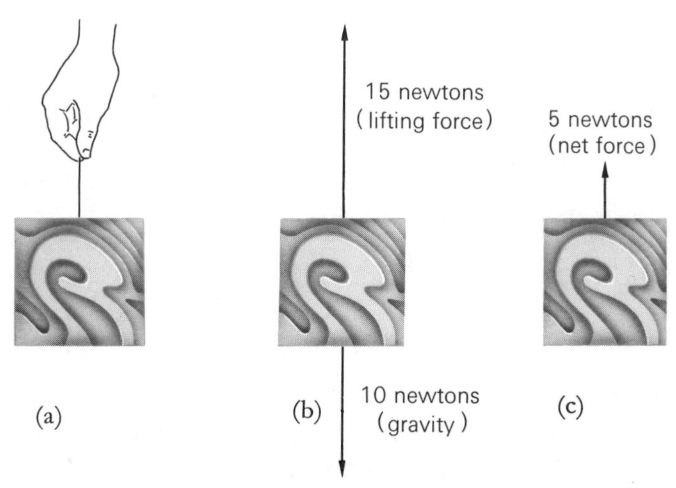

Figure 8–9. Two forces acting vertically on a block are equivalent to a single net force, in the sense that the same acceleration would occur in parts (b) and (c). (a) The block being lifted. (b) The forces acting on the block. (c) The net force on the block.

gravitational force pulling the wagon toward the earth; and $\vec{F}_U$, the upward push of the ground on the wagon. How is the motion of the wagon related to these forces?

We again appeal to our intuition about the way forces act. Consider a simpler example, like the one in Figure 8–9, in which a block is being lifted. Suppose the lifting force is 15 newtons and the pull of gravity is 10 newtons. One has the feeling that there is a "net force" upward of 5 newtons, in the sense that the motion produced by such a force would be the same as that produced by the combination of the lifting force and gravity.

In the more complicated situation of the boy pulling the wagon, where several forces are acting, we are also inclined to believe that there exists a net force, $\vec{F}_{\text{net}}$, that would produce the same motion as the combination of forces. This leads to a generalization of our force-acceleration relation to account for situations where more than one force might be acting:

$$\vec{F}_{\text{net}} = m\vec{a}$$

How is $\vec{F}_{\text{net}}$ related in general to the individual forces applied to the object? This matter has been studied experimentally. The conclusion, which seems applicable to all known forces, is just what one might guess. If a number of forces are acting on an object, the net force, $\vec{F}_{\text{net}}$, is the vector sum of the individual forces. This is sometimes called the *principle of superposition.*

In the example of the boy pulling the wagon, the motion of the wagon is described by:

$$\vec{F}_{\text{net}} = m\vec{a}$$

where

$$\vec{F}_{\text{net}} = \vec{F}_G + \vec{F}_B + \vec{F}_U + \vec{F}_f$$

This vector sum of the four forces of Figure 8–8(b) is shown in Figure 8–10. Notice that $\vec{F}_{\text{net}}$ is horizontal, and thus the cart must be accelerating in the horizontal direction.

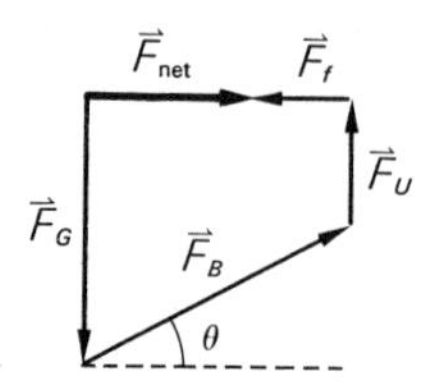

Figure 8–10. Construction of the vector $\vec{F}_{\text{net}} = \vec{F}_G + \vec{F}_B + \vec{F}_U + \vec{F}_f$. This construction is a simple generalization of the addition of two vectors discussed previously. Note that $\vec{F}_G + \vec{F}_B + \vec{F}_U = (\vec{F}_G + \vec{F}_B) + \vec{F}_U$ and $\vec{F}_G + \vec{F}_B + \vec{F}_U + \vec{F}_f = (\vec{F}_G + \vec{F}_B + \vec{F}_U) + \vec{F}_f$. The puzzled reader should use the vector law of addition for two vectors first to construct $\vec{F}_G + \vec{F}_B$, then $(\vec{F}_G + \vec{F}_B) + \vec{F}_U$, then $\vec{F}_{\text{net}} = (\vec{F}_G + \vec{F}_B + \vec{F}_U) + \vec{F}_f$. The figure shown will result.

Newton's Second Law

The relation $\vec{F} = m\vec{a}$, with $\vec{F}$ interpreted as the net force on an object, m its mass, and $\vec{a}$ its acceleration, is the essence of Law II in the *Principia*. Newton's second law is the keystone of his approach to investigating the interactions of objects in nature. As we have discussed, this equation is a definition. It defines the primitive means of measuring the net force on an object. One simply measures the acceleration of the object and multiplies by its mass.

A definition is a statement of logical equivalence. It can never be wrong. $\vec{F}$ always equals $m\vec{a}$ because that is the way we have defined it. One may well wonder then how $\vec{F} = m\vec{a}$ can have any physical content at all. This is actually a rather subtle issue that is difficult to discuss until one has applied $\vec{F} = m\vec{a}$ to some concrete physical situations and gets a feeling for how it is used. This is what we will do first. We shall then return to a discussion of the role of the second law. Our contention will be that $\vec{F} = m\vec{a}$ by itself does not have physical content. Rather it is important as an organizing principle—an instruction about how to comprehend nature. It says: to investigate forces, focus on accelerations.

Newton's Third Law

Newton's first two laws allow us, in principle, to describe the motion of a single object in terms of the forces acting on it. More often, however, we are interested in what happens when several bodies interact with each other. Consider, for example, the situation illustrated in Figure 8–11. Suppose that object 2 is attracted by object 1 with a force $\vec{F}_{21}$; the subscript implies the force on 2 due to 1. What can we say about the force $\vec{F}_{12}$ with which object 1 is attracted by object 2?

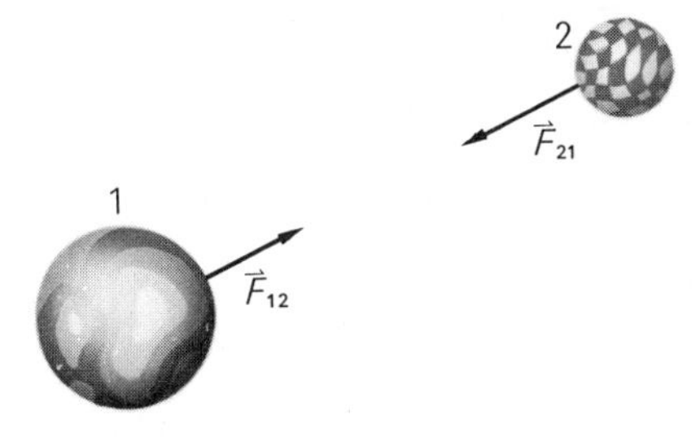

Figure 8–11. The forces between two objects. 2 pulls on 1 with $\vec{F}_{12}$; 1 pulls on 2 with $\vec{F}_{21}$.

A simple experiment can suggest an answer. Imagine object 2 to be a small magnet and object 1 a small piece of iron. Suppose each has the same mass. They are held apart on a smooth table and then released. What would we observe? Each object would accelerate toward the other. Thus each object must feel a force. Not only does the magnet attract the iron, but apparently the iron also attracts the magnet. In fact, careful measurements would show that the two objects meet at a point midway between their initial positions, allowing us to conclude that their accelerations were equal and opposite. Since each has the same mass, we may further conclude that the forces are equal and opposite:

$$\vec{F}_{12} = -\vec{F}_{21}$$

In the *Principia*, Newton generalized this result as Law III:

> *To every action there is always opposed an equal reaction or, the mutual actions of two bodies upon each other are always equal, and directed to contrary parts.*

In its support, he gave several arguments from common experience:

> If you press a stone with your finger, the finger is also pressed by the stone. If a horse draws a stone tied to a rope, the horse (if I may so say) will be equally drawn back towards the stone; for the distended rope, by the same endeavor to relax or unbend itself, will draw the horse as much towards the stone as it does the stone towards the horse . . .[3]

The third law is indispensable to the Newtonian program of analyzing the motion of a system of objects interacting with each other. However, we shall focus on such simple examples that, with one minor exception, we shall not have further recourse to this law.

Suggested Reading

Arons, A. *Development of Concepts of Physics*. Reading, Mass.: Addison-Wesley, 1965. Chapter 5 includes a careful discussion of the concept of force and Newton's second law.

Manuel, F. *A Portrait of Isaac Newton*. Cambridge: Harvard University Press, 1968. An engrossing biography of Newton, with particular emphasis on his psychological makeup.

Newton, I. *Principia*. Translated by F. Cajori. Berkeley: University of California Press, 1934. Newton's major work. The proofs are fairly intricate, but many of his commentaries are quite readable.

Toulmin, S. *Foresight and Understanding*. Bloomington: Indiana University Press, 1961. Chapter 3 is a particularly good discussion of the motions seen as natural by Aristotle, Galileo, and Newton.

Questions

1. We asserted on page 190 that a single constant force applied to an object causes that object to move with a constant acceleration. How might one determine experimentally whether the motion of such an object is in fact uniformly accelerated?
2. Refer again to Question 15 on page 102. Which of the two representations of the planetary data would be most likely to interest a Newtonian physicist? Why?
3. We have seen that Newton defined the magnitude of a force as proportional to the acceleration this force imparts to some standard object. Suppose that he had instead chosen to define the magnitude of a force as proportional to the square of the acceleration.
   (a) Would such a definition be logically admissible?
   (b) Would this definition be consistent with Newton's first law?
   (c) Would the force, defined in this manner, exerted by a spring whose extension is kept constant, still be a constant force?
   (d) Would the force exerted by two identical springs still be twice the force exerted by one of the springs alone?
4. Suppose that two identical objects, each of mass m, are attached together to form a larger, compound object. It is asserted that this compound object has a mass $2m$. What sort of experiment might be performed to investigate this assertion?

5. A stone, released from rest in a viscous medium such as oil, accelerates downward until it achieves a constant terminal velocity, which it maintains for the remainder of its descent.

   (a) Sketch the stone and indicate with arrows the forces acting on it.
   (b) What can you conclude about the net force on the stone during the constant-velocity phase of its motion? What can you therefore conclude about the relative magnitudes of the forces you cited in part (a)?
   (c) What can you conclude about the net force on the stone during the accelerated phase of its motion? What can you therefore conclude about the relative magnitudes of the forces you cited in part (a)?

6. As Galileo observed, all objects fall in a vacuum with the same constant acceleration. For example, consider three stones with masses of 1 kg, 2 kg, and 3 kg respectively. Each will fall toward the earth with an acceleration of 9.8 m/sec^2.

   (a) What is the gravitational force on each of these stones?
   (b) Can you generalize the result of part (a) to guess how the gravitational force experienced by an object depends on its mass?

7. Consider the boy and wagon shown in Figure 8–8. If the wagon is moving with constant velocity, what can you conclude about the net force on the wagon? What change in the magnitude of the frictional force shown in Figures 8–8 and 8–10 would bring about this net force?

8. A crate of mass 50 kg is being pulled along the floor at a constant velocity of 1 m/sec by a 200-newton force directed horizontally as shown.

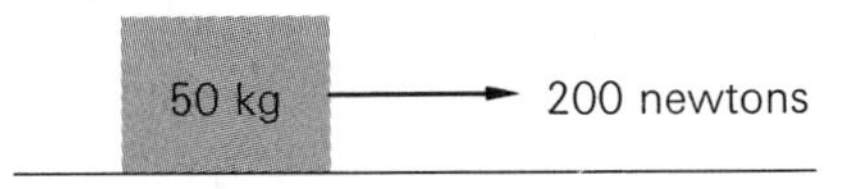

   (a) Indicate by drawing arrows the directions of the other three forces acting on the crate.

(b) What must be the net force $\vec{F}_{net}$ acting on the crate?
(c) Construct the vector sum of the force vectors, i.e., draw a diagram analogous to Figure 8–10. [Recall the result of part (b).]
(d) Determine the magnitude of the frictional force acting on the crate by referring to the diagram of part (c).
(e) What is the gravitational force on the crate? (Hint: What would its acceleration be if only the gravitational force were acting on it?)
(f) What is the magnitude of the force the floor exerts on the crate?

9. A child of mass 20 kg slides with constant acceleration down a playground slide. Suppose he travels down the length of the slide, 2 m, in 1 sec.
   (a) What is the direction and magnitude of his acceleration during this motion?
   (b) What is the direction and magnitude of the net force he experiences?
   (c) Discuss qualitatively the forces that combine to give this net force. Sketch the child on the slide and indicate the directions of the various forces on the child by drawing arrows.

10. A heavy box rests on the floor. If a horizontal force of 100 newtons is applied to it, it does not move. If a horizontal force of 200 newtons is applied to it, it still does not move. If a horizontal force of 300 newtons is applied to it, it moves with constant acceleration across the floor. What can you conclude about the magnitudes of the frictional forces on the box when the 100-, 200-, and 300-newton forces are applied? What would be the magnitude of the frictional force on the box if there were no other horizontal forces acting on it? What general conclusion can you draw from these observations about the magnitude of the frictional force acting on the box? Speculate on a mechanism responsible for the frictional force that would give rise to this behavior.

11. A 1-kg ball released near the surface of the earth falls with a constant acceleration of 9.8 m/sec^2.

(a) What is the gravitational force on the ball?
(b) Newton's third law implies that the ball exerts an equal and opposite force on the earth. How great an acceleration will the earth experience due to the action of this force? (The mass of the earth is 6×10^{24} kg.)
(c) How far will the earth "fall" in the one second immediately following the release of the ball? Could this motion be detected? Do you believe that the earth really does move toward a falling ball? Why or why not?

12. According to the atomic theory of matter a macroscopic object is composed of myriads of atoms. The matter holds together because of an attractive (electrical) force between the atoms. Assuming that this attractive force obeys Newton's third law,

(a) What can you conclude about the relation between the force one atom exerts on a second atom and the force the second exerts on the first?
(b) For this pair of atoms, what is the vector sum of their forces of mutual attraction?
(c) What can you therefore conclude about the vector sum of all the attractive forces between the roughly 10^{23} atoms in a chunk of macroscopic matter?
(d) What does the observation that macroscopic objects do not spontaneously accelerate in the absence of external forces imply about the validity of Newton's third law?

13. A student who had just finished reading Chapter 8 explained why he could no longer pick up this book: "If I pull the book up with a certain force, then according to Newton's third law the book will pull down on me with an equal and opposite force. Consequently I will be unable to lift the book." Comment on this argument. Can he pick up the book?

CHAPTER NINE

A Matter of Some Gravity

The lasting significance of the Newtonian approach lies in its success with the problems of terrestrial and celestial motion. Newton was able to interpret the results of Kepler and Galileo as specific applications of general principles, valid everywhere in the universe, making no distinction between sublunary and superlunary regions. He accomplished this by focusing on the forces that cause motion.

Newton's first law states that an object will move in a straight line with constant speed if no net force is acting on it. In the Newtonian world view, this motion is natural and needs no explanation. However, the speed of an apple falling to the earth is not constant and the elliptical paths of the moon orbiting the earth and the planets revolving about the sun are not straight lines. From the Newtonian perspective these motions need to be explained.

The second law directs our search for an explanation. It suggests that to investigate the force on the apple or the moon or a planet, one must observe its acceleration. We have already discussed the uniformly accelerated motion of a falling object. Let us now consider the acceleration of an orbiting object such as the moon or a planet. Since the description of an elliptical orbit is mathematically complex, we shall limit our analysis to the special case of an object moving with constant speed in a circular path. While this will only approximate the actual accelerations of orbiting planets in Kepler's solar system, the approximation is a very good one since the orbits are nearly circular.

The Acceleration Vector in Uniform Circular Motion

It may seem peculiar to say that an object moving in a circular path with constant speed is accelerating. But acceleration, as we have defined it, is a vector specifying the change of the velocity vector with time. Even though the magnitude of the object's velocity vector remains constant in uniform circular motion, its direction is changing. Hence the velocity vector is changing with time and the object is accelerating.

If the definition of the acceleration vector,

$$\vec{a} = \underset{\Delta t \to 0}{\text{limit}} \frac{\overrightarrow{\Delta v}}{\Delta t}$$

is applied to an object moving in a circular path of radius R with a constant speed v, a very simple result is obtained. (See Figure 9–1.) At any given time, the acceleration vector $\vec{a}$ is directed from the position of the object toward the center of the circle. The magnitude of $\vec{a}$ remains constant throughout the motion and has the value:

$$\boxed{a = \frac{v^2}{R}} \tag{7}$$

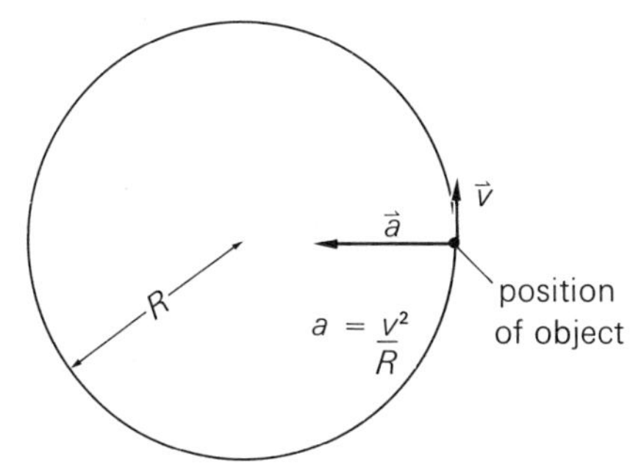

Figure 9–1. The acceleration vector in uniform circular motion has magnitude v^2/R and is directed from the position of the object toward the center of the circle.

The following remarks are intended to make this result seem plausible. A rigorous derivation is presented in Appendix 2.

Let us focus first on the direction of the acceleration vector. Consider a planet moving with a constant speed in a circular orbit about the sun. Figure 9–2(a) indicates the position and velocity of the planet at time t_i and at a later time $t_f = t_i + \Delta t$ (i.e., t_f is Δt seconds after t_i). The velocity vectors at these times are labeled $\vec{v}_i$ and $\vec{v}_f$ respectively. Since the planet is moving with constant speed, the vectors $\vec{v}_i$ and $\vec{v}_f$ are represented by arrows of the same length.

Figure 9–2. Construction of the change in velocity vector $\overrightarrow{\Delta v}$ corresponding to the time interval Δt. Recall that a vector is uniquely specified by its magnitude and direction. Its location on the page is quite arbitrary. Thus the corresponding arrows in (a) and (b) represent the same vectors.

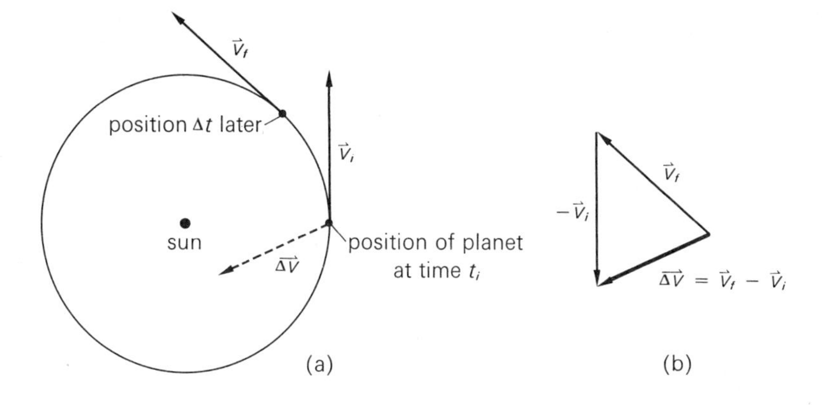

In Figure 9–2(b) we construct the vector $\overrightarrow{\Delta v} = \vec{v}_f - \vec{v}_i$. Notice that $\overrightarrow{\Delta v}$ points into the circle from the position of the planet at time t_i. It points in the general direction of the sun, though not precisely toward it. The average acceleration of the planet during the time interval Δt is $\overrightarrow{\Delta v}/\Delta t$. Since Δt is just a number, the direction of the average acceleration vector is the direction of $\overrightarrow{\Delta v}$. Hence the average acceleration vector during the time interval Δt points into the circle, in the general direction of the sun, from the position of the planet at time t_i.

This average acceleration in the time interval Δt following t_i approximates the instantaneous acceleration at t_i. The shorter the time interval, the better the approximation. Thus, to determine the direction of the instantaneous acceleration at t_i, we should consider a shorter time interval Δt than the one just discussed. This is illustrated in Figure 9–3(a).

Once again, we construct $\overrightarrow{\Delta v} = \vec{v}_f - \vec{v}_i$ [Figure 9–3(b)]. Notice that $\overrightarrow{\Delta v}$, and hence the average acceleration vector during this shorter time interval, point more nearly in the direction of the center of the circle. It should now be plausible that if we chose Δt very much smaller yet, the direction of $\overrightarrow{\Delta v}$ would be precisely toward the center of the circle. Consequently, that is the direction of the instantaneous acceleration vector $\vec{a}$ at time t_i, as indicated in Figure 9–1.

Now let us focus on the magnitude of the acceleration vector $\vec{a}$ in uniform circular motion. In Appendix 2, we prove rigorously

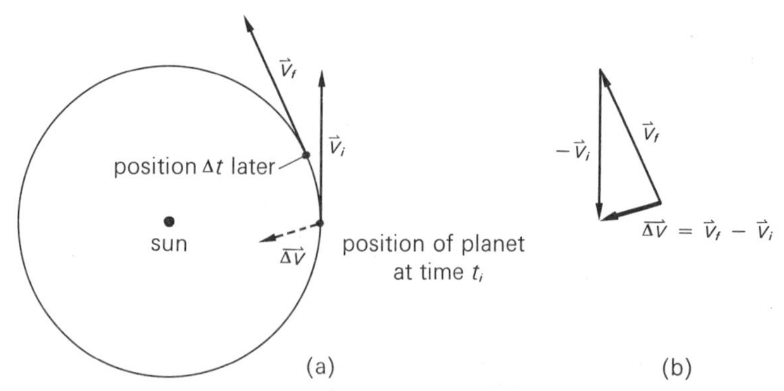

Figure 9–3. Construction of the change in velocity vector $\overrightarrow{\Delta v}$, corresponding to a shorter time interval Δt.

that $\vec{a}$ has the constant magnitude $a = v^2/R$. Let us forego this rigor here and simply demonstrate that this relation makes sense intuitively.

First, it is easy to see that the magnitude of the acceleration is proportional to the square of the speed. Consider two objects, 1 and 2, moving uniformly about identical circles of radius R, as shown in Figure 9–4. Suppose that object 1 moves with twice the speed of object 2. Let us calculate the average accelerations of the objects during the time it takes them to move from point A to point B.

The magnitude of the average acceleration is $\Delta v/\Delta t$. The construction of the change in velocity vector $\overrightarrow{\Delta v}$ for each object is also shown in Figure 9–4. Notice that, because of the double length of the velocity vectors of object 1, $\overrightarrow{\Delta v}_1$ is twice as large as $\overrightarrow{\Delta v}_2$. In addition, because object 1 is traveling twice as fast as object 2, the time it takes object 1 to move from A to B, Δt_1, is one-half as long as the time it takes object 2 to move from A to B, Δt_2. If $\Delta v_1 = 2\Delta v_2$ and $\Delta t_1 = \Delta t_2/2$, it follows that $\Delta v_1/\Delta t_1 = 4\Delta v_2/\Delta t_2$. In other words, if the speed of an object moving uniformly in a circle doubles, its average acceleration is four times as great.

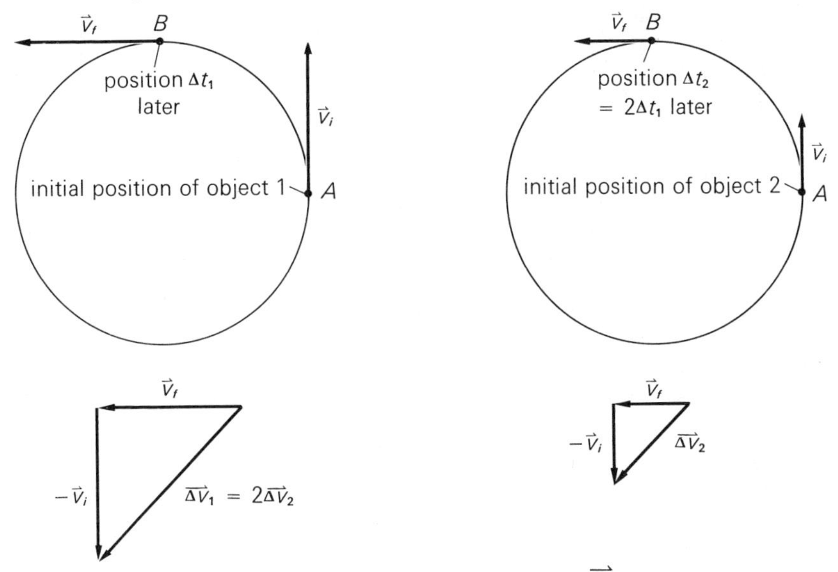

Figure 9–4. The change in velocity vector $\overrightarrow{\Delta v}$ is twice as large for the object moving with twice the speed.

The same result would be obtained no matter where we took point B. In particular, it would remain valid as we let point B get closer and closer to point A, i.e., as we let the time intervals approach zero. Hence the instantaneous acceleration of the object moving twice as fast is four times as great. If this argument is repeated for an object moving three times as fast as another, we find that the magnitude of its acceleration is nine times as large. This suggests that the magnitude of the acceleration of an object moving uniformly in a circle is proportional to the square of its speed.

In like manner we see that the magnitude of the acceleration vector is inversely proportional to the radius of the circle. Figure 9–5 shows two objects, 1 and 2, moving with the same speed around circles of radii $2R$ and R respectively. Let us calculate the average accelerations of the objects during the time it takes them to move from points A_1 and A_2 to points B_1 and B_2 respectively. The construction of $\overrightarrow{\Delta v}$ for each object is also shown in Figure 9–5, with the result that $\overrightarrow{\Delta v}$ is the same for the two objects. However, the distance from A_1 to B_1 is twice the distance from A_2 to B_2. (Recall that the circumference of a

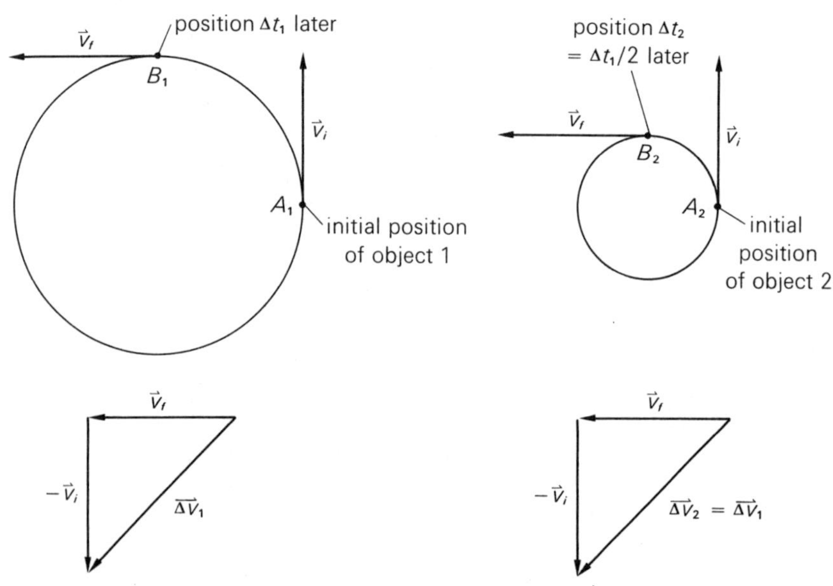

Figure 9–5. The change in velocity vectors $\overrightarrow{\Delta v}$ are the same for the two objects.

circle is proportional to its radius.) Consequently, the time Δt_1 that object 1 takes to move from A_1 to B_1 is twice as long as the time Δt_2 that object 2 takes to move from A_2 to B_2.

If $\Delta v_1 = \Delta v_2$ and $\Delta t_1 = 2\Delta t_2$, it follows that $\Delta v_1/\Delta t_1 = \frac{1}{2}(\Delta v_2/\Delta t_2)$. In other words, an object moving with the same speed as another on a circle of twice the radius experiences one-half the average acceleration.

The same result would be obtained if we moved points B_1 and B_2 closer and closer to A_1 and A_2 respectively, i.e., as the time intervals approached zero. Hence the instantaneous acceleration of the object moving on the circle of twice the radius would be one-half as large. If this argument is repeated for an object moving on a circle whose radius is three times as large, we find that the magnitude of its acceleration is one-third as great. This suggests that the magnitude of the acceleration of an object moving uniformly in a circle is inversely proportional to the radius of the circle.

It should now seem plausible that the acceleration vector in uniform circular motion has magnitude v^2/R and is directed from the position of the object toward the center of the circle. As we shall see, this is a key element in Newton's analysis of celestial motion.

The Hypothesis of Gravity

Consider the motion of the moon. It travels in an approximately circular orbit about the earth with approximately constant speed. Our analysis of uniform circular motion allows us to conclude that the moon therefore has a nearly constant acceleration directed toward the center of the earth. In the Newtonian framework, such an acceleration implies a force. The key question, according to Newton, is: "What is the nature of this force attracting the moon toward the center of the earth?"

That Newton discovered an answer to this question while sitting beneath an apple tree has become one of the legends of our culture. In fact, the story of the apple may have been true. While there is no mention of it in Newton's own writings, an early biographer reports that, years after the event,

> we went into the garden and drank tea under the shade of some apple trees . . . he told me he was just in the same situation as when formerly the notion of gravitation came into his mind. It was occasion'd by the fall of an apple, as he sat in a contemplative mood.[1]

It was an inspired notion. The force known as gravity binds men to the earth and pulls apples to the ground. Could this same mundane force extend far enough into space to attract not only apples but also the moon? This idea is particularly appealing because, in Newton's view, the apple and the moon have a very important element in common. Both accelerate toward the center of the earth. (See Figure 9–6.) Perhaps the same force is responsible for both accelerations.

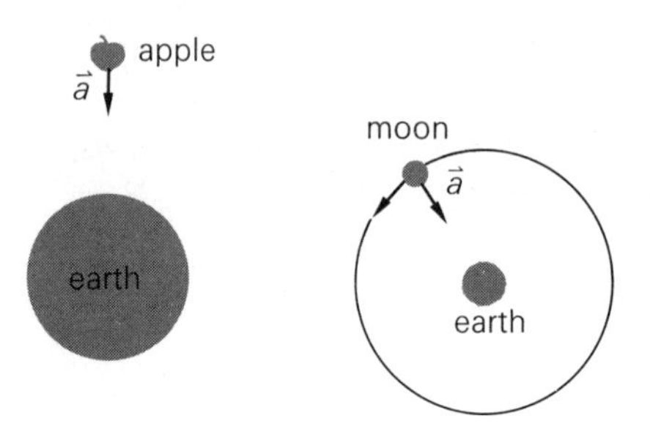

Figure 9–6. The acceleration vectors of both apple and moon are directed toward the center of the earth.

This leap of the imagination can be extended even further. In the heliocentric model of the universe, each planet moves with nearly constant speed around the sun in an elliptical orbit that can be roughly approximated by a circle (Figure 9–7). The acceleration of each planet, and hence the force on each planet, is at all times directed toward the sun. Could the force responsible for these accelerations be of the same essential nature as the force that pulls the apple and the moon toward the earth?

To convert this speculation into a meaningful theory, Newton followed a line of reasoning that involves two steps. The first step can be described as follows: $\vec{F} = m\vec{a}$ gives a prescription for measuring forces. That is, measuring the acceleration of an object gives the magnitude and direction of the force acting on the object. This result alone, however, is not very interesting;

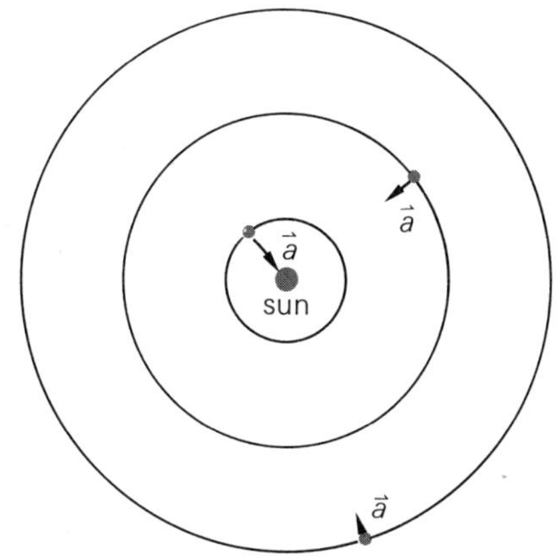

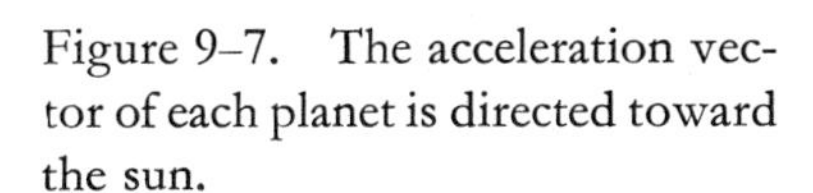
Figure 9–7. The acceleration vector of each planet is directed toward the sun.

numbers by themselves give little understanding. Of more interest—indeed of central importance to the entire Newtonian scheme—is an investigation of how a force depends on the object and on its surroundings. We may call this dependence a *force law*. Newton's first step was to guess a force law to describe the gravitational force.

An example should help clarify this procedure. We have already seen how, in principle, one might investigate the force of a spring: Suppose the spring has a length l_0 when relaxed. Attach to it an object of known mass m and pull on the spring, stretching its length to $l_0 + l$ (Figure 9–8). Continue to pull in such a way that the spring's length remains $l_0 + l$ as the object moves. Now measure the object's acceleration a. The force exerted by the spring corresponding to this extension l can then be calculated using $F = ma$. The force corresponding to other

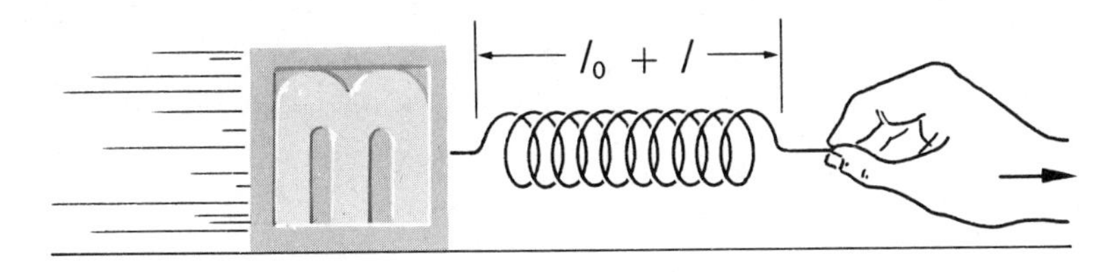

Figure 9–8. A spring, stretched a distance l, pulls a block. The relaxed length of the spring is l_0.

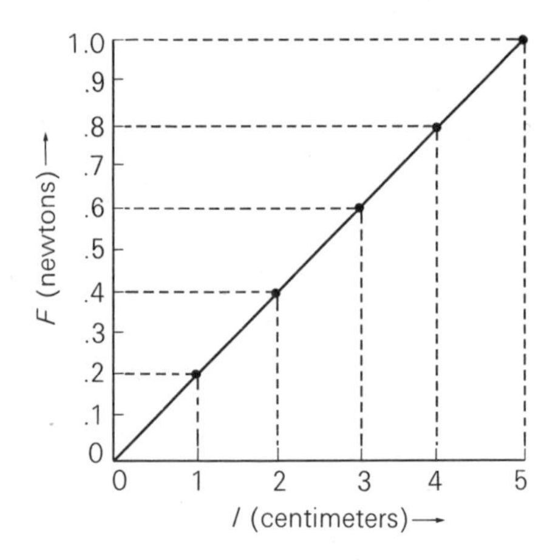

Figure 9–9. The graph of the force F exerted by a spring vs. its extension l appears to be a straight line.

extensions can be obtained in the same manner. Results for a typical spring are represented schematically in Figure 9–9.

Now the question is, what is the force law? That is, how does the force depend on the extension of the spring? In this case, the graph suggests a particularly simple dependence: the force F is proportional to the extension l. The force law for a spring thus seems to be:

$$F = kl$$

where k is a proportionality constant, the slope of the F vs. l graph of Figure 9–9.*

The discovery of the force law for gravity is necessarily much less direct, but the basic idea is the same. One observes the motion of objects in nature, in particular their accelerations, and ferrets out which characteristics (or *parameters*) of the objects and their environment influence the motion. One then attempts to guess the detailed mathematical dependence of the force on each of these parameters. Logically speaking, this inductive process of guessing the force law is circular. The force law is guessed by observing the motion of objects, and the only "proof" of the law's validity lies in its ability to predict such motion correctly.

The second step in the Newtonian scheme involves using the force law to explain observed motion. This is a deductive process.

* This law was first stated by Newton's contemporary, Robert Hooke, in 1678.

For example, having guessed a gravitational force law, one can then use $\vec{F} = m\vec{a}$ to deduce how objects will move in a wide variety of circumstances, from the surface of the earth to the heavens, from a ball rolling down an inclined plane to a comet sweeping around the sun. In the sense that one simple mathematical expression for the gravitational force can account for such a large number of diverse phenomena, Newton's law of gravity can be said to constitute an explanation. Of course, one cannot know *a priori* that it will be possible to find a simple force law that will account for the observed motion of terrestrial and celestial bodies. In fact Newton did find such a law, and this remarkable result is the ultimate success (and vindication) of his scheme.

Guessing the Gravitational Force Law

Figure 9–10 illustrates the simplest example of gravitational attraction. Object 1 of mass m_1 attracts object 2 of mass m_2, which is a distance r away. For example, the two objects might be the earth and the moon. In Newtonian terms, the question is how to describe the attractive force between them.

Presumably the force will depend on the characteristics of the objects (perhaps their masses or the material of which they are made) and on the geometry (for example, the distance between

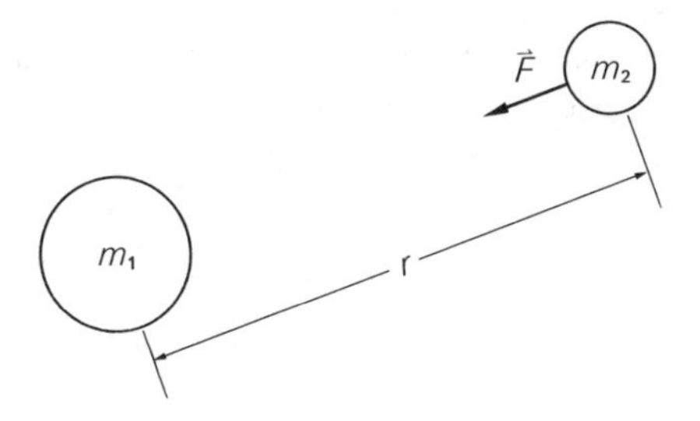

Figure 9–10. An object of mass m_2 experiences a gravitational force $\vec{F}$ caused by the presence of an object of mass m_1.

the objects). Let us consider how the force depends on these parameters:

1. The mass of the attracting object.
2. The mass of the attracted object.
3. The nature of the material composing the objects.
4. The distance between the objects.

One might consider other parameters—the relative orientation of the objects in space, their color, or the time of day. However, the four we have chosen will suffice to illustrate the process of guessing the force law. To determine the dependence of the force on any one of these parameters, naturally we must consider examples of motion in which the other parameters are fixed.

Clues from Falling Bodies

Let us begin by asking what we can learn about the gravitational force from observations of objects near the surface of the earth. Consider a free-fall experiment. Suppose a number of objects of the same material but different mass are dropped from the same position above the surface of the earth. The acceleration of each object is measured just after release.

In this experiment, three of the four parameters we have chosen to consider do not vary. In each case, the material composing the object is the same. In addition, the mass of the attracting body, the earth, does not change. Finally, if each measurement is made at the same position above the surface of the earth, the distance of each object from the matter composing the earth is the same. Thus the only parameter that varies is the mass of the falling object. In this way, we can determine how the force depends on the mass of the object on which the force is acting.

This falling body experiment is, of course, just the one Galileo had performed. He found that the acceleration of each falling object is the same. Thus the acceleration apparently does not depend on the mass of the object. But according to the

Newtonian scheme the acceleration is related to the force by $a = F/m$. Hence, a body of mass m near the surface of the earth experiences a gravitational force F_{grav} characterized by the property:

$$a = \frac{F_{\text{grav}}}{m} = 980 \text{ cm/sec}^2 \text{ independent of } m$$

But how can F_{grav}/m be the same for all values of m? This is possible only if F_{grav} is proportional to m—if the way the gravitational force depends on m is:

$$F_{\text{grav}} = Km$$

where K, in the falling body experiment, is constant. If the gravitational force does depend on m in this way, then

$$a = \frac{F_{\text{grav}}}{m} = \frac{Km}{m} = K$$

independent of m. This is consistent with Galileo's observation that all objects fall with the same constant acceleration.

To investigate how the gravitational force depends on the composition of the object, let us consider a similar experiment. Suppose a number of objects of different material but the same mass are observed to fall. For example, Figure 9–11 represents the fall of wood, iron, and lead spheres, each with mass m. Let F_W, F_I, and F_L be the gravitational forces experienced by the wood, iron, and lead spheres respectively. Once more, Galileo's experiments provide the result. Each object falls with the same

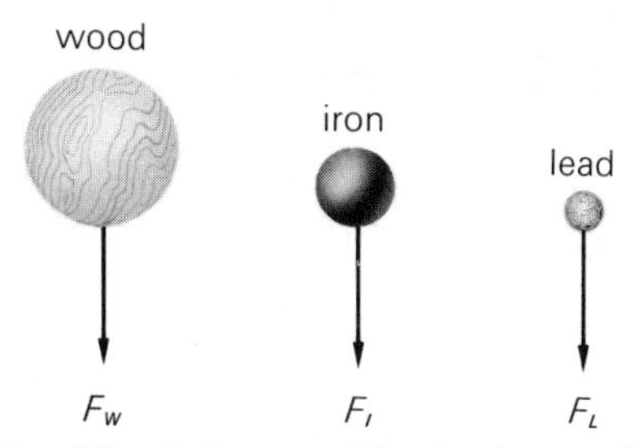

Figure 9–11. Wood, iron, and lead spheres of equal mass experience gravitational attractions F_W, F_I, and F_L, respectively.

constant acceleration, 980 cm/sec^2. What does this imply about the dependence of the gravitational force on the composition of the object?

Again, the acceleration of an object is related to the force it experiences by $a = F/m$. Hence, the result that each sphere of mass m falls with the same constant acceleration, 980 cm/sec^2, allows us to write:

$$a = \frac{F_W}{m} = \frac{F_I}{m} = \frac{F_L}{m} = 980 \text{ cm/sec}^2$$

This implies that $F_W = F_I = F_L$. In other words, the gravitational force does not depend on the composition of the object.

One caution is in order. These conclusions—that the gravitational force is proportional to the mass of the attracted object and independent of its composition—are inferred on the basis of experiment. They cannot, however, be regarded as absolute truths. It is possible to determine with modern equipment that the times of fall of different objects in a vacuum are the same to within perhaps .0001 second. This is quite suggestive of the conclusion that the times of fall do not depend at all on the mass or composition of the objects. However, no experimental result can be absolutely precise. It is always possible that there is some slight difference in the times, smaller than presently available equipment can measure. Hence our conclusion derives from considerations broader than experiment. In this case, the conclusion seems the simplest one we can reach consistent with present experimental knowledge.

Clues from the Motion of the Moon

In the eyes of Copernicus or Kepler or Galileo, the motion of the moon had nothing in common with that of a falling stone. From the revolutionary perspective of Isaac Newton, however, they are very much the same. In the *Principia* Newton presented a simple argument to demonstrate this unity. Consider the tall mountain in Figure 9–12. Suppose a man throws a stone horizontally as indicated by path *A*. To Newton, this projectile

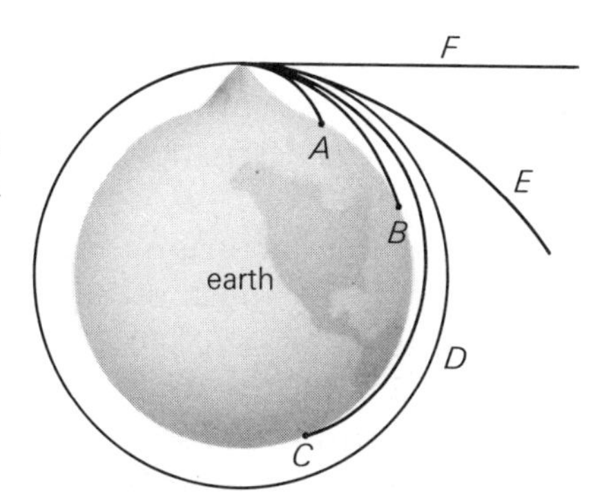

Figure 9–12. Trajectories of stones thrown from a mountaintop with increasing horizontal velocities.

motion could be considered a combination of the constant-velocity motion in the horizontal direction that would have occurred had the earth not been present (path *F*), and the free fall toward the center of the earth caused by the pull of gravity.

Suppose another stone is thrown even harder, as indicated by path *B*. It will have a longer range, both because of its increased horizontal velocity and because the spherical shape of the earth means that in effect the earth is curving out from under it. This latter effect is even more apparent in path *C*. However, if a stone is thrown extremely hard, it will follow a path like *E*. Although this stone continually falls from *F*, the path it would follow in the absence of gravity, its great initial horizontal velocity carries it away from the earth faster than it can fall toward the earth. Now imagine a stone thrown harder than that in path *C*, but not as hard as that in *E*. It should be clear that for the right initial horizontal velocity, path *D* will result. On this path, the stone falls toward the center of the earth throughout its motion just like any projectile, but because of the curvature of the earth, it never reaches the ground.

This is how Newton viewed the moon. It is nothing but a gigantic projectile. At some time in the distant past, the moon had somehow been projected horizontally with the proper initial velocity. Experiencing the same kind of attraction toward the earth as any stone, the moon falls toward the earth and hence orbits in a nearly circular path.

From this point of view, the force responsible for the circular motion of the moon is the same gravitational force responsible for the constant acceleration of a falling stone. We can measure the gravitational force attracting objects at the surface of the earth. By also measuring the force attracting the distant moon, perhaps we can learn how the gravitational force varies with distance.

To determine the magnitude of the force attracting the moon to the earth, we must measure the moon's acceleration as it "falls" in its orbit. We have found that an object moving with speed v in a circular path of radius r experiences an acceleration $a = v^2/r$ toward the center of the circle. This expression can be cast in a more convenient form by expressing the speed v in terms of the radius r and the period T (the time taken to complete one orbit). Since the circumference of the orbit is $2\pi r$ and the speed is the distance traveled divided by the time required to travel that distance, we may write

$$v = \frac{2\pi r}{T}$$

Therefore we have

$$a = \frac{v^2}{r} = \frac{1}{r}\left(\frac{2\pi r}{T}\right)^2 = \frac{4\pi^2 r}{T^2} \qquad (8)$$

Let us calculate this acceleration using the data available to Newton: $r = 60$ earth radii, or 223,000 miles, and $T = 27.3$ days. If these numbers are substituted in equation (8) and the result is converted to units of cm/sec², we find:

$$a_{\text{moon}} = .26 \text{ cm/sec}^2$$

This acceleration is much smaller than that of objects at the surface of the earth. One might wonder if the acceleration decreases with distance in any simple manner. Suppose we try a number of simple dependences—$1/r$, $1/r^2$, and $1/r^3$, for example. If the acceleration corresponding to a distance of one earth radius is 980 cm/sec² and the acceleration were inversely proportional to the distance, then the acceleration at the moon's distance of 60 earth radii would be:

$$1/r: \quad a = \frac{980 \text{ cm/sec}^2}{60} = 16.4 \text{ cm/sec}^2$$

Similarly, if the acceleration were inversely proportional to the square of the distance or the cube of the distance, the acceleration of the moon would be expected to be:

$$1/r^2: \quad a = \frac{980 \text{ cm/sec}^2}{(60)^2} = .27 \text{ cm/sec}^2$$

$$1/r^3: \quad a = \frac{980 \text{ cm/sec}^2}{(60)^3} = .0045 \text{ cm/sec}^2$$

It is clear from these figures that the measured acceleration of the moon, .26 cm/sec^2, is in poor agreement with the acceleration calculated assuming a $1/r$ or a $1/r^3$ dependence. However, it is in good agreement with a $1/r^2$ dependence. Thus we, with Newton, may hypothesize that the acceleration and hence the gravitational force (since $F = ma$) decrease as $1/r^2$.

Notice, however, that this hypothesis has not been proved. The measured acceleration, .26 cm/sec^2, does not even agree precisely with the acceleration calculated on the assumption of a $1/r^2$ dependence, .27 cm/sec^2. The difference may arise from uncertainties in the data or approximations in the calculation (the distance from the earth to the moon and the radius of the earth were not known very precisely, and the path of the moon is not quite the circular one we have assumed). However, from these numbers alone, all we can logically conclude is that the gravitational force seems much closer to $1/r^2$ than to $1/r$ or $1/r^3$. To assume it to be precisely $1/r^2$ is a matter of aesthetics, not experiment.

The Gravitational Force Law

On the basis of observations of falling bodies and of the orbiting moon, the magnitude of the gravitational force attracting an object to the earth appears proportional to the mass of the object and inversely proportional to the square of its distance from the center of the earth. How does the force depend on the one remaining parameter, the mass of the attracting body—the earth in these examples?

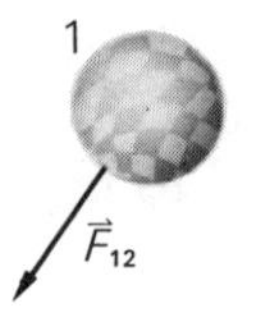

Figure 9–13. Objects 1 and 2 experience equal and opposite gravitational forces.

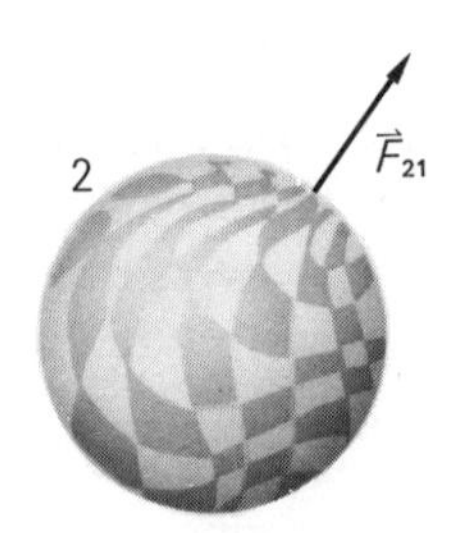

Figure 9–13 illustrates again the prototype gravitational situation. The object labeled 2 might be the earth and that labeled 1 might be the moon. $\vec{F}_{12}$ is the gravitational force with which object 2 pulls on object 1. We wish to discover how $\vec{F}_{12}$ depends on the remaining parameter, the mass of the attracting body 2. For this we turn to Newton's third law.

According to this law, if object 2 exerts a force $\vec{F}_{12}$ on object 1, then object 1 exerts an equal and opposite force $\vec{F}_{21}$ on object 2. In particular, the magnitudes of these forces are equal:

$$F_{12} = F_{21}$$

However, we have concluded that the gravitational force is proportional to the mass of the attracted object. Thus, F_{12} is proportional to m_1 and F_{21} is proportional to m_2. This can be reconciled with the requirement that $F_{12} = F_{21}$ provided that the gravitational force experienced by an object is proportional to the mass of the attracting object as well as to the mass of the object itself. This is easily seen: We have concluded that F_{12} is proportional to m_1 and inversely proportional to r^2. If F_{12} is also proportional to m_2, then its full dependence on m_1, m_2, and r is:

$$F_{12} = \frac{Gm_1 m_2}{r^2} \tag{9}$$

where G is a proportionality constant. If equation (9) is the form of the force law, then the gravitational force object 2 experiences is:

$$F_{21} = \frac{Gm_2m_1}{r^2}$$

which is equal to F_{12}, as the third law requires.

Thus we have arrived at a gravitational force law that Newton believed to describe the attraction between any two bits of matter in the universe. The mathematical expression

$$\boxed{F_{\text{grav}} = \frac{Gm_1m_2}{r^2}}$$

singles out the parameters on which this force appears to depend and indicates how it depends on them. The law was guessed from observations of nature. The fact that the gravitational force seems to depend so simply on so few parameters is a vindication of the Newtonian program. Newton's concept of force—measuring pushes or pulls by the accelerations they produce—appears fruitful.

An Explanation of Celestial Motion

Let us now turn to the second stage in the Newtonian program. Having guessed a gravitational force law, we can use it to explain numerous motions observed in nature.

Kepler in his three laws had pointed out the key features of planetary motion. A major task for Newton was to show that Kepler's laws were a necessary consequence of the gravitational force law. His success was the first major triumph of the Newtonian world view.

The easiest of Kepler's laws to account for was the second, the equal-area law. Newton demonstrated that any object subject to a central force (i.e., a force whose direction is along the line from the object on which the force is acting to the object responsible for the force) will obey the equal-area law. A geometric proof similar to Newton's is presented in Appendix 3. Since the gravitational force is a central force, Kepler's second law is a direct consequence.

Kepler's first law, that the path of each planet is an ellipse with the sun at a focus, was shown by Newton to follow directly from the inverse-square distance dependence of the gravitational force.* He further demonstrated that any other force law would lead to departures from a perfect re-entrant ellipse. For example, Figure 9–14 shows an orbit resulting from a hypothetical $1/r$ force: The orbit is roughly elliptical but does not retrace itself—it is said to *precess*. Thus, the observation that the planetary orbits are ellipses that do not precess to any significant degree is a strong argument in favor of the $1/r^2$ force law.**

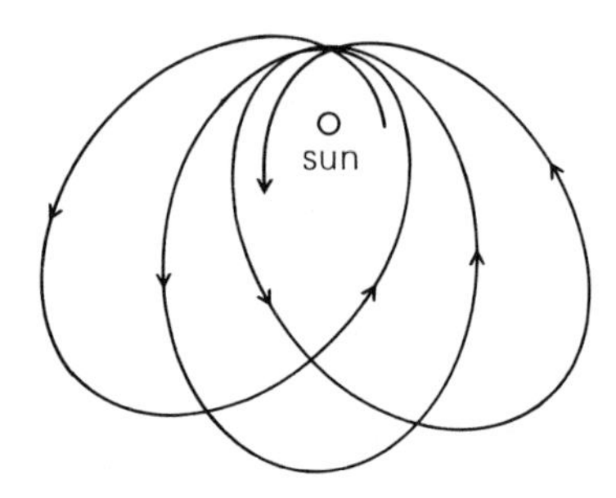

Figure 9–14. A planetary orbit resulting from a $1/r$ force. It does not retrace itself.

We shall not attempt to reproduce Newton's proof of these statements. It involves only geometry, and in principle is not difficult. It would, however, require a discussion of the geometric properties of the ellipse, which would take us too far afield.

* Other open paths, such as hyperbolas and parabolas, can also result from a $1/r^2$ law, but here we are clearly interested only in closed paths, around the sun. Hyperbolic and parabolic paths prove appropriate to objects from outside the solar system, like many comets.

** Newton's contemporaries first learned of these results from a visit paid to him in 1684 by the astronomer Halley (the discoverer of the comet). There was at that time much discussion in London about the nature of the trajectories followed by bodies subject to various hypothetical kinds of forces. Halley asked Newton, then living a withdrawn life in Cambridge, what sort of orbit would result from an inverse-square force law. Newton answered immediately that it would be an ellipse, and said that he had found the result years earlier. Pressed by Halley for the derivation, Newton could neither remember it nor find the papers where he had performed the calculation. Fortunately he was able to reproduce the proof shortly thereafter.

Finally, Newton showed that the gravitational force law implies Kepler's third law: if an object is attracted to a point by a $1/r^2$ force, the square of the period of its motion in the resulting ellipse is proportional to the cube of the average distance, where "average" distance denotes half the larger axis of the ellipse. Again we omit a discussion of the general proof because of its geometric complexity; however, we can give a proof for the special case of circular motion, which is a good approximation for the planets.

Let m_p be the mass of a planet, and m_s be the mass of the sun. Let r be the distance between the planet and the sun, and a_p be the acceleration of the planet in its circular orbit. We write the force law as

$$F_{\text{grav}} = \frac{Gm_pm_s}{r^2}$$

This force acting on the planet will bring about an acceleration a_p, which is related to F_{grav} by Newton's second law:

$$F_{\text{grav}} = m_pa_p$$

Equating these two expressions for F_{grav}, we find

$$\frac{Gm_sm_p}{r^2} = m_pa_p$$

But we know that for uniform circular motion the acceleration can be written [equation (8)]

$$a_p = \frac{4\pi^2r}{T^2}$$

which, in turn, can be substituted in the previous equation:

$$\frac{Gm_sm_p}{r^2} = m_p\frac{4\pi^2r}{T^2}$$

Rearranging, we have

$$T^2 = \left(\frac{4\pi^2}{Gm_s}\right)r^3 \tag{10}$$

This is Kepler's third law. The square of the period of a planet is proportional to the cube of its distance from the sun.

An Explanation of Terrestrial Motion

The gravitational force law also explains Galileo's observations of freely falling objects. Of course, we used one of these observations—that all objects exhibit the same acceleration independent of their mass and composition—to guess the force law in the first place. But we can show that the gravitational force law implies not only this observation but also the observation that the objects fall with a *constant* acceleration.

A calculation of the attraction between a small object near the surface of the earth and the extended mass of the earth would at first glance appear extremely complicated. After all, such an object is very close to some of the matter that makes up the earth, but very far from other matter of the earth. Since the force depends on distance, it might seem that the application of the gravitational force law to the motion of falling objects would be computationally intractable.

The problem was difficult even for Newton. However, he intuitively sensed a remarkably simple solution. He noted that for every bit of matter in the earth that attracts an object, there is a symmetrically placed identical bit of matter that also attracts it, as shown in Figure 9–15. The vector sum of these two forces is directed toward the center of the earth. Thus Newton argued that the net force from all the bits of matter in the earth should be directed toward the center. But what is the magnitude of that net force?

Here Newton's intuition suggested an answer. He speculated that the force attracting an object to the earth is equivalent to the force the object would experience if the earth were replaced by a particle having the earth's mass and located at the position of the earth's center (Figure 9–16).

Believing this to be true, Newton used it repeatedly as he analyzed the motions of gravitationally interacting bodies, in the hope—even faith—that he or someone else would eventually establish its validity. His doubts on the matter, however, were a major factor in causing him to withhold publication of the

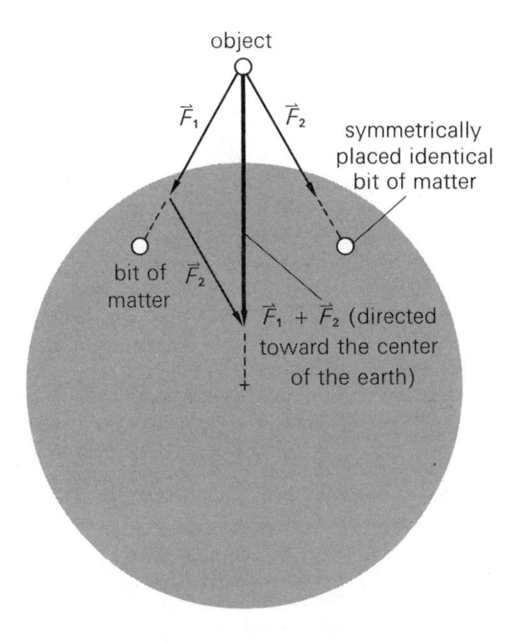

Figure 9–15. The net gravitational force on an object is directed toward the center of the earth.

Figure 9–16. Newton's speculation. In calculating the gravitational force, the earth may be replaced by a particle of mass M_e located at the center of the earth.

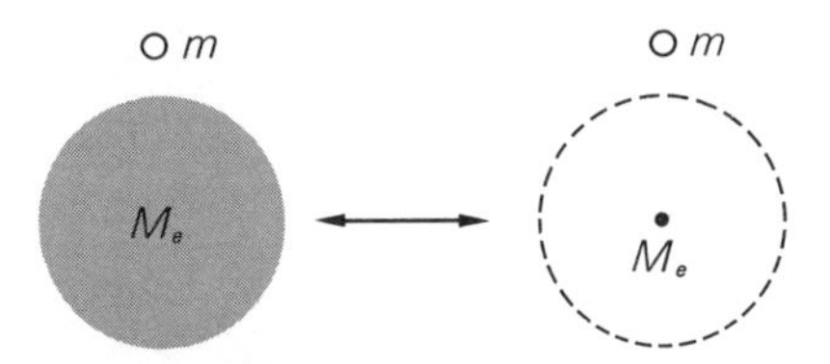

Principia for twenty years. During this time, he created the branch of mathematics known as the integral calculus and was able to complete the proof.

We see in this another example of the intuition and guesswork that is present in any creative science and that underlies even the strictly logical presentation of the *Principia*. One might make a distinction between "public" science, theories and conclusions presented in their most logical, ordered, and elegant form, and "private" science, the imaginative, intuitive, groping process of developing a theory. As noted earlier, we have very little information or insight into Newton's thought as he created his physics. The cold logic of the *Principia* is only public science.

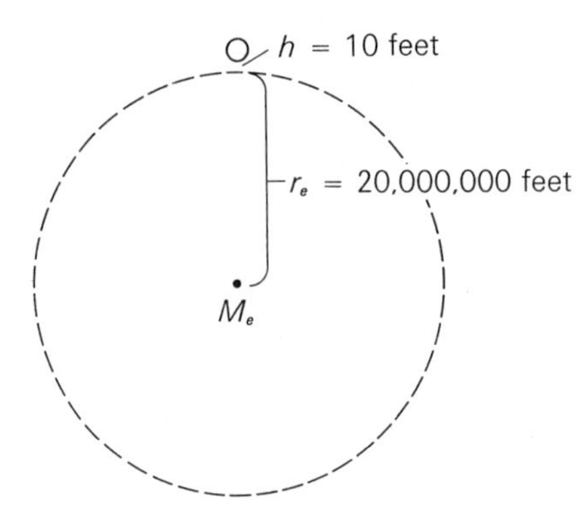

Figure 9–17. The distance of fall of an object is negligible in comparison with the radius of the earth.

Let us return to Galileo's observations of freely falling objects. Consider a stone of mass m released from a height h, say 10 feet, above the surface of the earth, as shown in Figure 9–17. What does the gravitational force law predict about its motion? According to Newton's second law, this motion is described by $F_{\text{grav}} = ma$. Using Newton's observation that for purposes of analysis the earth may be replaced by a particle with the earth's mass M_e located at the position of the center of the earth, we may write:

$$F_{\text{grav}} = \frac{GM_e m}{(r_e + h)^2}$$

and therefore

$$ma = \frac{GM_e m}{(r_e + h)^2}$$

where a is the acceleration of the stone and r_e is the radius of the earth. Notice that r_e is approximately equal to 4000 miles, or about 20 million feet, and h is less than 10 feet throughout the entire fall of the stone. Consequently, to a very, very good approximation $r_e + h$ can be taken equal to r_e. This means that the last equation above can be approximated by:

$$a = \frac{GM_e}{r_e^2} \qquad (11)$$

Notice that the mass of the stone has canceled out of the equation. Since G, M_e, and r_e are all constant, we see that the gravitational force law predicts that a falling object will move with a constant acceleration, and furthermore that this acceleration is independent of the mass of the object. Thus Newton was able to account for Galileo's results.

In this chapter we have shown how the gravitational force law, a simple mathematical formula, can be said to explain both the celestial and terrestrial motions that had puzzled man for centuries. This was a truly remarkable accomplishment, but it was just the beginning of the role the Newtonian approach would play in organizing scientific inquiry for the next three centuries.

Suggested Reading

Holton, G. *Introduction to Concepts and Theories in Physical Science.* Reading, Mass.: Addison-Wesley, 1952. Chapter 11 discusses Newton's gravitational force law and its application to the solar system.

Cooper, L. *An Introduction to the Meaning and Structure of Physics.* New York: Harper & Row, 1968. A briefer treatment of the same material is contained in Chapter 6.

Butterfield, H. *The Origins of Modern Science.* London: Bell, 1949. Chapter 8 is a qualitative discussion of Newton's theory of gravitation and its historical foundations.

Gillispie, C. *The Edge of Objectivity.* Princeton: Princeton University Press, 1960. Chapter 4 treats several aspects of Newton's work, including gravitation and optics.

Questions

1. Does the acceptance of the Newtonian world view necessitate acceptance of the heliocentric rather than the geocentric model as the way the universe really is? Explain.
2. We have noted that the gravitational force on an object apparently does not depend on the material composing it. Can you think of a force that does depend on the material composing the object on which the force acts?
3. Find the direction and magnitude of the acceleration of the car moving around a circular track in the exercise on page 160.
4. A car of mass 1000 kg races with a constant speed of 30

m/sec around the track illustrated, which has semicircular ends of radius 90 m, and straight sides. During which part of its motion is a net force acting on the car? Find the magnitude and direction of this force. What is the origin of this force?

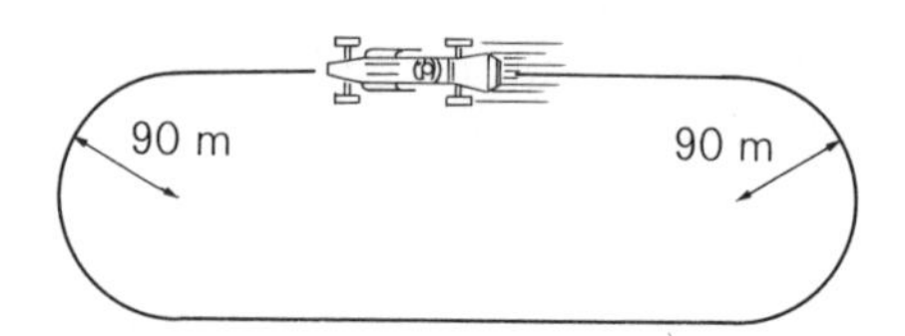

5. A ball of mass 100 g on the end of a 1-m string is whirled in a horizontal circle. (For simplicity, we neglect the downward force of gravity in this problem.) If it is whirled at a rate of four revolutions/sec:

   (a) Determine the magnitude of the acceleration of the ball.
   (b) Determine the force the string exerts on the ball.

6. Repeat the calculation of Figure 9–4, page 211, for the case of one object moving with three times the speed of the other object.
7. Repeat the calculation of Figure 9–5, page 212, for the case in which the radii of the circles are R and $3R$.
8. When asked if a car moving around a circular track with constant speed is accelerating, most people would say that it is not. Devise an alternative definition of the concept of acceleration to agree with the above response. Why would this concept of acceleration prove less useful in physics than the one we have adopted? (Can you think of any specific examples where the definition we have adopted is more apt for the description of nature than your alternative?)
9. Which would you guess is larger, the gravitational force attracting the moon to the earth, or the gravitational force attracting the moon to the sun? Explain the rationale for your conjecture. Now calculate which is larger. (You may wish to know that the mass of the sun is approximately 3×10^5 times the mass of the earth, and the distance from

the moon to the sun is about 400 times the distance from the moon to the earth.) How can you reconcile the result of this calculation with your conjecture?

10. Assume for simplicity that the earth and the moon are made of the same uniform material. The radius of the earth is roughly four times the radius of the moon.

    (a) Suppose that Galileo's Leaning Tower experiment is performed by an astronaut from atop his lunar module on the surface of the moon. What acceleration would the astronaut measure for free fall on the moon? (Hint: First determine how the masses of the earth and the moon are related. Then relate the acceleration on the moon to that on the earth, which you know to be 9.8 m/sec^2.)
    (b) How much would a 200-pound astronaut weigh on the moon? (A person's weight is the gravitational force he experiences.)

11. In a hypothetical universe, suppose that planets move about a sun in circular orbits under the influence of a gravitational force inversely proportional to the cube of the distance between planet and sun. Following Newton's analysis, deduce the form of the period-vs.-distance law (analogous to Kepler's third law) for that solar system.

12. In a hypothetical universe, the gravitational force law is

$$F = \frac{Gm_1m_2}{r^n}$$

where n is a positive integer ($n = 1$, or 2, or 3, or . . .). Consider a system of four planets orbiting a star in this universe.

    (a) Show that the analogue of Kepler's third law for this planetary system is:

$$T^2 = Kr^{n+1}$$

    where K is a constant.
    (b) Suppose the period and distance from the star are measured for each planet and the data obtained are

those listed in Question 13, page 101. What could you conclude about the gravitational force law in this universe?

13. Consider a hypothetical universe in which there is discovered a star with three planets, A, B, and C, moving around it in circular orbits. The radii of the planetary orbits and their respective periods are given in the following table:

Planet	Distance from Star (million miles)	Period (years)
A	1	1
B	2	2
C	3	3

On further investigation, it is found that on one of the planets, which has mass M and radius R, all objects released from rest fall vertically with the same uniform acceleration k. Assuming that Newton's three laws of motion are appropriate to this universe, discuss the nature of the gravitational force law. In particular,

(a) How does the gravitational force F on a particle depend on its mass m? Explain.
(b) How does the gravitational force F between two particles depend on the distance r between the particles? Explain.
(c) Express the gravitational force F between two point particles of mass m_1 and m_2 separated by a distance r in terms of m_1, m_2, r, M, R, and k. Explain your reasoning.

14. Suppose the gravitational force law had turned out to be very complicated, perhaps

$$F = \frac{A}{r^6} + \frac{Bm_1m_2}{r^{7/8}} + \frac{Cr^{1/2}}{m_1 + m_2} + \frac{Dr^{1/4}}{(m_1 + m_2)^2}$$

Would this prove the Newtonian approach wrong? Explain.

CHAPTER TEN

Newton's Legacy

With his theory of gravitation, Newton was able to link what had been two quite distinct strands of thought, explaining at once the observed motions of the planets in the heavens and objects on the earth. Others before Newton had grappled with the phenomena of celestial and terrestrial motion; what is striking is the diversity of their perceptions of the "problem of celestial motion" and the "problem of terrestrial motion." *A priori*, one could not know what questions would open the most fruitful line of inquiry.

To appreciate Newton's historical role, we must recognize the profound ambiguities confronting him. In a scientific revolution —a major shift in the conceptual scheme men use to understand their world—it is not clear which observations are significant, what grouping of these observations is the most pregnant, and what sort of explanation should be regarded as satisfactory. Viewed in this perspective, Newton's primary contribution was not the hypothesis of a gravitational interaction that explains celestial and terrestrial motion. Rather, the way he looked at the problem, the questions he asked, and the way he organized his inquiry are Newton's most significant legacy. As we shall demonstrate in this chapter, the conceptual framework Newton suggested has proved so successful that it has in an important sense *directed* inquiry in physics for the past three hundred years. In so doing it has also exerted a significant influence on our civilization.

Forces and Motion

Let us briefly review the Newtonian approach. As we have seen, it involves two steps. The first is the discovery of a force law. One attempts to observe the motion of an object in a simple situation—for example, a planet subject to the gravitational force of the sun. By determining the acceleration of the object and noting how it depends on the position and other parameters, one can apply $\vec{F} = m\vec{a}$ and try to obtain the force law.

The second step is to use this force law to predict the motion of other objects under other conditions. Thus, from the gravitational force law Newton could deduce Kepler's laws describing the motion of all the planets. We followed some of these deductions for particularly simple situations (circular planetary orbits, for example). However, it is important to realize that in principle one can always calculate the exact path of an object subject to known forces. This can be seen as follows.

Suppose we know the initial position and velocity of an object of mass m that experiences a known force $\vec{F}$. Figure 10–1 illustrates the situation. For example, $\vec{F}$ might be the gravitational force depending on the inverse square of the distance from the sun located at the origin. If the force were not present, the object would move in the direction of its instantaneous velocity $\vec{v}$. However, because of the force, the object experiences an acceleration given by $\vec{a} = \vec{F}/m$. This means that the velocity $\vec{v}$ changes, generally in both magnitude and direction. The position

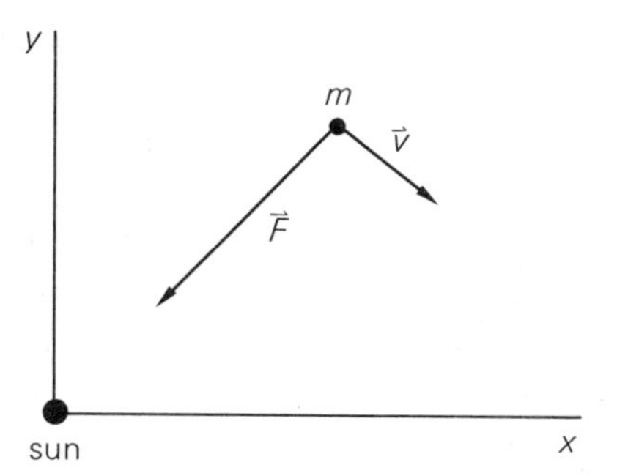

Figure 10–1. The object of mass m is moving with velocity $\vec{v}$. The force $\vec{F}$ acting on the mass brings about an acceleration of the object, which will lead to a change in $\vec{v}$.

of the object changes accordingly, and this means that the force may change. This force in turn produces a further acceleration, and corresponding further changes in velocity and position, leading to a still different force, and so it goes. The process is well defined mathematically, and the techniques of the calculus can be used to calculate the precise position of the object at any later time.

Planetary Perturbations

An important early success of the Newtonian techniques came in their application to the details of planetary orbits. In Newton's time, the accuracy of astronomical measurements had improved enough for astronomers to discover many small departures from ellipticity in the orbits of the planets.* Kepler's ellipses were not quite right. What then of Newton's theory, which predicted elliptical orbits?

In fact, Newton's prediction had been made only in the idealized circumstance of a single planet orbiting about a sun in an otherwise empty universe. In the solar system, however, there are other planets, and Newton showed that the departures from ellipticity, or *perturbations*, are caused by the gravitational

* The telescope had been invented in Holland at the end of the sixteenth century and was given wide publicity by Galileo, who exhibited to an incredulous world the moons of Jupiter and the phases of Venus. By Newton's time, the telescope was in general use.

forces between the planets. When one planet passes sufficiently close to another, it may be pulled slightly away from its elliptical orbit. These interplanetary forces are ordinarily very small compared with the force between a planet and the sun, since the masses of the planets are much smaller than the mass of the sun. However, when two planets pass near each other, this perturbing force is large enough to have an observable effect. "And hence arises a perturbation of the orbit of Saturn in every conjunction of this planet with Jupiter, so sensible, that astronomers are puzzled with it," Newton wrote.[1] He was able to calculate the perturbations in the motion of any given planet, obtaining results in good agreement with observations.

This story has a familiar ring. We discussed earlier how Ptolemy, faced with the departures of the planets from the ideal of uniform circular motion, had patched up the theory with epicycles. Here we see how Newton, faced with deviations from simple elliptical motion, also patched up his calculations by invoking gravitational forces from other planets. However, there is an important difference. Ptolemy's epicycles were *ad hoc* devices, with no justification other than effectiveness. But the interplanetary forces are a necessary consequence of the Newtonian theory, and their existence knits it into a tighter fabric.

The Discovery of Neptune

The success of the Newtonian method with the details of planetary orbits is dramatically shown in the story of the discovery of Neptune. In the late eighteenth century, a new planet, Uranus, had been discovered accidentally. Astronomers set out to determine its orbit and motion. However, even after the effects of all the other planets had been laboriously calculated, significant discrepancies remained between the observed motion and that predicted on the basis of all known gravitational attractions.

Two theoretical astronomers, John Adams in England and Urbain Leverrier in France, independently tried to resolve the

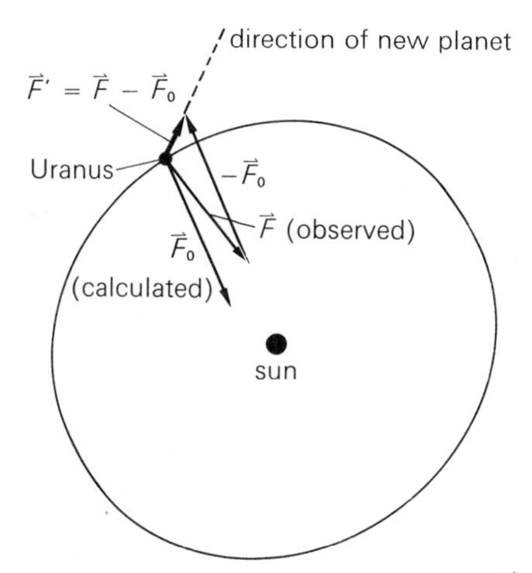

Figure 10–2. From the difference between $\vec{F}$, the net force on the planet calculated from its observed acceleration using $\vec{F} = m\vec{a}$, and $\vec{F}_0$, the vector sum of the gravitational forces due to the sun and each of the other known planets, the position of a new planet may be inferred.

difficulty by assuming the presence of another, undiscovered planet. Their approach is shown in Figure 10–2. From a knowledge of the observed motion of Uranus, its acceleration vector at any time can be calculated by taking the usual limit of $\overrightarrow{\Delta v}/\Delta t$. This acceleration vector can be substituted into $\vec{F} = m\vec{a}$ to determine the magnitude and direction of the net force $\vec{F}$ on Uranus. On the other hand, the gravitational force law can be used directly to find the forces exerted on Uranus by the sun and each of the other planets. We designate the vector sum of these forces by $\vec{F}_0$. If all forces have been taken into account, $\vec{F}_0$ should be identical with $\vec{F}$, the net force on Uranus. As Figure 10–2 shows, however, $\vec{F}_0$ was found to be somewhat different from $\vec{F}$. The difference, $\vec{F}' = \vec{F} - \vec{F}_0$, must arise from some other attracting object—a new planet. The direction of $\vec{F}'$ is in the direction of the planet and its magnitude indicates the distance of the planet from Uranus. By calculating $\vec{F}'$ at many points on the orbit of Uranus, the orbit of the hypothetical planet could be determined.

Adams made this calculation and asked the observatory at Greenwich to look for the new planet at a point he would specify. But Adams was a young man with no reputation, and the calculation was difficult, with many opportunities for mistakes. The observatory staff, preoccupied with other matters, gave the search low priority. Leverrier finished his calculation soon afterward and had better luck with the observatory at Berlin. In September of 1846, the planet Neptune was discovered very close to the position predicted by Leverrier.

When careful observations of Neptune's orbit were made, similar difficulties arose. These led in the same way to the discovery of Pluto in 1930.

Weighing the Earth

When we discussed the Newtonian gravitational force law, $F = Gm_1m_2/r^2$, we did not give the value of the gravitational force constant G. In the *Principia*, Newton never explicitly wrote an equation with this constant. He simply described the force as proportional to the product of the masses and inversely proportional to the square of their separation.

In fact, Newton did not need to know the value of G to explain terrestrial and celestial motion. Nor could he determine G by observing the motion of falling bodies on the earth or the motion of the planets about the sun. Recall the discussion of freely falling bodies in Chapter 9. Equation (11), obtained by writing $\vec{F} = m\vec{a}$ for an object near the surface of the earth, asserts that the object will fall with the constant acceleration

$$a = \frac{GM_e}{r_e^{\,2}} = 980 \text{ cm/sec}^2 \qquad (11)$$

Since r_e, the radius of the earth, can be measured, this equation enables us to calculate the combination GM_e, where M_e is the mass of the earth. However, this does not determine G by itself since the mass of the earth cannot be measured directly.

We encounter a similar problem if we attempt to calculate G from the planetary data. We found in equation (10) in Chapter 9 that if $\vec{F} = m\vec{a}$ is applied to a planet orbiting about the sun, its period T is related to its distance r from the sun by

$$T^2 = \left(\frac{4\pi^2}{Gm_s}\right)r^3 \qquad (10)$$

If we know the period of a planet and its distance from the sun, this equation enables us to calculate the combination Gm_s. Since m_s, the mass of the sun, cannot be measured directly, we cannot determine G itself from the planetary data.

There is an obvious way to measure G: Take two objects whose masses m_1 and m_2 are known, separate them by a distance r, and measure the force between them. One way to do this (in principle) is illustrated in Figure 10–3. Attach the object of mass m_1 to the end of a previously calibrated spring balance. Let it rest with the spring at its relaxed length l_0. Then bring up the other mass to a distance r. The first mass will be attracted, stretching the spring by an amount l (for which the corresponding force F has been measured by an acceleration experiment). Substituting the result into $F = Gm_1m_2/r^2$ allows one to solve for G.

The trouble with this technique is the extremely small magnitude of the gravitational force between laboratory-sized chunks

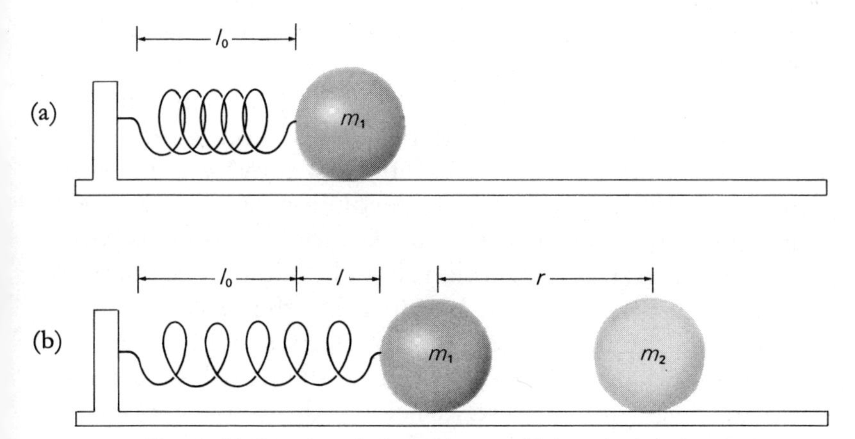

Figure 10–3. A technique that could in principle be used to measure the gravitational force. (a) A mass at rest, attached to a relaxed spring. (b) The gravitational attraction of another mass stretches the spring.

of matter. It was simply not technically feasible to measure this force in Newton's time, or for some time afterward. Only late in the nineteenth century did such an experiment first succeed. Henry Cavendish in England measured G by an extremely delicate experiment similar in principle to the one just discussed, and found a value $G = 6.7 \times 10^{-11}$ newton m^2/kg^2. The constant by itself doesn't tell us much, but in equation (11) it enables us to calculate the value of the mass of the earth. The result is:

$$M_e = 6.0 \times 10^{24} \text{ kg}$$

More impressive than the actual number is the simple realization that it can be found at all. Cavendish showed obvious pride in his accomplishment by entitling the paper in which he described it "Weighing the Earth"—in effect what his experiment allowed him to do.

Gravity Outside the Solar System

Considerable evidence has accumulated to indicate that the gravitational force is not limited to the solar system. For example, *double stars* have been observed in our galaxy. They consist of two stars revolving about one another. Figure 10–4 indicates the relative motion of the two stars in the double-star system known as ξ-Boötis. The cross in the figure represents the position of one of the stars and the circled dots indicate the positions of the other star relative to the first at five-year intervals beginning in 1830. One complete revolution of this star takes nearly 150 years. Nevertheless, if Newton's gravity is what attracts these two stars, the motion should exhibit the regularities summarized in Kepler's laws. In particular, the orbit should be an ellipse and the line joining the two stars should sweep out equal areas in equal times. The data of Figure 10–4 do show these regularities (see the exercises below); this is strong evidence for the existence of the gravitational interaction outside our solar system.

Even farther from the earth, whole galaxies of stars have been discovered to rotate about other galaxies. The data are not

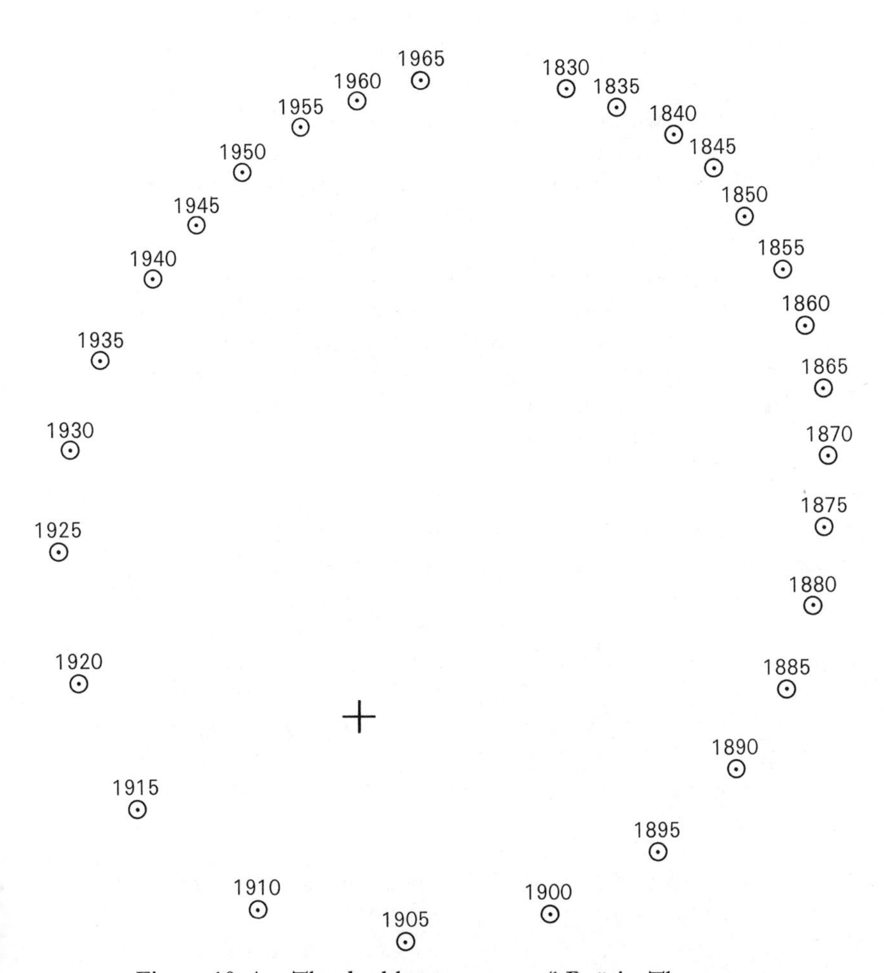

Figure 10–4. The double-star system ξ-Boötis. The cross represents the position of one of the stars, while the circled dots represent the positions of the other star relative to the first at five-year intervals. The circles represent rough estimates of the uncertainties in the determinations of positions. (Data courtesy of Roger B. Carr and Robert T. Mathews, Carleton College Astronomy Department.)

accurate enough to allow detailed comparison with Kepler's laws, but one cannot resist the assumption that here, too, gravity is present. Newton's assertion of the universality of the gravitational force appears justified.

Using two pins, a string, and a pencil, try to construct an ellipse to fit the data of Figure 10–4. (See Figure 4–14 for a description of how an ellipse may be constructed.) It should be noted that each data point is actually the average of several observations collected from the published literature. Each dot on the graph represents a calculated average position; the size of the circle around it is a rough estimate of the scatter of the individual observations. That is, the circle represents the uncertainty in each point. Are the data consistent with an ellipse?

Using the ellipse constructed in the above exercise, test the applicability of Kepler's second (equal-area) law by comparing the areas corresponding to five-year intervals at three or four different parts of the orbit. Of course the areas will not be precisely equal. What then can you say about the applicability of the equal-area law to ξ-Boötis? (A statement like "My areas were close enough" is not very informative. How close should they be? In answering this question use the uncertainties in the points to estimate uncertainties in the areas.)

Construct the velocity vectors of the orbiting star at a number of times. Using the definition

$$\vec{a}_{\text{ave}} = \frac{\vec{v}_f - \vec{v}_i}{t_f - t_i}$$

construct the average acceleration vectors for a number of time intervals. According to Newton's second law, $\vec{F} = m\vec{a}$, the force on the orbiting star is in the direction of the acceleration of the star. By examining the acceleration vectors you have constructed at various points of the orbit, what can you say about the force?

Parallax

In Chapter 4 we discussed one of the strongest arguments against the heliocentric system—the absence of any observed parallax in the positions of the stars. If the earth moves, the appearance of the stars ought to be somewhat different at different times of the year—that is, at different points in the earth's orbit. In the absence of any parallax, Copernicus had to assume that the stars were so far away that the effect was too small to be observed. With each improvement in observational accuracy, astronomers searched for parallax. Each failure to observe it forced them to assume a still greater distance to the stars. Not until 1838 was this effect finally discovered. It amounted to less than one second of arc even for the nearest stars.

The long series of attempts to detect a stellar parallax, and the eventual acceptance of the heliocentric model long before such parallax had been observed, are a useful counter-example to the naïve picture of the scientific method discussed in Chapter 1. From the time of Copernicus to the eighteenth century, the basic observation (no parallax) did not change, yet the interpretation and significance of this observation were altered in the light of changing theories of the heavens. Once a "proof" of the geocentric model, parallax finally came to be viewed as merely an indication of the great distance to the stars from the earth.

Artificial Satellites

In discussing the achievements of Newtonian gravitation, we can hardly omit the spectacular feats of recent years involving man-made heavenly bodies. Innumerable earth satellites have been placed in orbit in the past two decades, men have traveled to the moon, and voyages to the planets are being considered. Man is reaching out into the universe in a way Newton never dreamed of.

The theory used in the calculation and control of satellite motion is entirely Newtonian. To place an earth satellite in orbit, one fires it sufficiently high above the earth so that atmospheric friction will be small, and then gives it a large enough speed parallel to the earth's surface. In accordance with $\vec{F} = m\vec{a}$ and the gravitational force law, the satellite will then revolve about the earth like the moon, "falling" in its orbit under the pull of the earth's gravity.

Applying to earth satellites the analysis Newton used for the moon and planets, we may predict that, to the extent that the earth is a uniform sphere, all earth satellites should obey Kepler's laws. Each should move in an elliptical orbit with the earth's center at a focus, and the areas it sweeps out in equal time intervals should be the same. The squares of the periods of such satellites should be proportional to the cubes of their distances from the earth's center. Observations of actual earth satellite motion show that Kepler's laws apply to a good approximation. However, small departures from perfect elliptical orbits can be seen. These are attributed to concentrations of mass within the earth and deviation from a spherical shape. In fact, the motion of artificial satellites has been used to determine more accurately the shape and internal structure of the earth. Such is our faith in Newtonian physics.

In like manner, Newton's physics is used in the planning and execution of voyages to the moon. One first decides on the path he wishes a lunar vehicle to follow, then calculates the acceleration needed at each point of the path to keep the vehicle on course, and finally arranges to fire the engines at such a rate and in such a direction that their force, when added to the gravitational forces from the earth and the moon, produces the necessary acceleration.

The successful orbiting of earth satellites and the landing of men on the moon are often called triumphs of science. This accolade, however, can betray a misunderstanding of the distinction between science and technology. Consider the first trip to the moon. Leaving aside the question of what man may learn by examining the moon firsthand, little in the voyage itself contributed to an understanding of the physical world. In the three hundred years of their existence, the basic principles of Newtonian physics have become so well established and so successful in their domain that no one doubted their applicability to extra-terrestrial voyages. Only powerful enough engines and sufficiently accurate guidance systems were required. When a satellite first orbited the earth, it was a triumph for the designers and manufacturers of the engines and guidance systems, but not for science. The first trip to the moon converted no one to Newtonian physics and discouraged no Aristotelians. We understood no more about the nature of motion after the voyage ended.

$\vec{F} = m\vec{a}$: *An Organizing Principle*

In the preceding sections we have applied the Newtonian conceptual scheme to several aspects of the problem of celestial motion. The key element in this scheme is Newton's second law, and we can now assess its role in physics. While it is referred to as a "law," $\vec{F} = m\vec{a}$ is in fact a definition, and hence has no physical content by itself. Essentially it is a prescription for how to go about studying the interactions of matter.

It was not at all certain *a priori* that such a study would lead to a successful theory of celestial or terrestrial motion. As we have seen, Ptolemy's epicycles, Kepler's perfect solids, and Descartes' vortices represent quite different conceptions with which men have tried to understand observations of the planets. Newton's $\vec{F} = m\vec{a}$ provides a radically different conceptual context. It is of fundamental importance as an *organizing principle*. It singles out the relevant questions to ask and the kinds of concepts that are appropriate to use in answering them.

We can illustrate this role of physical theories in organizing observations of nature by contrasting the Aristotelian and Newtonian views of projectile motion. To the Aristotelians, Earth, Water, Air, and Fire each had a natural motion. Movement of a stone toward the center of the earth, its natural place, required no further explanation. Movement not directed toward the natural place had to be explained in the Aristotelian context. We have seen that the motion of a projectile can be viewed as a composite of horizontal motion with constant speed and vertical motion with constant acceleration toward the center of the earth. To an Aristotelian, the vertical fall of the stone was natural, requiring no further explanation. What the Aristotelian philosophers sought to explain was the horizontal component. Various theories, including those of impetus, were proposed to explain why a stone did not simply seek its natural place after it had been projected.

The concept of natural motion, explained simply as an innate tendency of an object to move in a certain manner, may sound unscientific to modern ears. But is it really so different from the Newtonian concepts in current use? When a Newtonian physicist views projectile motion, what questions does he ask? Does he inquire about the horizontal, constant-speed component of the projectile? Not at all. That is the motion the projectile would have had if no force had been acting on it; it requires no further explanation. It is the Newtonian concept of natural motion. To a Newtonian, it is the vertical, accelerated motion which must be explained. We understand projectile motion as a combination of constant acceleration in the vertical direction caused by the gravitational force of the earth and constant speed in the horizontal direction, its natural, inertial motion.

Thus the Aristotelian and Newtonian world views lead to different conceptions of the significant facts about projectile motion. A physical theory does more than explain; it singles out the observations that need to be explained.

Let us consider the sense in which Newton's theory explains planetary motion. The Newtonian scheme focuses on the interaction between objects. It asks about the nature of the forces that cause objects to move the way they do. This is very much in the spirit of Descartes' attempt to describe how the planets are made to move about the sun and quite different from Ptolemy's mathematical fitting of planetary orbits or Kepler's geometric scheme to account for the number of planets and their distance from the sun. On the other hand, Descartes and Newton differed radically in what they considered an acceptable response to an inquiry about the nature of an interaction. Descartes sought a physical mechanism to explain how one object could influence another. Newton was content with a mathematical formula.

Perhaps "content" is too strong a word to describe Newton's attitude. He recognized that in one sense, his theory of gravity explains everything. Yet in another, it explains nothing. For it has not told us what gravity is. Newton could only write the mathematical formula Gm_1m_2/r^2. But what mechanism is responsible for this attraction? Why the inverse square law? Newton did not know. The question concerned him deeply. He made many attempts to explain gravity as due to some sort of "aether" pervading all of space and pushing gravitating bodies together, or as due to a vortex model like that of Descartes. But in the end he was unsuccessful. He concluded the *Principia* with these words:

> But hitherto I have not been able to discover the cause of those properties of gravity from phenomena, and I frame no hypotheses . . . To us it is enough that gravity does really exist, and act according to the laws which we have explained, and abundantly serves to account for all the motions of the celestial bodies . . .[2]

This inability to explain gravity accounted for much early resistance to the Newtonian approach. The followers of Descartes in particular felt strongly that their vortex model provided such an explanation and was therefore a better theory. For more than fifty years after the publication of the *Principia*, Descartes' physics remained preeminent in continental Europe. In the mid-eighteenth century, Voltaire, the French novelist and philosopher, wrote:

> A Frenchman who arrives in London finds a great change in philosophy as in everything else. He left the world full, he finds it empty. In Paris one sees the Universe composed of vortices of subtle matter. In London one sees nothing of this. In Paris it is the pressure of the moon that causes the flux of the sea; in England it is the sea which gravitates toward the moon. With your Cartesians, everything is done by an impulsion that nobody understands; with Mr. Newton, it is by an attraction, the cause of which is not better known.[3]

In the end, the Newtonian approach did eclipse that of Descartes, on the strength of its far greater pragmatic accomplishments. It left the question of the cause of gravity unanswered. Gradually men stopped asking the question, and the Newtonian success seemed complete.

A World of Law

The eventual acceptance of the *Principia* in England brought Newton fame and adulation rare in any era. His leadership among English scientists was acknowledged with the presidency of the Royal Society, the chief governing body of English science, which he ruled in dictatorial fashion for the last twenty-five years of his life. The powers of patronage he exercised by virtue of his position and prestige were wielded to advance the cause of the new physics. University posts were filled with his disciples, careful not to displease their great benefactor. Newton, a supremely original man in his youth, seems in his later years to have repressed the originality of others.

Little of this, however, was apparent to the British public, to whom Newton was the genius who had discovered the laws of the heavens. He was honored with the position of Master of the Mint in 1700, and five years later was knighted by Queen Anne. As the years passed, more and more honors attested to the recognition of his achievements not only by his scientific colleagues, but by the entire English intellectual world.

At least part of the reason for this enthusiastic reception stems from the general mood of the day. When the *Principia* appeared in 1687, England had been embroiled in a long period of intense religious and civil strife. Disputations between Catholics, Anglicans, and Puritans had torn both the universities and the country at large, leading to the civil war of 1643–51. With the restoration of the monarchy in 1660, the country was exhausted, intellectually as well as physically, from the confused and passionate issues that had divided it.

In this atmosphere, Newton's system of the world was welcome. Not only was it a monumental intellectual feat in itself, but it was refreshingly remote from the subtle and complex controversies of theology and politics. One could discuss Newton's work without exhausting emotional arguments. The very style of the *Principia* contributed to this reception. Its Euclidean progression of definitions, theorems, and corollaries seemed to give its results the objective certainty of geometry. Through the power of Reason, Newton appeared to have discovered Truth. According to his contemporary Alexander Pope:

> Nature and Nature's laws lay hid in night:
> God said, "Let Newton be!" and all was light.[4]

In the *Principia*, it seemed that Newton had not merely found a way of understanding nature; he had discovered the very laws by which nature operates. The belief that nature is governed by mathematical laws is an important part of Newton's legacy. As the successes of Newtonian physics became ever more numerous and impressive, not only in England but soon over the entire Western world, this belief strengthened. A rule of natural law governing not only the physical world but even the affairs of men became an irresistible idea. Furthermore, Newton's example suggested that man's mind could discover these laws. John Locke in philosophy and Adam Smith in economics, among others, attempted to construct theories of human affairs in which laws of nature were central. To this day the use of natural laws to explain events even in fields like history and aesthetics remains an important element of our culture.

Determinism

Closely related to the idea of natural law is that of determinism, which lies at the heart of Newtonian physics. We have already seen how, given the initial conditions and the force laws, one can use Newtonian analysis to predict the entire future motion of an object. The same result applies to a collection of interacting objects—that is, to the universe. If the positions and velocities of all objects at one instant are known, the forces on each object can be calculated, and their ensuing motion can be determined unambiguously. Hence only one predetermined future can evolve.

Newton himself hesitated to attribute such a complete determinism to his system. The intervention of God in the functioning of the physical world seemed inevitable to him. For example, he was concerned about the eventual consequences of planetary perturbations. It appeared probable to him that the accumulation of these perturbations over the years would lead to instabilities in the whole planetary system, and that the planets and sun would eventually collide. The fact that the solar system had survived for such a long time seemed to point to the occasional participation of God in its operation, correcting the planetary motions to prevent catastrophes. As Newton was a deeply religious man, who saw in the order of the heavens an indication of the greatness of God, the need to invoke divine assistance seems to have pleased rather than disappointed him.

At the end of the eighteenth century, however, the French mathematician Laplace demonstrated that the solar system should remain stable in spite of the perturbations. To Laplace, the world seemed to be nothing but a huge machine, subject to a complete determinism. He wrote:

> If an intelligence, for one given instant, recognizes all the forces which animate Nature, and the respective positions of the things which compose it, and if that intelligence is also sufficiently vast to subject these data to analysis, it will comprehend in one formula the movements of the largest bodies of the universe as well as those of the minutest atom: nothing will be uncertain to it, and the future as well as the past will be present to its vision. The human mind offers in the perfection which it has been able to give to astronomy, a modest example of such an intelligence.[5]

This view has obvious bearing on a number of philosophical and religious questions such as the place of God in the universe and the possibility of human freedom. It lay at the heart of the nineteenth-century conflict between science and religion. Although Newtonian determinism has been abandoned by twentieth-century physics in favor of the probabilistic approach of quantum mechanics, the extent to which the macroscopic world is deterministic is still debated among physicists and philosophers.

The Electrostatic Interaction

The success of Newton and his followers in using the gravitational interaction to explain phenomena in the heavens and on the earth soon led to the application of the Newtonian approach to other areas of physics. Other interactions were known, and it was natural to study them from a Newtonian point of view. For two centuries after Newton's death, most physicists were occupied either with determining what force laws describe these interactions, or with using these force laws to explain, in Newtonian terms, the structure and properties of matter. In the following sections we shall briefly examine several of the more significant of these investigations. Each involved notable intellectual accomplishments by a number of people. Central to all these efforts, however, was the debt each man owed to the work of Newton, who established the conceptual framework within which they worked.

From ancient times it had been known that when certain materials are rubbed with certain other materials, electrical effects are produced. For example, suppose a balloon is rubbed with a woolen cloth. If the cloth is then brought near the balloon, the two attract each other, as illustrated in Figure 10–5(a). Now suppose two balloons are rubbed with the cloth and then suspended as illustrated in Figure 10–5(b). They repel each other. The forces involved are relatively strong, far stronger than the virtually undetectable gravitational forces between the balloons or between the balloon and cloth.

Early explanations of these effects involved something called *electric charge*, which is transferred from one object to another when they are rubbed together. Benjamin Franklin, who performed a number of electrical experiments in the 1740s, called the charge on the balloons *negative*, and that on the wool *positive*.

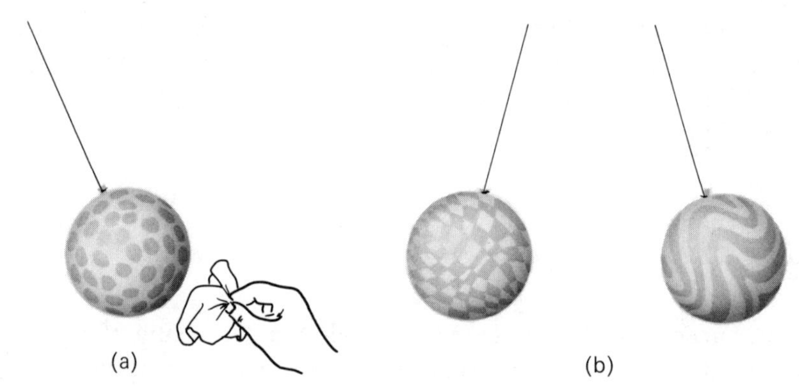

Figure 10–5. Examples of the electrostatic force. (a) A balloon and a woolen cloth with which it has been rubbed attract each other. (b) Two balloons rubbed with wool repel each other.

Using his terminology, we can account for the experimental observations by assuming that like charges repel and unlike charges attract.

In 1785, a century after the publication of the *Principia*, the French physicist Charles de Coulomb performed a series of experiments to investigate the nature of this *electrostatic* interaction. The idea behind these experiments is indicated schematically in Figure 10–6. (In practice, Coulomb's experiments were somewhat more sophisticated in their technical details, but this picture suggests the essential idea.) Two objects, with charges q_1 and q_2, are placed a distance r apart.* The elongation of a

Figure 10–6. A technique that could in principle be used to measure the electrostatic force. The extension of the spring indicates the magnitude of the force.

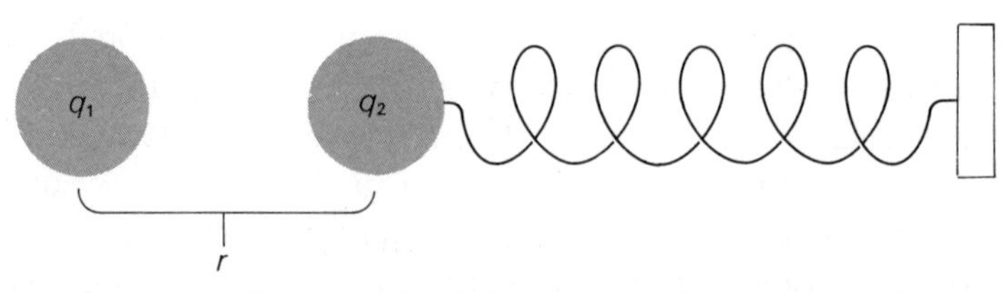

* Here and in what follows we avoid the question of how one can determine the magnitude of the electrical charges. Interested readers are referred to the books listed at the end of the chapter.

calibrated spring is used to determine the electrostatic force corresponding to this separation r. The measurement is then repeated for various separations and for different values of the charges.

On the basis of his experiments, Coulomb proposed the following force law:

$$F_{\text{elect}} = \frac{kq_1q_2}{r^2}$$

The similarity of this electrostatic force law to the gravitational force law is striking. Both forces decrease in inverse proportion to the square of the separation between the interacting objects. In the electrostatic case, however, each charge may be either positive or negative, resulting in a force that can be either attractive or repulsive.

It is significant that just as in the case of gravity, observations of electrostatic phenomena can be accounted for by a very simple and elegant force law. The Newtonian concept of force seems an apt way to describe the electrostatic interaction as well. Just as with gravity, however, no mechanism that would account for the existence of the electrostatic force is apparent. In the Newtonian context, none is required. The force law is a sufficient explanation.

The Electromagnetic Interaction

Magnetism has already been alluded to in Chapter 4, where we noted that Kepler was influenced by Gilbert's studies of magnetism in suggesting a possible motive force for the planets. As in the case of electricity, there seemed to be two kinds of magnetic "charge," which were called north and south poles. Like poles were found to repel each other, while unlike poles attract. Gilbert was the first to suggest that the earth itself was a large magnet with two poles, thus accounting for the behavior of compass needles, which are merely small magnets.

One difference between electricity and magnetism is the apparent impossibility of separating north and south poles. A bar magnet broken in half results in two smaller magnets, each with a north pole and a south pole, as shown in Figure

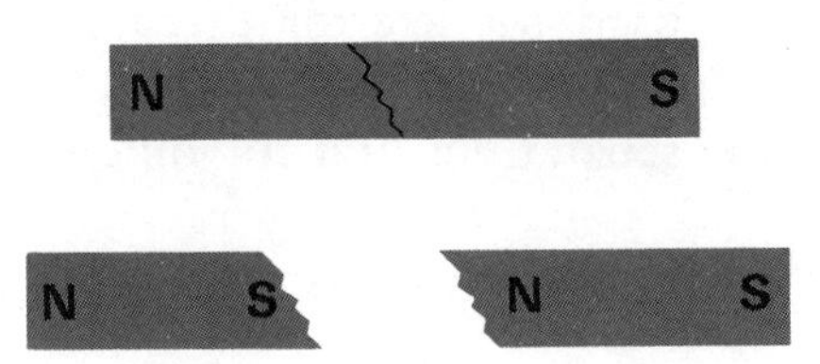

Figure 10–7. The north and south poles of a magnet cannot be isolated.

10–7. This phenomenon makes the discovery of a magnetic force law more difficult than in the electrical case, since it means that one cannot measure directly the force between two isolated magnetic poles. Nevertheless, the English physicist Michell in 1750 interpreted his measurements of forces between magnets in terms of a simple force law:

$$F_{\text{mag}} = \frac{bp_1p_2}{r^2}$$

where b is a constant of proportionality and p_1 and p_2 are magnetic pole strengths. However, as we shall soon see, this force law is apparently less fundamental than those we have written to describe gravity and electricity.

Reports of iron objects being magnetized after they had been struck by lightning prompted physicists in the eighteenth century to seek a connection between electrical and magnetic effects. It became evident, however, that a stationary magnet and a stationary electrically charged object neither attract nor repel each other. The first connection between the two interactions was discovered early in the nineteenth century by the Danish physicist Hans Christian Oersted.

An Italian contemporary of Oersted, Alessandro Volta, had found that a flow of charge could be produced by connecting a wire to the opposite ends of a series of zinc and copper plates immersed in acid—a battery. Oersted found that a wire carrying a battery-produced current would deflect a small magnet, and shortly thereafter, he confirmed the existence of the converse effect: a magnet will deflect a current-carrying wire. Apparently there is an interaction between a magnet and *moving* electric charge.

Only a week after hearing of Oersted's discovery, André-Marie Ampère announced to the French Academy that a current-carrying wire interacts not only with a magnet, but also with another current-carrying wire. Thus there are magnetic effects involving only moving charges and no magnets at all. In fact, Ampère found that a circular loop of wire carrying a current duplicates in all essential respects the properties of an ordinary bar magnet. This led him to propose that all magnets derive their properties from an internal, microscopic circulation of electric charge. From this point of view, the magnetic force law involving pole strengths that Michell had found loses its fundamental importance. The magnetic force can be more fundamentally described as a consequence of the interaction of moving electric charges.

With the acceptance of Ampère's model of magnetic effects, all magnetism came to be viewed as a particular aspect of electricity. In general, two electric charges are now thought to interact through both electric and magnetic forces. When either charge is at rest, the interaction is purely electrostatic. When both charges are moving, there is also a magnetic interaction, which depends on both the magnitudes and the velocities of the charges. It is possible to express both electric and magnetic effects in terms of a single *electromagnetic* interaction, for which a simple Newtonian force law can be written.

The Properties of Matter

Thus far we have ignored many of the forces that are most common in everyday life. A fundamental understanding of such forces as the pull of an extended spring, the tension in a taut rope, the blow of a steel hammer, or the pressure of a compressed gas was achieved through the study of the structure and properties of bulk matter. The behavior of such matter has come to be attributed to the action of the electrostatic force on the atomic level. Thus these other forces are now viewed as less fundamental than gravity and electromagnetism. They can all be explained by models based on electrostatic interactions between atoms.

Consider first the properties of a gas. Several empirical gas laws relating pressure, volume, and temperature had been discovered before Newton's time, but an understanding of these laws awaited the development of the atomic theory of matter in the nineteenth century. In the model then developed, the so-called kinetic theory, a gas is pictured as an incredibly large number (10^{23} or more) of molecules. Each molecule can be thought of as a tiny billiard ball caroming about, colliding with other molecules of the gas and with the walls of the container (Figure 10–8).

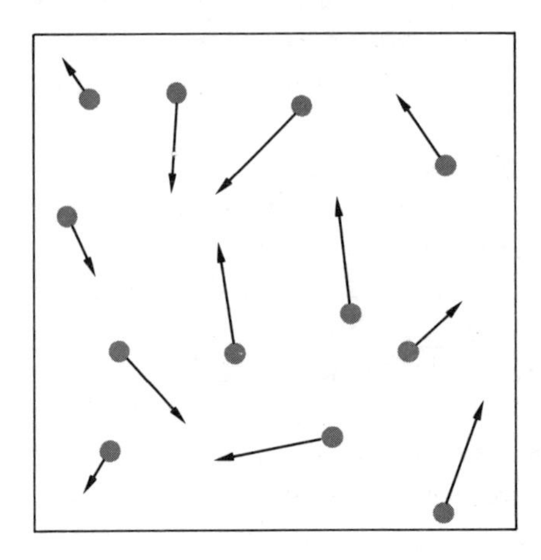

Figure 10–8. A gas in a container, as represented in the kinetic theory model. The dots represent molecules and the arrows represent their velocities.

This model explains the high compressibility of a gas by the vast amount of empty space between the molecules, and the supposition that the molecules interact only when they collide. The pressure of a gas is attributed to the collisions of the molecules with the container walls. When the temperature increases, the molecules are assumed to move faster. This leads to more frequent and more violent collisions with the walls, which explains why an increase in temperature leads to an increase in pressure. A decrease in volume also increases the frequency of collisions with the walls and thus is accompanied by an increase in pressure.

While the motion of each molecule could in principle be calculated by using $\vec{F} = m\vec{a}$, in practice of course it would not be possible to carry out such a calculation for 10^{23} interacting particles, even with the largest computer imaginable. However, as the Swiss physicist Daniel Bernoulli and others found in the nineteenth century, it is not necessary to do this to obtain quantitative relations between the temperature, pressure, and volume of a gas from the kinetic theory model. The very fact that there are so many molecules in a gas allows one to treat the collection statistically instead of following the motion of individual molecules. Elementary calculations based on this model lead to predictions that agree with the observed gas laws.

We now understand the billiard-ball-like interaction between colliding gas molecules to be an electrostatic repulsion. The molecules have no net charge, and hence when they are far apart they do not interact electrostatically. When they collide, however, their negatively charged electrons overlap and repel each other. Thus the kinetic theory provides a successful Newtonian model of a gas without requiring the introduction of any new force laws.

A highly successful atomic model of solids has also been developed. It pictures a crystalline solid as a regular array of positively charged *ions* (atoms that have lost some electrons) held together by spring-like forces as indicated in Figure 10–9. These forces can be understood to result from electrostatic interactions among the ions and between the ions and the myriads of electrons wandering about through the crystal.

Figure 10–9. A simple model of a crystalline solid. The black dots represent ions, and the springs represent the electrostatic forces between them.

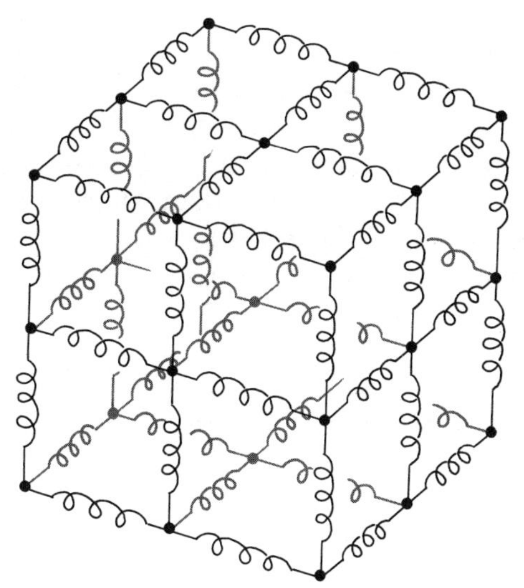

This model can account for such properties as the conduction of electricity and heat by solid substances. Using the model one can also explain the pull of a macroscopic spring or the tension in a rope. When the spring or the rope is stretched its atoms are pulled farther apart and the electrostatic forces (the tiny springs of the model) resist this separation. The ability of a floor to sustain a great weight is easily understood if we think of the floor as a gigantic innerspring mattress, as depicted in Figure 10–9. When we step on the floor, it is like stepping on a very stiff mattress. It depresses slightly under our feet, pushing together the atoms. The electrostatic "springs" resist this compression. Another common macroscopic force, friction, experienced when two solids rub against one another, can be understood in terms of an electrostatic attraction between the atoms of the adjacent surfaces of the solids.

That all non-gravitational forces between macroscopic bodies are explainable in terms of electromagnetic interactions on the atomic level, is in many ways the most significant result of the two centuries of physics following the death of Newton. The approach he introduced led to a striking conclusion: All motion, at least on the macroscopic level, can apparently be explained using just two forces, gravity and electromagnetism, each describable by a simple force law.*

This is no mean accomplishment. The conceptual structure that underlies it is Newton's legacy to physics, and, through physics, to the contemporary world. That the key to physics should be interactions, that the strength of these interactions should be measured by the accelerations they produce, and that motion could be understood in terms of mathematical force laws describing these interactions, was a program no more obviously destined for success than those of Aristotle or Descartes. Yet Newton's laws led to a simple, compelling, and widely applicable description of nature. This was the ultimate measure of their success.

* More recent studies of the particles that make up the nuclei of atoms indicate that their interactions cannot be attributed to gravity and electromagnetism alone. Two new interactions, the "strong" and the "weak" nuclear forces, have been introduced. For a further discussion of the "strong" interaction, see Chapter 16.

Suggested Reading

Butterfield, H. *The Origins of Modern Science*. London: Bell, 1949. Chapter 9 discusses the impact of Newtonian physics on eighteenth-century thought.

Gillispie, C. *The Edge of Objectivity*. Princeton: Princeton University Press, 1960. Chapter 5 is a similar study.

Randall, J. H., Jr. *The Making of the Modern Mind*. Boston: Houghton Mifflin, 1940. Chapter 11 is another well-known treatment of the same subject.

Buchdahl, G. *The Image of Newton and Locke in the Age of Reason*. New York: Sheed & Ward, 1961. This short book includes a large section of quotations from eighteenth-century writers illustrating their reaction to Newtonian science.

Arons, A. *Development of Concepts of Physics*. Reading, Mass.: Addison-Wesley, 1965. Chapter 7 is a good, non-mathematical discussion of the phenomena underlying the force laws we have discussed.

Questions

1. Consider the three laws of physics listed below. Logically speaking, are they similar in character? That is, do they play similar roles in physics? Explain.

   (a) Kepler's first law: The planets move in ellipses with the sun at one focus.
   (b) Newton's second law: $\vec{F} = m\vec{a}$.
   (c) Coulomb's electrostatic force law: The electrostatic force between two objects is proportional to the product of their charges and inversely proportional to the square of the separation between them.

2. How might you define a revolution in physics? Consider the events and ideas we have discussed from Greek times to Newton. Identify any revolutions that may have occurred during this period, and discuss their relative importance, both to the development of physics and to society in general.

3. Cavendish entitled his paper "Weighing the Earth." In fact, his determination of the gravitational force constant G also enabled him to weigh the sun.

   (a) How was this possible? Explain how the mass of the sun may be determined once G is known.
   (b) Use the value of $G = 6.7 \times 10^{-11}$ newton m^2/kg^2 to calculate the mass of the sun.
   (c) Does a knowledge of G also enable one to calculate the mass of the moon and the mass of the other planets?

4. In 1772, the German astronomer Johann Bode discovered a striking regularity in the planetary data. He noted that the average distance R of each of the planets from the sun, expressed in astronomical units (see footnote to table on page 92), could apparently be represented by the formula

$$R = 0.3 \times 2^{(n-2)} + 0.4$$

   where n is an integer, 1, 2, 3, etc., provided that in the case of $n = 1$ the first term is set equal to zero.

   (a) Use this relation to predict the radii corresponding to the integers 1 through 7, and compare these values with the observed radii listed in Table 4–1 on page 92. If this correspondence is taken seriously, what new observations are suggested?
   (b) Since there was no apparent explanation for the form of Bode's law, it was met with skepticism by most astronomers, who tended to view the correspondence found in (a) as a coincidence. Not long afterward, however, several planetoids were discovered between Mars and Jupiter, all having an average distance from the sun of 2.77 a.u. In addition, the distance from the sun to Uranus (discovered about the same time) proved to be 19.2 a.u. Speculate on the effect these discoveries might have had on the acceptance of Bode's law.
   (c) The eventual discoveries of Neptune and Pluto provided two more tests of Bode's law. Their distances from the sun were 30.1 and 39.5 a.u., respectively. What conclusions about the law might be drawn from these discoveries?

5. As is well known from the extensive television and radio coverage of manned orbiting satellites, it takes roughly an hour and a half for a low-altitude earth satellite to complete a full circular orbit. Show that this is just what Newton's law of gravitation predicts. (Hint: You could either calculate the period directly using Cavendish's value of G, or you could relate the period of the satellite to that of the moon, about 28 days at a distance of 60 earth radii, or you could relate the period of the satellite to the gravitational acceleration of a freely falling object at the surface of the earth.) It is convenient to approximate the radius of the low-altitude satellite's orbit as simply the radius of the earth.
6. Suppose a satellite is launched so that it completes a circular orbit about the earth's equator with a period of 24 hours. Since it takes precisely the same time to complete an orbit as the earth takes to complete a rotation, it will remain directly above a point on the equator. Such synchronous satellites would prove very useful as communications relays. Three of them, appropriately spaced around the equator, can see virtually every point on the earth's surface. Determine the radius of the orbit in which such a satellite must be placed. (Hint: the period of the moon, whose orbital radius is about 60 earth radii, is about 28 days.)
7. A tiny new planet has reportedly been observed in the solar system. It takes one month to complete a circular orbit about the sun. What is the radius of its orbit? Speculate on why the planet remained unobserved for so long.
8. Consider a satellite in a low-altitude circular orbit about a spherical body of radius R. Suppose the body is made of uniform material with density ρ. (The density of an object with uniform composition is simply its mass divided by its volume.)

   (a) Express the mass of the body in terms of ρ and R.
   (b) Use the result of part (a) and equation (10), page 241 (interpreting the m_s in that equation to be the mass of the body) to express the period of the low-altitude satellite in terms of the density ρ. (Approximate the

radius of the satellite's orbit by the radius of the body, R.)

(c) If the density of the material composing the moon is the same as that of the material composing the earth, how would the period of a low-altitude satellite around the moon compare with the period of a low-altitude satellite around the earth? [Hint: use the result of part (b).] How does your result compare with the experience of astronauts who have orbited the earth and moon?

9. Objects released from rest in an orbiting earth satellite are observed to float in the satellite. Explain this phenomenon in Newtonian terms.

10. The two stars of a particular double-star system each have the same mass m. The figure shows the observed motions of these stars. Each moves with the same speed v in a circle about the midpoint of the line joining the two stars.

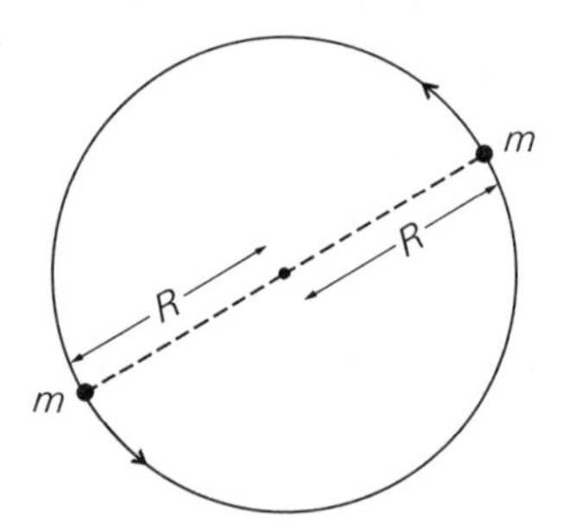

(a) Argue that this motion is consistent with Newton's laws.

(b) Show that the period T of either star is related to R, the radius of its circular orbit, by

$$T^2 = \frac{16\pi^2}{Gm} R^3$$

11. A spy satellite orbits about the earth taking pictures. Once a roll of film has been exposed, a tiny door in the bottom of the satellite opens and the can of film is allowed to drop out and fall to the earth, where it is recovered and developed. What is wrong with the physics of this scheme?

CHAPTER ELEVEN

It All Depends on Your Frame of Reference

Having seen how Newton's laws are applied, we are now in a position to evaluate their meaning more critically. In the process we hope to show that these laws are far more than merely a device for predicting results of experiments. They provide a pattern for organizing our observations of the world.

We begin by restating the first and second laws:

I. *An isolated body moves with a constant velocity.*

II. $\vec{F}_{\text{net}} = m\vec{a}$.

The first statement is a slightly more succinct version of Newton's phrasing: "Every body continues in its state of rest or of uniform motion in a straight line unless it is compelled to change that state by forces impressed on it."

There seems to be a serious logical difficulty in Law I. How do we know that a body is isolated, i.e., that no forces are acting on it? An obvious way is to see whether it moves with a constant velocity. Sir Arthur Eddington, the twentieth-century British cosmologist and philosopher, was struck by this apparent circularity. He suggested that one could formulate the first law as: "Every body continues in a state of rest or of uniform motion in a straight line except in so far as it doesn't."[1]

In fact, while in a certain sense the first law is indeed circular, it is not empty. The resolution of this difficulty provides a good example of the groping way scientific concepts—like all others—are conceived. The most difficult task in creating a physical theory is formulating the concepts in terms of which the theory is to be expressed. Further, only after the concepts have been woven together in a nice neat package that agrees with observations can we be sure the concepts are of value. Indeed it is this ultimate pragmatic test that determines their value.

So it is with the notion of an isolated body. We have no positive *a priori* way of knowing when an object has no forces acting on it. As a first guess, we might try to move any matter that might interact with it very far away and then assume the body is isolated. On the basis of this assumption we investigate interactions. We bring one bit of matter at a time (e.g., a charged object) near the previously isolated body and observe the accelerations produced. In this way we might investigate the electrostatic force or study other interactions.*

This is the program that has been carried out. Interactions have been investigated on the assumption that if all matter is moved far enough away, its effect can be disregarded. As we noted in Chapter 10, the striking result is that all macroscopic natural phenomena can be described by $\vec{F} = m\vec{a}$ with only two different force laws:

1. Gravity
2. Electromagnetism

* Of course, one might object that we cannot move the earth very far away, so the gravitational force is always present. This does complicate matters, but it is possible to get around the problem. For example, we can arrange the experiments so that the force under investigation is horizontal. The horizontal accelerations that provide information about the force are not affected by the vertical gravitational force.

What is more, we find that both these forces decrease rapidly with distance. Hence, in this circular manner our initial assumption about isolated bodies—that if matter is moved far enough away it will exert a negligible force—appears justified. The program is self-consistent.

Another aspect of the relation between Newton's first and second laws seems puzzling. According to the second law, the acceleration of an object is related to the force it experiences by $\vec{F} = m\vec{a}$. Consider an isolated body. For such a body $\vec{F} = 0$. Hence we find $\vec{a} = 0$. Thus the second law predicts that an isolated body will not accelerate. In other words, it will move with a constant velocity. But that is the content of the first law. It seems that the first law is only a special case of the second! If that is the case, why did Newton bother to write it as a separate law?

In fact, as we shall demonstrate, the first law has an independent and essential purpose. It specifies the kind of reference frame to be used in applying the second law. We discussed the concept of a reference frame in Chapter 5 in connection with our description of the motion of a falling apple. There our choice of a reference frame seemed straightforward and obvious. It is not always so. To understand the role of Newton's first law, we must examine the consequences of different choices of reference frames.

Events

The mathematical description of the motion of an object entails choosing a frame of reference and specifying the position of the object with respect to this reference frame at each instant of time. For example, Figure 11–1 depicts an object moving in two dimensions. The path of the object and its position at times t_1, t_2, and t_3 are shown.

We can make this description more vivid by introducing the concept of an *event*. Consider a still simpler example. A car is moving at a speed of 2 m/sec in the $+x$ direction. The car passes the origin of the reference frame at $t = 0$. Suppose the car

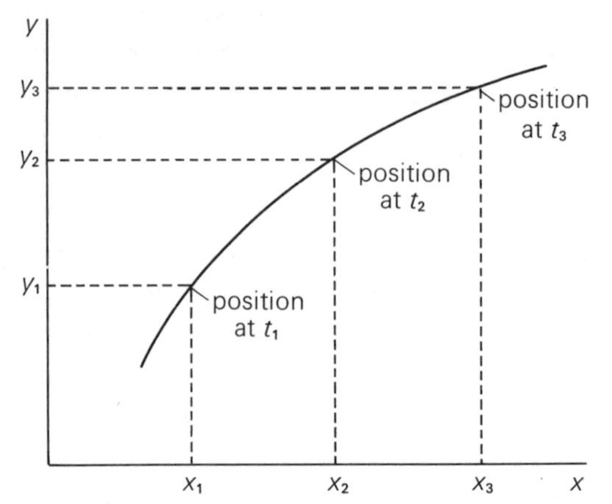

Figure 11–1. The motion of an object in two dimensions can be specified by indicating its position with respect to a reference frame at each instant of time.

backfires every second beginning at $t = 0$. The position and time of each backfire (represented by a "pop") are shown in Figure 11–2. We shall refer to each pop as an event. In general, an event is an occurrence that can be described by specifying its

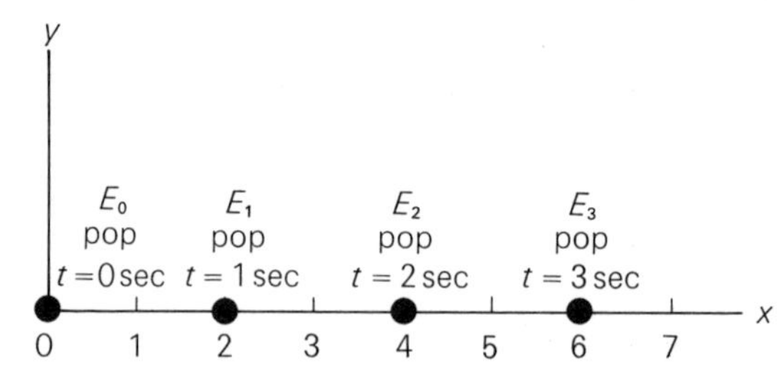

Figure 11–2. A sequence of events: the backfirings of a moving car.

position and time. Thus the successive pops, which we label E_0, E_1, E_2, etc., are described in Figure 11–3.

Clearly, an event need not be as startling as a backfire. In the motion of Figure 11–1, for example, occurrences we may call events are: object passes through point x_1, y_1 at time t_1; object passes through point x_2, y_2 at time t_2, etc. In this sense, the motion of any object is a succession of events.

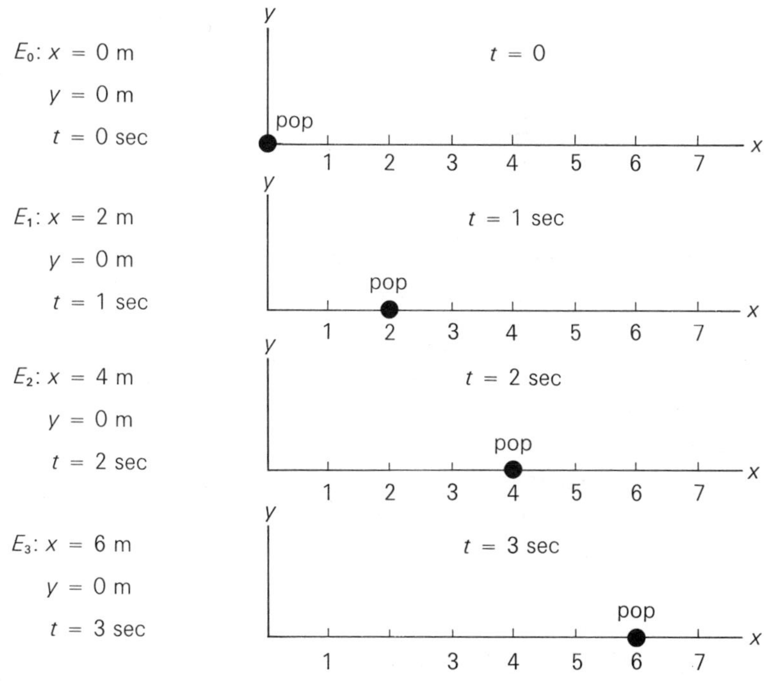

Figure 11–3. The positions and times of the events in Figure 11–2.

Choice of Reference Frame

Of course, there is nothing sacred about the reference frames we have chosen. The concept of a reference frame is something we have created, and the particular frame we use for the description of a given motion is our choice, not nature's. To specify position, we might equally well use a frame with a different origin [Figure 11–4(a)] or one whose axes are oriented in a different manner [Figure 11–4(b)].

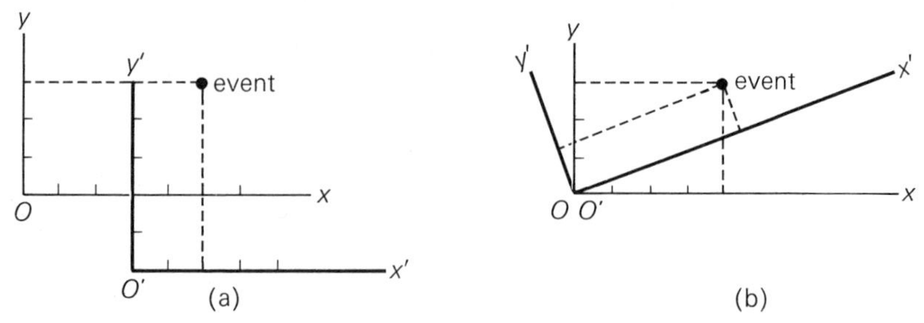

Figure 11–4. Alternative frames of reference for the description of an event. (a) indicates two frames with different origins but parallel x- and y-axes, while (b) indicates two frames with the same origin but a different orientation of their axes.

More relevant to our discussion of Newton's laws is a reference frame that is moving relative to our original frame. Figure 11–5 shows a particularly simple case of one reference frame moving relative to another. The S' frame is moving relative to

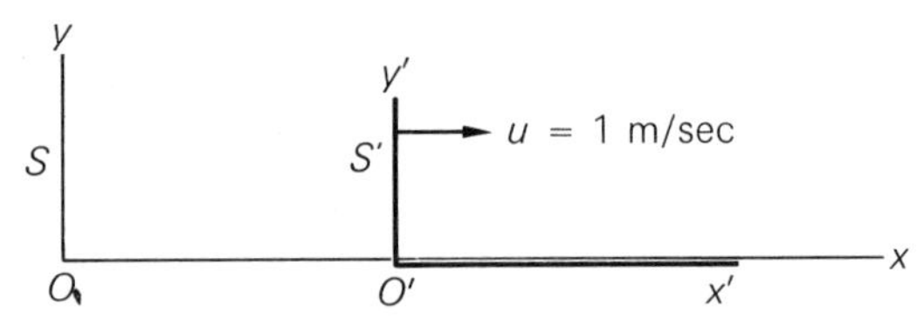

Figure 11–5. The S' frame moves relative to the S frame with a velocity of 1 m/sec in the $+x$ direction.

our original frame, the S frame, with a speed of 1 m/sec in the $+x$ direction.

What is the relation between the corresponding descriptions of events and the motion of objects in the two frames of reference? Let us consider the motion of the backfiring car of Figures 11–2 and 11–3. Suppose that the origins of the S and S' frames coincide at time $t = 0$. Figure 11–6 shows the position of the backfiring car and the S' reference frame at one-second intervals. It is clear from this figure that the first pop E_0 takes place when the car is located at coordinates $x' = 0$, $y' = 0$ in the S' frame. The second pop E_1 occurs when the car's co-

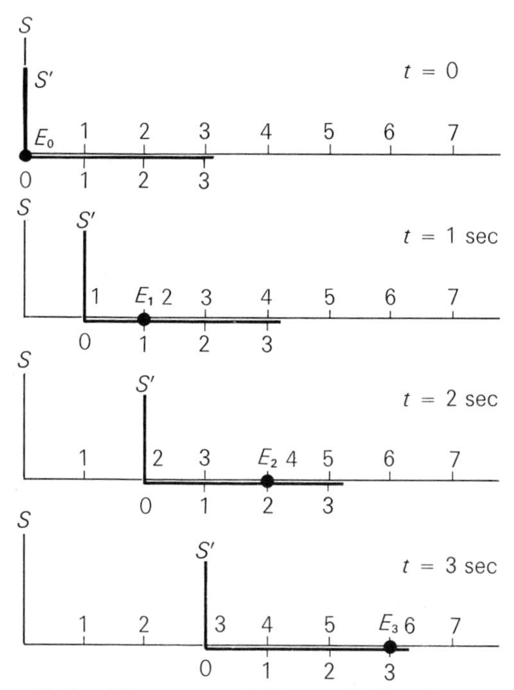

Figure 11–6. The events of Figure 11–3 and corresponding positions of the S' frame.

ordinates are $x' = 1$ m, $y' = 0$, etc. Figure 11–7 illustrates the events relative to the S' frame. In other words, with respect to the S' frame, the events are described by:

$$
\begin{aligned}
E_0 : x' &= 0 \text{ m} \\
y' &= 0 \text{ m} \\
t' &= 0 \text{ sec} \\
E_1 : x' &= 1 \text{ m} \\
y' &= 0 \text{ m} \\
t' &= 1 \text{ sec} \\
E_2 : x' &= 2 \text{ m} \\
y' &= 0 \text{ m} \\
t' &= 2 \text{ sec} \\
E_3 : x' &= 3 \text{ m} \\
y' &= 0 \text{ m} \\
t' &= 3 \text{ sec}
\end{aligned}
$$

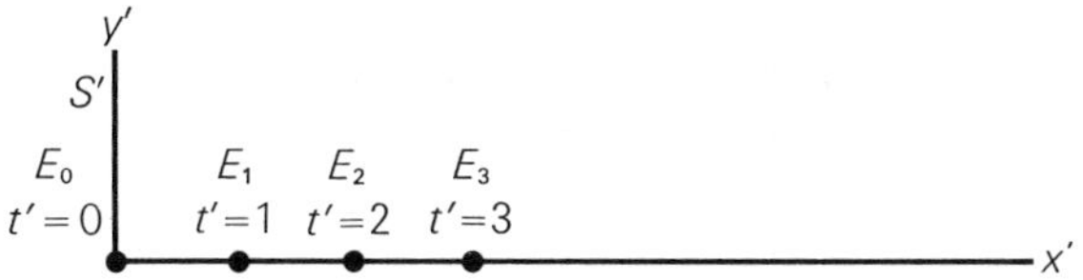

Figure 11–7. The sequence of events E_0, E_1, E_2, and E_3 described with respect to the S' frame.

Notice that we have assigned to each event the same time that we used to describe the event with respect to the S frame. There is no obvious reason why the time of an event should depend in any way on the reference frame we choose to specify its position.

A comparison of Figures 11–2 and 11–7 indicates a simple relation between the motion of the car as measured in these two reference frames. With respect to the S frame the car is moving with a constant velocity of 2 m/sec in the $+x$ direction. With respect to the S' frame, it also moves with a constant velocity, 1 m/sec in the $+x'$ direction.

> Consider one frame of reference that is accelerating relative to another. As shown in Figure 11–8, the S' frame is moving relative to the S frame with a uniform acceleration of 1 m/sec² in the $+x$ direction. Suppose that prior to time $t = 0$ the origins coincide. At $t = 0$, S' begins to accelerate. In Figure 11–9, indicate the position of the S' frame at times $t = 1$, 2, and 3 sec. [Cf. equation (4), page 139.]

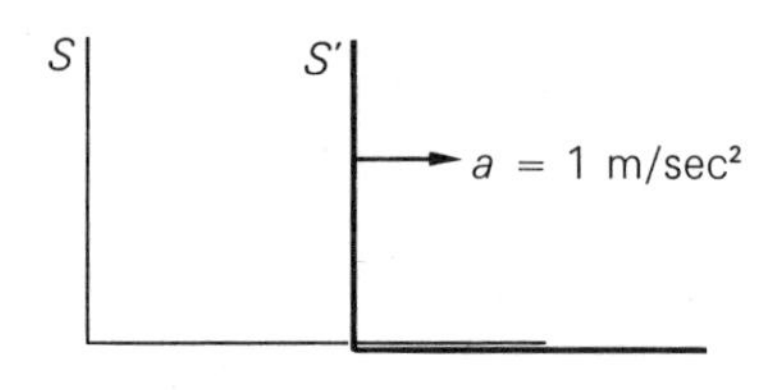

Figure 11–8. The S' frame accelerates relative to the S frame with an acceleration of 1 m/sec² in the $+x$ direction.

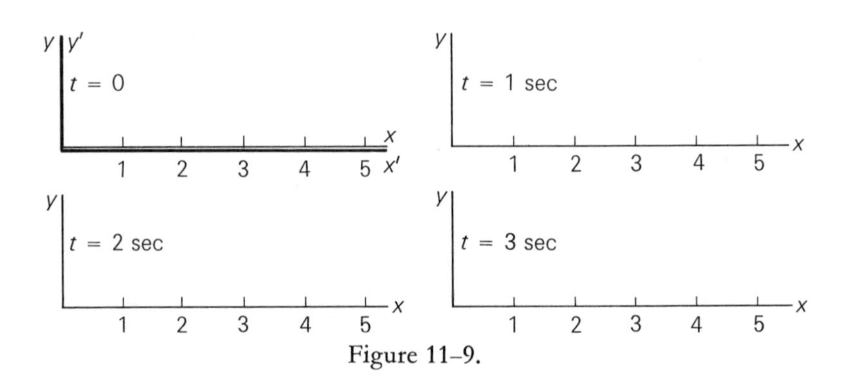

Figure 11–9.

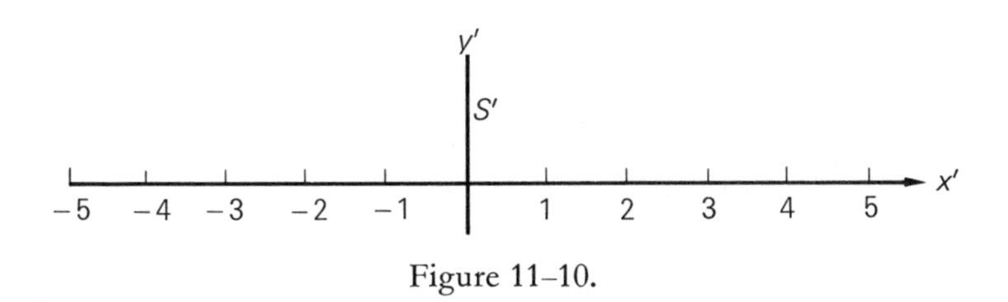

Figure 11–10.

Consider an object that remains at rest at the origin of the S frame. In Figures 11–9 and 11–10, indicate the position of the object at times $t = 0$, 1, 2, and 3 sec with respect to the S' frame.

From this exercise it should be clear that although the object is at rest with respect to the S frame, it accelerates with respect to the S' frame.

Inertial Reference Frames

We are now prepared to reexamine Newton's first law. Consider an isolated object at rest at the origin of a reference frame S, as shown in Figure 11–11(a). So far, so good for the first law. The isolated object remains at rest. But what about the motion of this isolated object as measured in the S' frame, which is accelerating relative to S? This is depicted in Figure 11–11(b). As the above exercise illustrated, the isolated object accelerates relative to the S' frame and hence the first law, as we have stated it, is *wrong*. An isolated object does not move with a constant velocity relative to *every* reference frame.

Similarly, $\vec{F} = m\vec{a}$, with $\vec{F}$ restricted to the two macroscopic force laws, is not valid in an arbitrary reference frame. Consider a freely falling object of mass m near the surface of the earth. The force acting on the object is the gravitational force. Applying $\vec{F} = m\vec{a}$ we find:

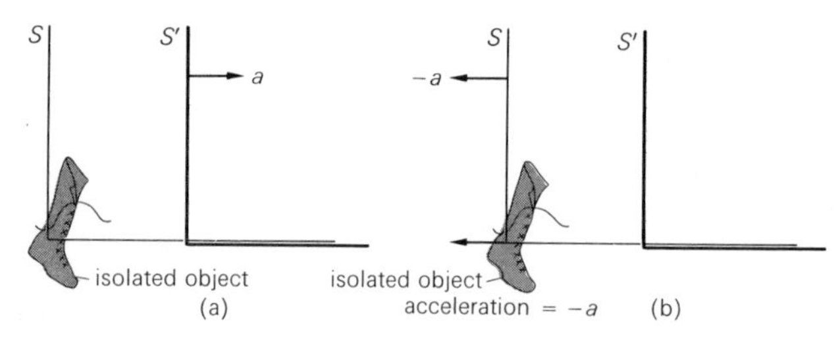

Figure 11–11. The motion of an isolated object as measured in two reference frames accelerating relative to each other. (a) The isolated object remains at rest relative to the S frame. (b) The isolated object accelerates relative to the S' frame.

$$\frac{GmM_e}{r_e^2} = ma, \qquad \text{or} \qquad a = \frac{GM_e}{r_e^2} = 980 \text{ cm/sec}^2$$

where M_e is the mass of the earth and r_e is the radius of the earth. This is precisely the acceleration actually measured with respect to a reference frame fixed to the earth such as the S frame in Figure 11–12. However, what if we had chosen as our reference frame the S' frame, which is moving toward the earth with a uniform acceleration of 980 cm/sec^2? For example, the S' frame might be attached to an elevator whose cable has snapped. The object of mass m is at rest in this frame of reference. Therefore, in the S' frame its acceleration is 0, not $GM_e/r_e^2 = 980$ cm/sec^2. Hence its motion does not satisfy $\vec{F} = m\vec{a}$ if $\vec{F}$ is restricted to gravitational and electromagnetic forces.

We have found that neither the first law nor the second law is valid for every reference frame. However, there is a restricted class of reference frames in which isolated bodies do move with a constant velocity. Experimental observations indicate that the frame of reference attached to the surface of the earth is very nearly such a frame. These reference frames in which Newton's first law (sometimes known as the law of inertia) is valid are known as *inertial reference frames*.

The key point is that it is also in these inertial frames that $\vec{F} = m\vec{a}$, with $\vec{F}$ drawn from the two basic laws, gravitational and electromagnetic, correctly predicts the motion of macroscopic objects. Hence, Newton's first law does play an important

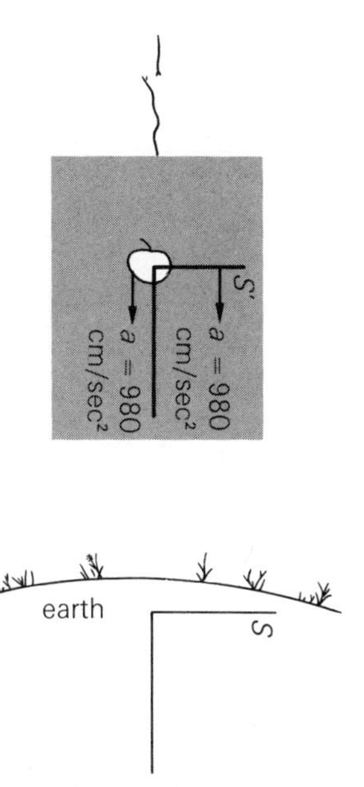

Figure 11–12. Two reference frames near the earth's surface. S is fixed to the earth; S' is fixed to a falling elevator. A freely falling object accelerates relative to the S frame, but remains at rest relative to the S' frame.

role in the Newtonian scheme. It singles out the inertial reference frames, the frames in which the $\vec{F} = m\vec{a}$ program of investigating interactions leads to such a small number of compellingly simple force laws.

Relation between Inertial Frames

If one of these inertial reference frames is known, we can specify how to find all of the others: they are all the reference frames moving at constant velocity relative to the first. To see this, suppose we designate an inertial reference frame by the symbol S, and let S' be a reference frame with coordinate axes parallel to those of S and moving with speed u relative to S in the $+x$ direction (see Figure 11–13). We shall demonstrate that S' is also an inertial reference frame.

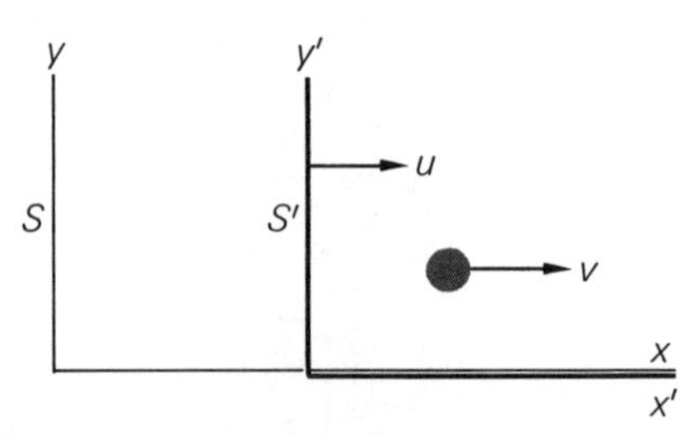

Figure 11–13. An object and an S' reference frame move relative to the S reference frame with constant velocities v and u, respectively, in the $+x$ direction.

Consider an isolated object moving relative to S with constant velocity v. For simplicity, suppose v is in the $+x$ direction, as shown in the figure. Then the velocity v' of the object with respect to S' will be $v' = v - u$.

This may be easier to visualize in terms of Figure 11–14, where S is a reference frame fixed to a road, the isolated object is car B coasting frictionlessly with speed v down the road, and S' is a reference frame fixed to car A, which is moving with speed u down the road. Then v', the speed of car B as viewed from car A, is $v' = v - u$. (For example, if the speed of B relative to the road is 50 miles/hour and the speed of A relative to the road is 30 miles/hour, then the speed of B relative to A is 20 miles/hour.)

More generally, if the velocity of the S' reference frame relative to the S frame is $\vec{u}$ and the velocity of the isolated object

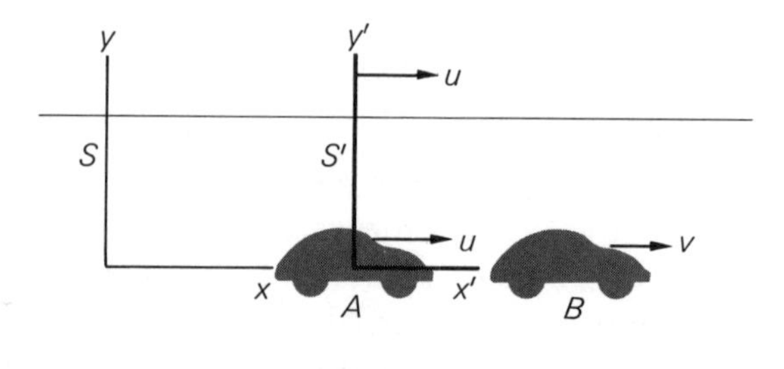

Figure 11–14. Car B and the S' reference frame, fixed to car A, both move with constant speed relative to the S reference frame, fixed to the road.

with respect to S is $\vec{v}$, then the velocity $\vec{v}'$ of the isolated object with respect to S' is $\vec{v}' = \vec{v} - \vec{u}$. Consider a man swimming in a moving stream. We take the S frame to be fixed to the bank of the stream. The stream flows with velocity $\vec{u}$ in the $+x$ direction, and we take the S' frame to be moving with the water [Figure 11–15(a)]. Suppose the swimmer swims in the y' direction—that is, $\vec{v}'$ is at right angles to the flow of the water. As he swims across the stream, he is at the same time swept downstream by the motion of the water. It should be clear that his net velocity with respect to the banks of the stream (that is, $\vec{v}$) will be that shown in Figure 11–15(b). Note that our assertion above, $\vec{v}' = \vec{v} - \vec{u}$, agrees with this figure. It should also be clear that the same result will hold whatever directions of $\vec{v}$, $\vec{v}'$, and $\vec{u}$ we choose. Whatever the motion with respect to the water, we have to add the water's motion to it to get the motion with respect to the shore.

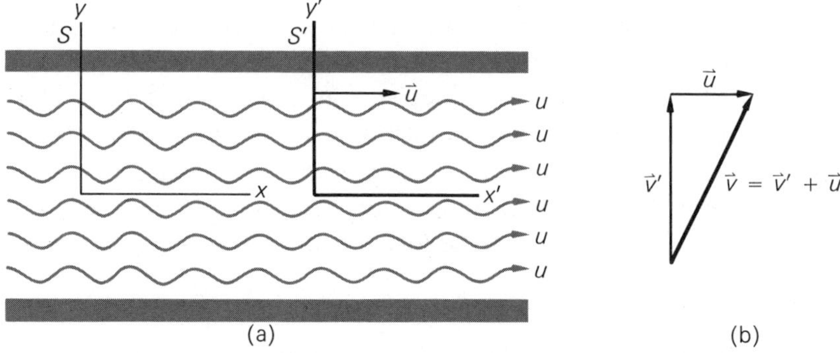

Figure 11–15. The motion of a swimmer in a moving stream illustrates the relation $\vec{v} = \vec{v}' + \vec{u}$, which implies $\vec{v}' = \vec{v} - \vec{u}$. (a) The S reference frame is fixed to the bank. The S' reference frame moves with the water with velocity $\vec{u}$ relative to the S frame. (b) The velocity of a swimmer relative to the shore, $\vec{v}$, is the vector sum of his velocity relative to the water, $\vec{v}'$, and the velocity of the water relative to the shore, $\vec{u}$.

This is all we need to prove our assertion that any reference frame S' moving with constant velocity $\vec{u}$ relative to the inertial reference frame S is also an inertial reference frame: Consider any isolated object. Its velocity $\vec{v}$ relative to S is constant since

S is an inertial reference frame. We have assumed $\vec{u}$ to be a constant. Since $\vec{v}$ and $\vec{u}$ are constant velocities, then so is $\vec{v}' = \vec{v} - \vec{u}$. Hence, the velocity of any isolated body relative to S' is also constant. Since we have defined an inertial reference frame as one in which an isolated body moves with constant velocity, we have therefore demonstrated that S' is an inertial reference frame.

> Show by a similar argument that any reference frame that is not moving with a constant velocity relative to an inertial frame is not an inertial reference frame.

The Principle of Galilean Relativity

We have seen that the reason for the central role inertial reference frames play in Newtonian mechanics is that the common force laws are expressed most simply in these frames. We shall now show that each of these inertial frames is equivalent to any other for the description of motion in Newtonian mechanics. This means that the equation of motion of a body, $\vec{F} = m\vec{a}$, takes exactly the same form in every inertial reference frame. This striking observation is known as the *principle of Galilean relativity*.

The validity of this principle is based on the result that the forces encountered in Newtonian mechanics, such as the fundamental gravitational force and the elastic forces of springs, strings, and table tops, apparently do not depend on the velocities of the interacting objects. To the extent that this is true, the net force $\vec{F}$ on an object is the same in all inertial reference frames. Similarly, since inertial reference frames move at a constant velocity with respect to one another, the acceleration (change of velocity per unit time) of an object will be the same in every inertial frame. To show this, let us consider the following example.

The S reference frame in Figure 11–16 is fixed to the road. The S' reference frame is fixed to car A, which is traveling at speed 10 m/sec in the $+x$ direction. Both S and S' are inertial reference frames. At time t_0, a second car B is traveling with

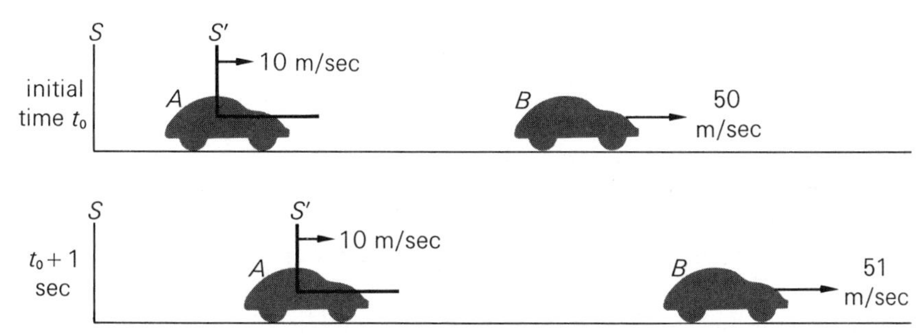

Figure 11–16. The speed of car B relative to the S frame is 50 m/sec at time t_0 and 51 m/sec 1 sec later. Relative to the S' frame, moving with a speed of 10 m/sec with respect to the S frame, the speed of car B is 40 m/sec at time t_0 and 41 m/sec 1 sec later.

speed 50 m/sec relative to S, also in the $+x$ direction. One second later, the speed in S of car B is measured to be 51 m/sec. Hence, the average acceleration during this one-second interval as measured in the S frame is:

$$\frac{51 \text{ m/sec} - 50 \text{ m/sec}}{1 \text{ sec}} = 1 \text{ m/sec}^2$$

Consider the same sequence of events measured in S'. At t_0 the speed of car B is 40 m/sec relative to the S' frame, while one second later the speed is 41 m/sec relative to S'. Hence the average acceleration in S' during the one-second interval is:

$$\frac{41 \text{ m/sec} - 40 \text{ m/sec}}{1 \text{ sec}} = 1 \text{ m/sec}^2$$

Thus the average acceleration is the same in the two inertial frames. If we took an arbitrarily small time interval the results would still be the same. Hence the instantaneous accelerations of an object measured in two inertial frames are equal.

The conclusion that $\vec{F}$ and $\vec{a}$ are the same in all inertial frames, combined with the observation that, at least for the moderate velocities attainable in pre-twentieth-century laboratories, the mass of a body is independent of its velocity, gives the result that $\vec{F} = m\vec{a}$ takes the same form in all inertial frames. This is the principle of Galilean relativity. We shall consider its full implications in the following sections.

The Galilean Transformation

Newton's laws hold in all inertial reference frames, but it is often more convenient to analyze a physical situation in one reference frame than in another. In particular, the most convenient frame for analysis may not be that of the laboratory where the observations are made. Consequently it is useful to derive a transformation to relate the description of events in one inertial reference frame to that in another. Recall that by an event we mean an occurrence, such as a collision of particles or a flash of light, whose description involves the specification of its spatial coordinates x, y, z, and its time t.

The *Galilean transformation* relates the space-time coordinates x, y, z, t of an event as measured in one inertial frame S to the space-time coordinates x', y', z', t' of the same event as measured in another inertial reference frame S', which, as we have seen, moves with a constant velocity relative to S.

Let us first consider the time. There is no obvious reason why the time t of an event measured in the S frame should be any different from the time t' of the event measured in the S' frame. In fact it is hard to imagine at this point what $t \neq t'$ might mean. Newton in the *Principia* alluded to an absolute time scale with respect to which all events could be specified: "Absolute, true, and mathematical time, of itself and from its own nature, flows equably without relation to anything external . . ."[2] A tacit assumption of Newtonian mechanics, so "obvious" that it was never explicitly stated, was that $t' = t$.

Now let us examine the spatial coordinates of an event as measured in two inertial frames. For simplicity, let us again assume that the coordinate axes of S and S' are parallel and that S' moves relative to S with speed u in the $+x$ direction (Figure 11–17). A standard convention is to call the time at which the two origins of the coordinate systems coincide $t = t' = 0$ [Figure 11–17(a)]. We are free to make this choice since we are

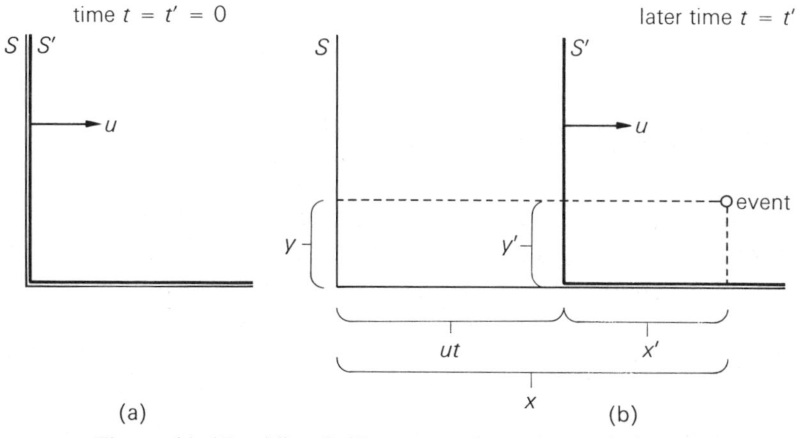

Figure 11–17. The Galilean transformation, relating the coordinates x, y, t of an event relative to the S frame and the coordinates x', y', t' of the event relative to the S' frame. We choose the time at which the origins of S and S' coincide to be $t = t' = 0$ [part (a)]. Note that $x' = x - ut$ and $y' = y$ [part (b)].

concerned only with time intervals (differences in the times at which two events occur).

Suppose an event takes place at time t at a position specified by coordinates x, y, z in the S frame. We wish to find the corresponding coordinates x', y', z' in S'. As can be seen in Figure 11–17(b) (where we have suppressed the z axis to simplify matters), at time t the origin of S' has moved a distance ut from the origin of S in the $+x$ direction. From the figure it is clear that the coordinate x' is related to the coordinate x by $x' = x - ut$, while $y' = y$. Hence the Galilean transformation can be read directly from Figure 11–17(b). In fact, it is the figure that should be remembered, not the equations:

Galilean transformation

$$x' = x - ut$$

$$y' = y$$

$$t' = t$$

Indistinguishability of Absolute Motion

We can now discuss an important implication of the principle of relativity. Consider two laboratories moving relative to each other. One might be on the ground and the other might be in a uniformly moving ship. We will define two inertial reference frames: S, which is at rest relative to the ground laboratory, and S', which is at rest relative to the ship. Suppose that we perform an experiment in Newtonian mechanics—i.e., one involving such forces as gravity, the pulls of springs and strings, friction, etc. The principle of Galilean relativity implies that there is no such experiment we could do whose result in the ground laboratory would be different from the result of the same experiment performed in the ship laboratory. Consequently, there exists no such experiment that would allow us to conclude that one of the laboratories was *really* moving and the other *really* at rest.

For example, the experiment might involve dropping a stone of mass m from rest. We wish to inquire whether the result of this experiment when performed in the ground laboratory will differ from the result obtained when it is performed in the ship. We know that in either laboratory the force is the same, namely a gravitational force GM_em/r_e^2 directed vertically downward. Substituting this force into Newton's second law, we obtain the equation of motion describing the falling stone in each experiment. As expected from the principle of relativity, in each laboratory $\vec{F} = m\vec{a}$ has the same form,

$$\frac{GM_em}{r_e^2} = ma$$

implying motion with the constant vertical acceleration:

$$a = \frac{GM_e}{r_e^2} = 980 \text{ cm/sec}^2$$

If a stone is released from rest in either laboratory, this equation predicts that it will fall vertically with an acceleration of 980 cm/sec^2. Since the result is the same in both laboratories, this experiment will not differentiate between the two reference frames S and S'. More generally, the fact that the gravitational force law does not depend on velocity implies that any experiment involving this interaction will give the same result in all inertial reference frames.

This result implies that we can rephrase the principle of Galilean relativity in a more striking fashion: *No experiment in Newtonian mechanics can distinguish one inertial frame from another.* Galileo seems to have been the first to enunciate such a principle. As we discussed in Chapter 7, he argued that the vertical path of falling stones does not necessarily imply that the earth is at rest. Citing observations of a stone falling on land and on a uniformly moving ship, Galileo concluded:

> . . . from seeing the rock always fall in the same place, nothing can be guessed about the motion or stability of the ship . . . From this experience it may be seen that, at most, the falling body might drop behind if it were made of light material and the air did not follow the ship's motion; but if the air were moving with equal speed, no imaginable difference could be found in this or in any other experiment you please . . .[3]

The principle of Galilean relativity, with its infinite set of inertial reference frames all equally well suited to the description of interactions, is a notable feature of Newtonian mechanics. Newton himself was well aware of it. In Corollary V after the statement of the laws of motion in the *Principia*, he says, "The motions of bodies included in a given space are the same among themselves, whether that space is at rest, or moves uniformly forwards in a straight line without any circular motion."[4]

However, aside from its importance in establishing that the earth could be moving, the relativity principle was a rather esoteric consequence of Newton's theory, and attracted little attention. Although the principle denies the possibility of establishing experimentally that a particular inertial reference frame is really at rest, it did not prevent Newton from imagining the existence of such a frame. In his words, "Absolute space, in its own nature, without relation to anything external, remains always similar and immovable."[5]

There was an obvious candidate for such a reference frame. Kepler's heliocentric universe consisted of planets orbiting about a fixed sun, surrounded by innumerable distant fixed stars. It was natural for Newton to assume that the stationary objects in this model, the sun and stars, were really at rest. The inertial reference frame in which this was true, often referred to as the *frame of the fixed stars*, tacitly assumed the role of "absolute space" in the minds of men for two hundred years after Newton.

Suggested Reading

Eisenbud, L. "On the Classical Laws of Motion." *American Journal of Physics*, **26** (1958): 144–159. This article is a careful, closely reasoned critical examination of Newton's laws. The first two sections are non-mathematical and deal with several of the issues raised in this chapter.

Cooper, L. *An Introduction to the Meaning and Structure of Physics*. New York: Harper & Row, 1968. Chapter 28 is a brief discussion of Galilean relativity and the Galilean transformation.

Questions

1. In each of the following situations, what would happen that would enable one to conclude that the reference frame specified is not an inertial frame? Explain.

   (a) A man is sitting on the seat of an automobile that suddenly brakes. (The reference frame is fixed to the automobile.)
   (b) A ball hangs on the end of a string in an accelerating railroad car. (The reference frame is fixed to the car.)
   (c) A ball hangs on the end of a string attached to the ceiling at the outer edge of a merry-go-round. (The reference frame rotates with the merry-go-round.)
   (d) A man stands on a scale in an elevator that starts at the third floor and stops at the ground floor. (The reference frame is fixed to the elevator.)

2. A hawk flies in a straight path at a speed of 15 m/sec in the $+x$ direction relative to the S frame, which is fixed to the ground. A dove is following the hawk, flying at a speed of 20 m/sec in the $+x$ direction. Suppose the S' frame is also moving relative to the S frame at a speed of 20 m/sec in the $+x$ direction.

   (a) Describe the motion of the dove relative to the S' frame.

   (b) Describe the motion of the hawk relative to the S' frame.

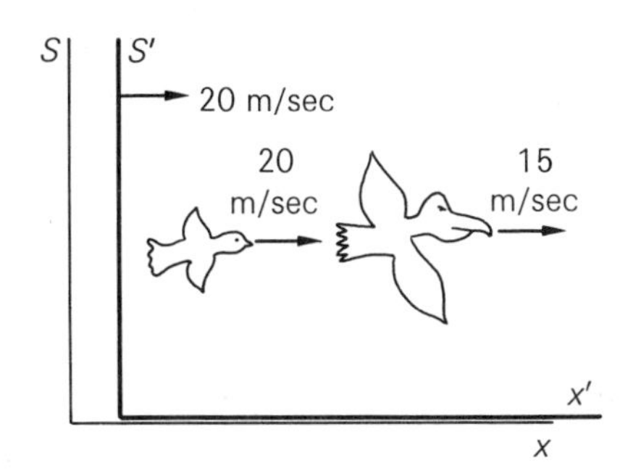

3. The water in a stream moves with speed u relative to the shore. A man is rowing a boat upstream with a speed w relative to the water. The man's hat falls out of the boat and floats downstream. A time T later the man, realizing his hat is gone, turns the boat around and rows (still with speed w relative to the water) after the hat.

   (a) Describe the motion of the hat and boat with respect to the frame of reference moving with the water. Use this description to determine how long after the hat fell in the water the man will reach the hat.

   (b) Describe the motion of the hat and boat with respect to a reference frame fixed to the land. Use this description to determine how long after the hat fell in the water the man will reach the hat. Compare with your result in part (a).

   (c) What is the moral of these two calculations?

4. Suppose the swimmer on page 277 wishes to swim directly across the stream to a point on the opposite shore. Draw a

vector diagram like that of Figure 11–15(b) showing the direction in which he must swim relative to the water. If the water speed is 1 mile/hour and he can swim $\sqrt{2}$ miles/hour relative to the water, what is this direction? What if he can only swim 1 mile/hour relative to the water?

5. The water in a stream moves with speed u relative to the shore. Suppose a swimmer, who can swim with speed w relative to the water, wishes to swim from point A to point B, a distance L, and then back to A.

   (a) While swimming from A to B what is his speed relative to the land (in terms of u and w)? How long will it take him to swim from A to B?

   (b) While returning to A, what is his speed relative to the land? How long will it take him to swim from B to A?

   (c) Show that the total time for the trip from A to B and back to A is

$$\frac{2L}{w}\frac{1}{(1 - u^2/w^2)}$$

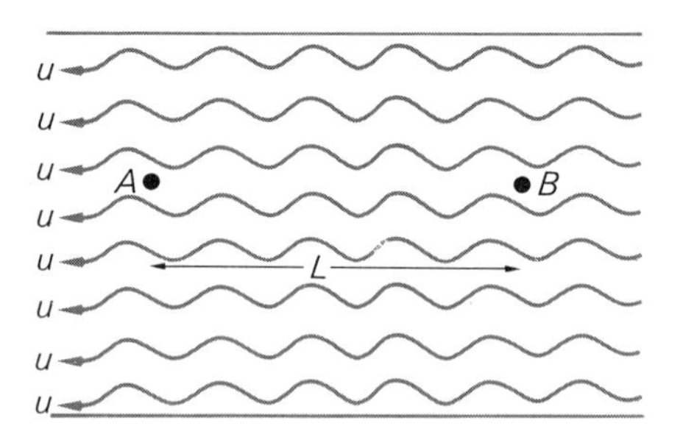

6. The water in a stream of width L moves with speed u relative to the shore. Suppose a swimmer, who can swim with speed w relative to the water, wishes to swim directly across the stream from C to D. Let $\vec{w}$ be the velocity of the swimmer relative to the water, $\vec{u}$ be the velocity of the water relative to the land, and $\vec{v}$ be the velocity of the swimmer relative to the land.

   (a) If the swimmer is to swim from C to D, what must be the direction of $\vec{v}$?

(b) What is the relation between $\vec{w}$, $\vec{u}$, and $\vec{v}$? Draw these velocity vectors, indicating their directions and showing that they satisfy this relationship.
(c) What is the speed of the swimmer relative to the land? I.e., what is the magnitude of $\vec{v}$, v, in terms of u and w?
(d) How long will it take him to swim from C to D (in terms of u, w, and L)?
(e) If he wishes to swim back from D to C, what will be the direction of his velocity $\vec{w}'$ relative to the water? How long will it take?
(f) Show that the total time for the trip from C to D and back to C is

$$\frac{2L}{w}\frac{1}{\sqrt{1 - u^2/w^2}}$$

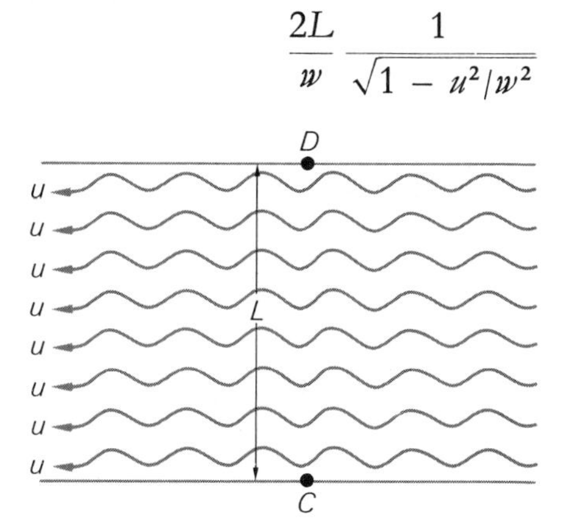

7. We found the Galilean transformation, which relates the coordinates of an event as measured in two reference frames whose relative motion is in the x direction. Draw a figure similar to Figure 11–17 for the case in which the S' frame moves in the $+y$ direction with speed u relative to the S frame, and deduce the corresponding Galilean transformation.

8. The S' frame moves relative to the S frame with a speed $u = 1$ m/sec in the $+x$ direction as shown in Figure 11–5. Consider the events E_0, E_1, E_2, and E_3 indicated in Figure 11–3. Use the Galilean transformation to calculate the position and time of each event relative to the S' frame. Compare your results with those given in Figure 11–7.

9. A car moves with a speed of 2 m/sec along the $+y$ axis of the

S reference frame, emitting a backfire every second. Let E_0 be the backfire emitted at $t = 0$, when the car is at the origin, E_1 the backfire at $t = 1$ sec, when the car is at $y = 2$ m, E_2 the backfire at $t = 2$ sec, and E_3 the backfire at $t = 3$ sec. Suppose the S' frame moves with a speed of 1 m/sec in the $+x$ direction relative to the S frame as shown in Figure 11–5. Let the origins of S and S' coincide at $t = t' = 0$.

(a) Draw a figure analogous to Figure 11–6, showing the position of the S' frame and the events at $t = 0, 1, 2,$ and 3 seconds.
(b) Draw a figure analogous to Figure 11–7, showing the events E_0, E_1, E_2, and E_3 with respect to the S' frame.
(c) Use the Galilean transformation to calculate the position of each event relative to the S' frame. Compare with the results you obtained in part (b).
(d) What is the velocity of the car relative to the S' frame? (Indicate both speed and direction.)

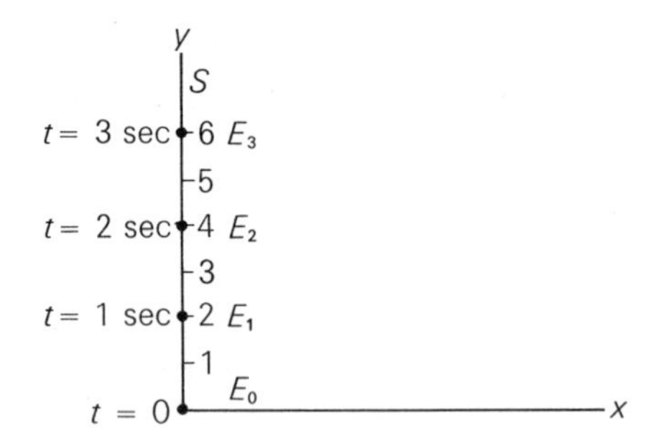

10. Consider again the cannonball shot from a cannon at the edge of a cliff. Equations (5) and (6) on page 167 specify the ball's x- and y-coordinates at time t with respect to an S reference frame fixed to the end of the cannon. Use the Galilean transformation to obtain the coordinates x' and y' of the ball at time $t' = t$ with respect to an S' reference frame moving with speed v_0 in the $+x$ direction, whose origin passes the origin of the S frame when the ball is fired at $t = t' = 0$. Describe in words the ball's path relative to the S' frame.

11. An object is located at rest at the origin of the S reference frame. The S' reference frame moves with speed u in the $+x$ direction relative to the S frame.

    (a) Describe in words the motion of the object relative to the S' frame.
    (b) Use the Galilean transformation to find the motion of the object relative to the S' frame and show that your result agrees with what you found in (a).

12. A ball released from rest at the surface of the earth falls with a constant acceleration of 9.8 m/sec^2. What is wrong with each of the following statements:

    (a) If the fall of this ball is viewed from another inertial reference frame S' which moves with a constant horizontal velocity relative to the earth, the ball appears to follow a parabolic trajectory. Hence the motion of the ball is different in two inertial reference frames, a violation of the principle of Galilean relativity.
    (b) If a ball is released from rest in a freely falling elevator, it remains at rest rather than falling. Hence the same experiment gives different results in two different reference frames, a violation of the principle of Galilean relativity.
    (c) If a ball is released from rest in a space ship far from the earth, but traveling with a constant velocity relative to the earth, it remains at rest rather than falling. Hence the same experiment gives different results in two different inertial reference frames, a violation of Galilean relativity.

THE THEORY OF RELATIVITY

Absolute, true, and mathematical time, of itself, and from its own nature, flows equably without relation to anything external . . .[1]

—Isaac Newton
Principia

A priori it is not at all necessary that the "times" . . . in different inertial frames agree with one another. One would have realized this long ago if, in the practical experience of everyday life, light did not appear (because of its high speed) as a means for the determination of absolute simultaneity.[2]

—Albert Einstein
"Autobiographical Notes"

If light signals do not travel with infinite speed, is there any way to determine absolutely whether two events occurring at widely separated positions are in fact simultaneous? What does this imply about the nature of time?

CHAPTER TWELVE

The Elusive Ether

The principle of Galilean relativity, asserting the impossibility of distinguishing between inertial reference frames within the scope of Newtonian mechanics, seemed to preclude the experimental determination of a reference frame that could be termed absolutely at rest. In the nineteenth century, however, new discoveries and theories in two areas of physics, optics and electromagnetism, seemed to indicate that in these fields the principle of Galilean relativity was not valid, and that consequently a way might be found after all to measure the absolute motion of a reference frame.

The Nature of Light

The history of men's views of the nature of light is long and fascinating. To the early Greeks, light was Fire, one of the four basic elements. Democritus later introduced the notion that light was composed of tiny particles that traveled from a luminous object to the eye. In the seventeenth century, Robert Hooke and particularly Christian Huygens, both contemporaries of Newton, proposed that light should be viewed instead as a wave phenomenon. Just as water waves travel from one point to another without the actual transport of water between the points, in their view, light involved the jiggling of a medium Huygens called "ether," without the actual transport of that medium from place to place.

Newton, however, favored the particle model of light. In support of this model, he observed that light travels in a straight line and propagates through a vacuum. If an object is placed in a beam of light, it casts a shadow. Similarly, if a sheet of paper with a hole is placed between a light source and screen as shown in Figure 12–1, the patch of light on the screen has the shape of the hole. In a lengthy dialogue with Huygens, Newton argued that this is characteristic of particles, not waves. The pattern on the screen is just what one would expect if light were a beam of bullet-like particles emitted by the source and absorbed or reflected by the paper. If light were a wave it would tend to bend around objects and thus not produce a sharp shadow. He also argued that the fact that one can see through an evacuated flask means that light cannot be a wave since the flask no longer contains a medium in which a wave could propagate.

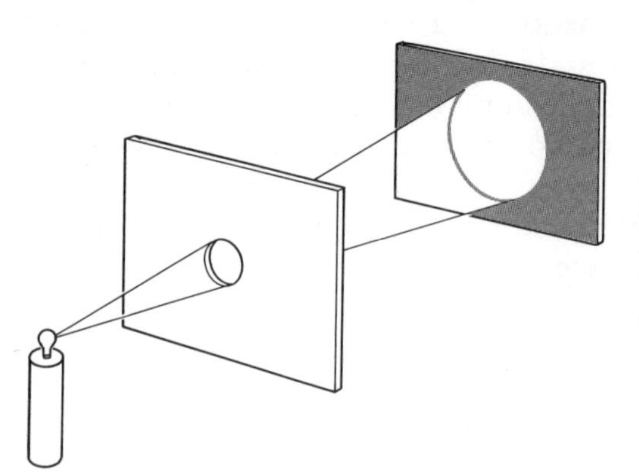

Figure 12–1. Evidence for the particle model of light. Light passing through a hole produces a sharp image of the hole.

Huygens countered by asserting that if light were a wave whose wavelength (the distance between successive crests) is much smaller than the dimensions of the object on which the light shines, the object would indeed appear to cast a sharp shadow. The bending of the light would be too small to be observed. Furthermore, he contended, the ether is not removed when air is evacuated from a flask. The ether is all-pervasive, occupying even the vast reaches of space between the earth and the stars. This view that an evacuated region of space actually contains ether appealed to the Cartesians and others who felt that the idea of a region of space containing nothing was meaningless.

The wave-particle controversy was not resolved until the beginning of the nineteenth century, when an Englishman, Thomas Young, and a Frenchman, Augustin Fresnel, demonstrated that light does bend around objects when they are extremely small. Similarly, when light passes through a tiny pinhole, it produces on a screen the pattern shown in Figure 12–2. This pattern is certainly not a sharp image of the hole. It is a series of concentric light and dark circles, known as a *diffraction pattern*, which can be explained straightforwardly by a wave model of light.

The subsequent discovery of many similar phenomena gave strong support to the wave model, and it soon gained universal acceptance. In buying the wave model, however, physicists also bought its peculiar medium, the ether. Since it had never been detected directly, and since it did not appear to disturb the motions of the planets, the ether was assumed to be massless,

Figure 12–2. Evidence for the wave model of light. The pattern found on the screen of Figure 12–1 when the hole is very small. (From D. H. Towne, *Wave Phenomena*, Addison-Wesley, Reading, Mass., 1967.)

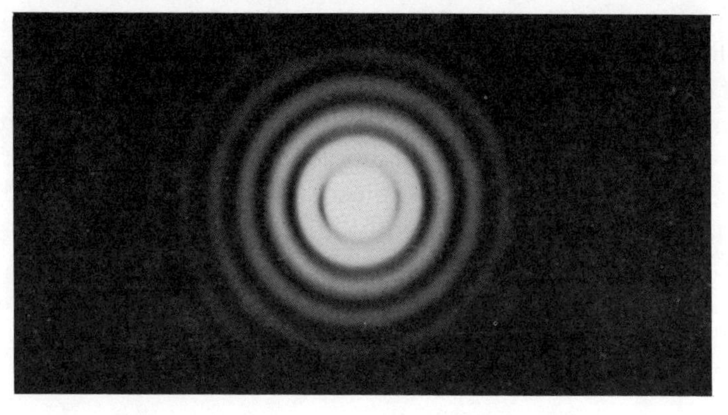

invisible, and all-pervasive. Its only *raison d'être* was to provide a medium for the propagation of light.

If one accepts the existence of the ether, then it appears that light, unlike the phenomena of Newtonian mechanics, can be most simply described in a single reference frame. This is the frame in which the ether is motionless—the *rest frame* of the ether. Thus the ether is a natural candidate for an absolute space in physics. The principle of Galilean relativity appears doomed.

As an illustration of this result, consider two inertial reference frames S and S', and suppose S is the rest frame of the ether. (See Figure 12–3.) Two pulses of light are shown, one traveling toward the right and one toward the left. (These might have been produced by setting off a flashbulb, for example.) Both

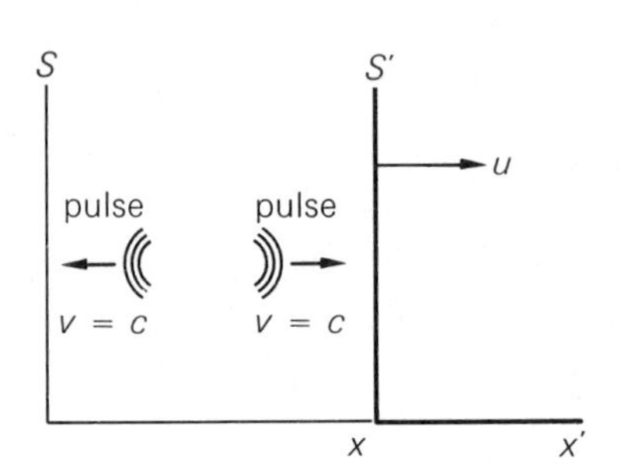

Figure 12–3. Two pulses of light moving in opposite directions. Relative to the S' frame, the pulses would have different speeds, $c + u$ and $c - u$.

pulses move relative to the ether (the S frame) with speed $v = c$.* From the velocity transformation of Chapter 11, we find that the speed of the pulse moving in the $+x'$ direction as measured in the S' frame would be $v' = v - u = c - u$; similarly, the pulse moving in the $-x'$ direction would have a speed $v' = c + u$. Hence the behavior of light in one inertial frame, S, is different from that in another, S'. In the former, the speed of light is the same in both directions, while in the latter it is faster one way than the other. Measurements of the velocity of light could thus be used to distinguish between inertial reference frames, and hence the principle of Galilean relativity did not seem valid for experiments with light.

* The speed of light in a vacuum is conventionally designated by the symbol c.

Electromagnetism

Another area of physics in which the principle of Galilean relativity did not seem valid was electromagnetism. We saw in Chapter 10 that electric charges at rest interact one way (electrostatically) while the same charges in motion interact quite differently (magnetically as well as electrostatically). Figure 12–4 illustrates a simple example. In part (a), two positively charged objects are shown at rest. Since they are at rest, the forces between them are wholly electrostatic; since the charges have the same sign, these electrostatic forces are repulsive.

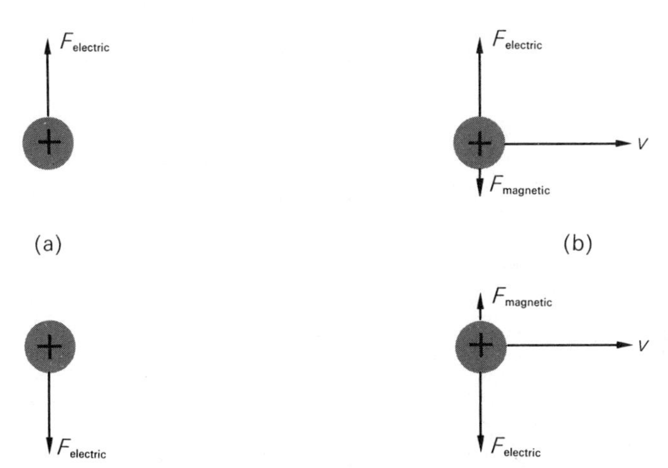

Figure 12–4. The net force between two moving charges is apparently less than the net force between the same two charges at rest. (a) Two identical charges at rest experience only repulsive electrostatic forces. (b) Two identical charges moving parallel to each other experience repulsive electrostatic and attractive magnetic forces.

In part (b), the two charged objects are moving toward the right with speed v. Here the electrostatic forces are still present, but there are also magnetic forces caused by the motion of the charges. The magnetic forces are found to be proportional to the speed of the charges, and, for charges of the same sign moving parallel to each other, these forces are attractive. Hence the net force on each object in (b) is less than that in (a).

This result implies that the principle of Galilean relativity is not applicable to electromagnetic phenomena. It suggests that charges at rest interact differently from charges in motion. But at rest or in motion with respect to what? Presumably, with respect to some inertial reference frame that could be considered absolutely at rest. Thus, this example suggests an experiment that could be used, at least in principle, to determine absolute space. Place two charges at rest relative to some inertial frame S (for example, the earth's surface) and measure the force between them. Repeat the experiment with the charges at rest in another inertial frame S' (for example, on a moving train). The frame in which the two charges at rest experience a smaller attractive magnetic force and hence larger net repulsive force is moving more slowly with respect to absolute space. Of all inertial frames, the one in which the two charges placed at rest experience the largest repulsive force is the frame that is at rest with respect to absolute space. Charges at rest in this frame experience no attractive magnetic force.

Further evidence for a relation between electromagnetic effects and an absolute space came later in the nineteenth century from the work of the Scottish theoretical physicist James Clerk Maxwell. Maxwell showed in 1865 that the equations describing electromagnetic interactions (now known as "Maxwell's equations") predict that the acceleration of charged particles should produce electromagnetic waves, oscillating electric and magnetic effects that propagate through space with a speed of approximately 3×10^8 m/sec—well known at the time to be the speed at which light waves travel. Maxwell showed these electromagnetic waves to have other properties, such as reflection, refraction, and polarization, which light also shared. He proposed that light is simply a type of electromagnetic wave. Maxwell's theory thus provided an unexpected consolidation of two areas of physics, light and electromagnetism, which had previously been considered quite distinct.

This synthesis of light and electromagnetism, one of the triumphs of nineteenth-century physics, immediately suggested that the special reference frame of light—the frame in which the ether is at rest—and the absolute space of electromagnetism were the same. Although Maxwell's theory made no particular mention of a medium for electromagnetic waves, physicists generally assumed that the ether was necessary as a medium, and that the speed of propagation specified by Maxwell's equations, 3×10^8 m/sec, should be interpreted as speed relative to the ether.

One further feature of Maxwell's theory is of interest here. His basic equations are written in terms of coordinates x, y, z, t which presumably should be associated with the absolute reference frame. If the Galilean transformation is used to express these equations in terms of the coordinates x', y', z', t' of some other relatively-moving inertial frame S', the resulting equations have a different, more complicated form. Thus the ether frame is special not only because it is absolutely at rest, but also because physics appears simpler if described in terms of the coordinates of that frame.

The Michelson-Morley Experiment

The existence of something as all-pervasive and yet as subtle as the ether was an intriguing idea. Many physicists attempted to detect this ether directly and to determine the motion of the earth and solar system relative to it. The most important of these experiments was carried out by two Americans, Albert A. Michelson and Edward W. Morley, in 1887.

The reasoning behind their experiment is straightforward. Consider the earth moving in its orbit about the sun with a speed of approximately 3×10^4 m/sec. (See Figure 12–5.) Regardless of the motion of the solar system relative to the ether, at some time of the year the earth should be moving relative to the ether with a speed of at least 3×10^4 m/sec.

The apparatus Michelson and Morley used to detect this motion is shown schematically in Figure 12–6. Let us assume for the sake of argument that the earth and hence the apparatus are moving relative to the ether with speed u. A pulse of light from

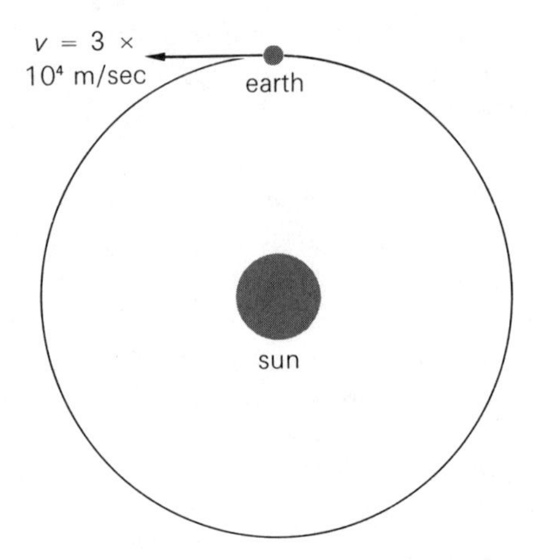

Figure 12–5. The motion of the earth about the sun. It seemed reasonable to assume that at some time of the year the earth should be moving relative to the ether with at least its orbital speed.

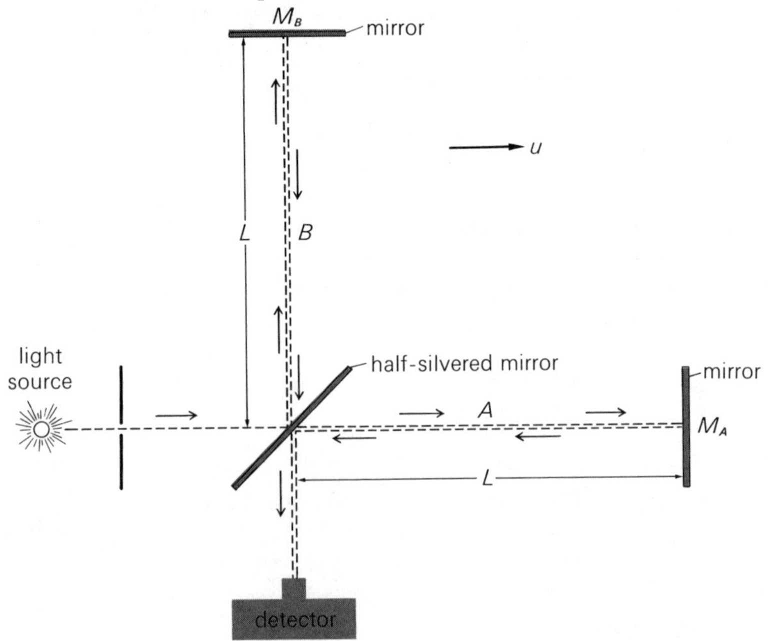

Figure 12–6. A schematic diagram of the Michelson-Morley apparatus.

the source strikes the half-silvered mirror, which has the property of transmitting half the light and reflecting half. The transmitted half travels along path A to a mirror M_A, and is reflected back along the same path to the half-silvered mirror, where part of it is reflected into the detector. The reflected half of the original pulse from the light source travels along path B to another mirror M_B, and is reflected back to the half-silvered mirror, through which part of it passes into the detector.

Assume that the apparatus is adjusted so that the distance from the half-silvered mirror to each of the other mirrors is the same, L. If the apparatus were at rest with respect to the ether, so that the speed of light was the same in all directions, then the light traveling along path A would arrive at the detector simultaneously with the light traveling along path B.

If the apparatus is moving with respect to the ether, however, we expect a different result. Let us analyze the experiment from the point of view of the reference frame in which the apparatus is at rest and the ether is streaming past with speed u to the left, as shown in Figure 12–7. Perhaps the easiest way to understand the effect of the relative motion of the apparatus and the ether is to think of the ether as a stream of water and the light pulse as a swimmer in that stream. The light pulse (swimmer) moves with speed c relative to the ether (stream).

Figure 12–7. In the reference frame in which the apparatus is at rest, the ether streams past with speed u.

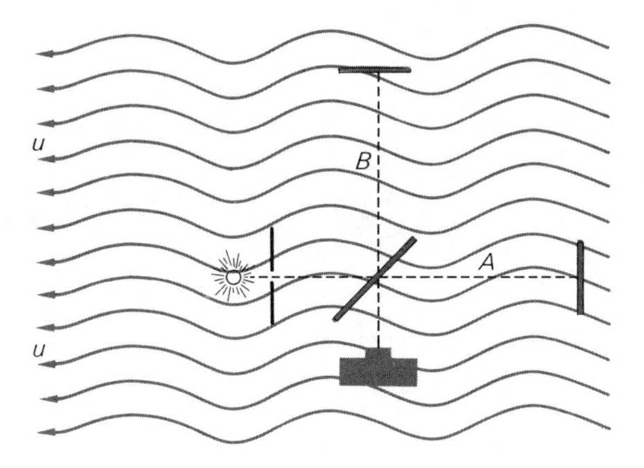

The half-silvered mirror separates the pulse into two pulses, one traveling along path *A*, the other along path *B*. Consider the pulse along path *A*. Going out toward the mirror, it is moving against the stream and consequently its net speed with respect to the apparatus is $c - u$. Coming back, it is moving with the stream and its net speed is $c + u$. Since both the outward and the return paths are of distance *L*, the time it takes for the entire trip is:

$$t_A = \text{time out} + \text{time back} = \frac{L}{c-u} + \frac{L}{c+u}$$

$$= \frac{L(c+u) + L(c-u)}{(c-u)(c+u)} = \frac{2Lc}{c^2 - u^2}$$

$$= \frac{2L}{c} \frac{1}{(1 - u^2/c^2)}$$

Now consider the pulse moving along path *B*. Here it is useful to think of a swimmer attempting to swim directly across the stream from point 1 on one shore to point 2 on the other shore, as shown in Figure 12–8(a). For his net velocity relative to the land to be directed along the line from 1 to 2, the swimmer must swim slightly upstream relative to the water. More technically, the velocity relative to the land, $\vec{v}'$, will be the vector

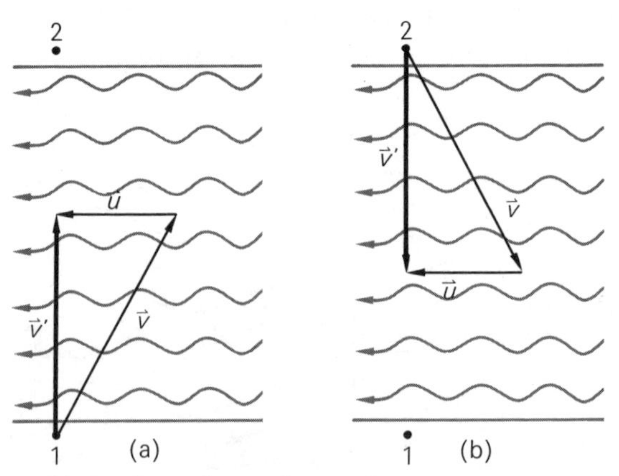

Figure 12–8. In each case for a swimmer to move between points 1 and 2 on the land with velocity $\vec{v}'$, he must swim with velocity $\vec{v}$ relative to the water. The water flows with velocity $\vec{u}$ relative to the land.

sum of the velocity relative to the water, $\vec{v}$, and the velocity of the water relative to the land, $\vec{u}$. From Figure 12–8 and the Pythagorean theorem we conclude that the net speed of the swimmer relative to the land will be $v' = \sqrt{v^2 - u^2}$. By exactly the same reasoning we can conclude that the light pulse whose speed is c relative to the ether stream will have a net velocity along path B of the Michelson-Morley apparatus that is the vector sum of its velocity relative to the ether plus the velocity of the ether relative to the apparatus. Thus the speed of the pulse on its outward trip along B is $\sqrt{c^2 - u^2}$.

On the return trip, a similar situation obtains, as indicated in Figure 12–8(b). Identical reasoning shows that the speed of the pulse on the return trip is also $v' = \sqrt{c^2 - u^2}$. Since the distances of the outward and the return trips are each L, the total time for the pulse traveling along path B is

$$t_B = \frac{L}{\sqrt{c^2 - u^2}} + \frac{L}{\sqrt{c^2 - u^2}} = \frac{2L}{\sqrt{c^2 - u^2}} = \frac{2L}{c} \frac{1}{\sqrt{1 - u^2/c^2}}$$

This is to be compared with the travel time of the pulse along path A:

$$t_A = \frac{2L}{c} \frac{1}{(1 - u^2/c^2)}$$

It is clear that these two times are related by:

$$t_B = \sqrt{1 - u^2/c^2}\, t_A \tag{12}$$

But if u is less than c, $1 - u^2/c^2$ is less than 1 and hence $\sqrt{1 - u^2/c^2}$ is less than 1. This implies that t_B is less than t_A. Hence the pulse that travels along path B should reach the detector before the pulse that travels along path A. By observing the time difference $t_A - t_B$ one could calculate u, the speed of the earth relative to the ether.*

* While this discussion correctly describes the Michelson-Morley apparatus and the essence of their measurement, in practice the time interval $t_A - t_B$ they expected is far too short to measure directly. Consequently an indirect technique for measuring the time interval is employed; it involves a continuous source of light waves rather than a pulse. The beams of light waves traveling along paths A and B are superposed at the detector, producing an "interference pattern." By studying this pattern of adjacent light and dark patches one can infer the difference in the times required for the light beams to traverse the two paths and hence establish the velocity of the earth relative to the ether.

According to this reasoning, the Michelson-Morley experiment should have provided a straightforward way to detect the ether. The time difference predicted from equation (12), assuming u to be at least as great as the earth's speed in its orbit, was large enough so that it should have been easily detected with the apparatus described above. However, when the experiment was first performed in 1881 by Michelson during a visit to Potsdam, Germany, only a small time difference was observed, just a fraction of that expected. Furthermore, the measured time difference was so small that it could easily be attributed to vibrations and strains in the delicate apparatus, having no connection with its motion through the ether.

To verify this surprising result, Michelson and Morley repeated the experiment in 1887 in Cleveland. The apparatus had been considerably improved, and still no significant time difference was detected. The only difference measured was "certainly less than the twentieth part . . . , and probably less that the fortieth part"[1] of the amount expected from equation (12). Such a small difference could be attributed to experimental uncertainties. Thus the Michelson-Morley experiment found no detectable motion of the earth relative to the ether.

Interpreting the Experiment

How could this negative result be interpreted? We have here a particularly useful example to compare with the popular view of the scientific process. A prediction had been made. Experiment showed the prediction to be wrong. How did physicists react? Did they simply abandon the foundations of physics that had been established with such effort and had dealt successfully with so many other aspects of the physical world?

Of course not. A widespread and natural reaction was to question whether the experiment had been properly performed and its results properly interpreted. Michelson and Morley themselves were disappointed by it. They had devised what seemed to be a straightforward way to detect the ether, but for some reason it had failed. In their discouragement over their negative result, they even neglected to follow up the original measurements with others they logically should have performed.

Michelson and Morley had estimated the expected time difference along the two paths of their apparatus by assuming the velocity of the earth relative to the ether to be the same as the velocity of the earth in its orbit around the sun. That is, for purposes of argument they assumed the sun to be at rest relative to the ether. But perhaps the entire solar system is moving relative to the ether. Then the earth's velocity with respect to the ether will be the vector sum of its velocity relative to the sun and the sun's velocity relative to the ether. "It is just possible," they wrote, "that the resultant velocity of the earth relative to the ether at the time of the observations was small, though the chances are much against it. The experiment will therefore be repeated at intervals of three months, and thus all uncertainty will be avoided."[2]

These three-month repetitions were required to rule out the unlikely possibility that the original experiment had been performed at that precise moment when the earth, in its motion relative to the sun, happened to be at rest relative to the ether. However, Michelson and Morley never carried out these additional measurements. They viewed their attempt to measure the earth's velocity relative to the ether as a failure.* Rather than repeat the experiment, they turned to other, unrelated projects that seemed more promising.

Others speculated about possible reasons for discounting the result of the experiment. In 1892, the Dutch physicist Hendrik Lorentz wrote, "I am totally at a loss to clear away this contradiction [to the accepted ether theory] . . . Can there be some point in the theory of Mr. Michelson's experiment which has as yet been overlooked?"[3] A few years later, the Irish physicist George Francis FitzGerald discussed the possibility of some

* Michelson's disappointment in this experiment is indicated by his comment fifteen years after the experiment was completed: "I think it will be admitted that the problem, by leading to the invention of the interferometer [the Michelson-Morley apparatus, which proved useful for many other purposes], more than compensated for the fact that this particular experiment gave a negative result." (Michelson, A. *Light Waves and Their Uses*. Chicago: University of Chicago Press, 1902, p. 159.)

unnoticed difficulty in the construction or behavior of the apparatus:

> I am fairly well satisfied with Michelson's and Morley's experiment. No doubt changes of temperature are going on, but . . . it is very unlikely that there could have been just that irregular distribution of warming and cooling . . . [that] the expansions and contractions should have exactly compensated for the effect they expected . . . It is however a not impossible coincidence. Another possibility might be more irregular gravitation to objects in the room . . . but this is still more improbable.[4]

In an attempt to clear up these uncertainties, Morley and D. C. Miller constructed several new versions of the Michelson-Morley apparatus during the period 1902–1906, using different materials from those of the original. With them they repeated the experiment at several times of the year and in several different locations. This series of experiments, however, only confirmed the earlier result.

The Ether Drag Hypothesis

The attempts to understand these negative Michelson-Morley results were much like the attempts of the Greeks to deal with the anomalies of planetary motion. Greek philosophy presupposed uniform circular motion, yet the planets were observed to retrogress. Rather than abandon uniform circular motion, the Greeks made elaborate attempts to patch up the geocentric model, augmenting it first with homocentric spheres and later with epicycles. Likewise, many ingenious suggestions were made to account for the Michelson-Morley result while still retaining the ether hypothesis.

Michelson and Morley felt the most likely inference to be drawn from their experiment was that, at least in the vicinity of the earth, there is no relative velocity of the earth and ether. Short of returning to a geocentric view of the universe, they could only assume that the earth moves relative to most of the ether in the universe, but that in its motion it drags the ether in its immediate vicinity along with it.

Such an assumption, however, conflicted with the accepted explanation of another observation. In 1728, the astronomer James Bradley, searching unsuccessfully for parallax, had found that during the course of a year, all the stars in the heavens appear to move in concert about tiny elliptical paths. As this is true of even the most distant stars, the effect, called *aberration*, is apparently not due to parallax. The explanation commonly accepted was that to view a particular star, an earthbound telescope has to be aimed slightly away from the direction of the star because the telescope is moving with respect to the ether.

To understand this effect, let us consider a star directly overhead, as shown in Figure 12–9. Suppose that at the time of year when the observation is made, the earth is moving to the right with speed u relative to the ether. While the light is traveling vertically with speed c with respect to the ether, the telescope is

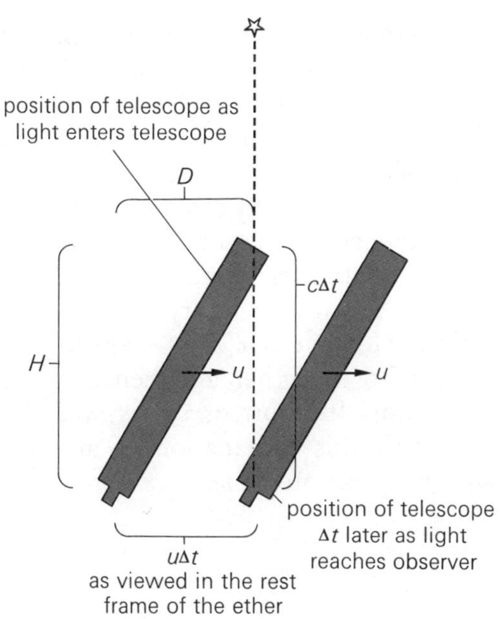

Figure 12–9. Aberration. The telescope, moving with speed u relative to the ether, must be tipped from the vertical so that light from a star directly overhead will be able to travel down the tube.

moving to the right with speed u. For the light to travel down the tube of the telescope to the observer, the telescope must be tipped slightly in the direction of its motion, as indicated in Figure 12–9. Let Δt be the time it takes the light to get from one end of the telescope to the other. During this time the light moves vertically a distance $H = c\Delta t$ while the base of the telescope moves to the right a distance $D = u\Delta t$. As can be seen from the figure, the ratio D/H is a measure of how much the telescope must be tipped. We find

$$\frac{D}{H} = \frac{u\Delta t}{c\Delta t} = \frac{u}{c}$$

Suppose we take the speed of the earth in its orbit, 3×10^4 m/sec, as an estimate of u. Recalling that $c = 3 \times 10^8$ m/sec, we calculate

$$\frac{D}{H} = \frac{u}{c} = \frac{3 \times 10^4}{3 \times 10^8} = 10^{-4}$$

Thus the tilt is extremely small. Suppose, for example, that H were 100 feet. Such a telescope must be tipped a distance $D = 10^{-4} \times 100$ ft, or about an eighth of an inch, for the light from the star to travel down the axis of the telescope. While small, this tilt is measurable. As the earth moves in its elliptical orbit about the sun, it continually changes its direction of motion, and the tilt of the telescope must be altered accordingly. The net effect is that each star appears to move in a tiny elliptical path.

This explanation is in satisfactory agreement with the observed aberration. However, it does require the telescope to move with respect to the ether and hence disagrees with the ether-drag hypothesis. In short, if we assume the existence of the ether and accept this explanation of aberration, then the ether-drag hypothesis is untenable.

Lorentz-FitzGerald Contraction

Another attempt to reconcile the existence of the ether with the null result of the Michelson-Morley experiment was the so-called *Lorentz-FitzGerald contraction*, first suggested in 1889 by FitzGerald:

I have read with much interest Messrs. Michelson and Morley's wonderfully delicate experiment attempting to decide the important question as to how far the ether is carried along by the earth. Their result seems opposed to other experiments [e.g., the observation of aberration] showing that the ether in the air can be carried along only to an inappreciable extent. I would suggest that almost the only hypothesis that can reconcile this opposition is that the length of material bodies changes, according as they are moving through the ether . . .[5]

That is, FitzGerald suggested that the reason the light pulse traveling along path A in Figure 12–6 did not take a longer time than that along path B is that the length of path A is not L but a shorter distance. If material objects shrink in the direction of their motion through the ether by just the right amount, the Michelson-Morley result, $t_A = t_B$, could be reconciled with the existence of the ether.

Figure 12–10 illustrates FitzGerald's hypothesis. Suppose the distance L is chosen to be one meter. To measure off this distance, we might lay a meter stick from the half-silvered mirror to each of the reflecting mirrors, M_A and M_B. Suppose that path A is along the direction of motion of the earth through the ether. According to FitzGerald's hypothesis, the meter stick lying along this path will shrink. Thus, although we measure path A to be one meter, its length is really less than a meter because the meter stick has shrunk due to its motion through

Figure 12–10. The effect of the Lorentz-FitzGerald contraction on the Michelson-Morley apparatus.

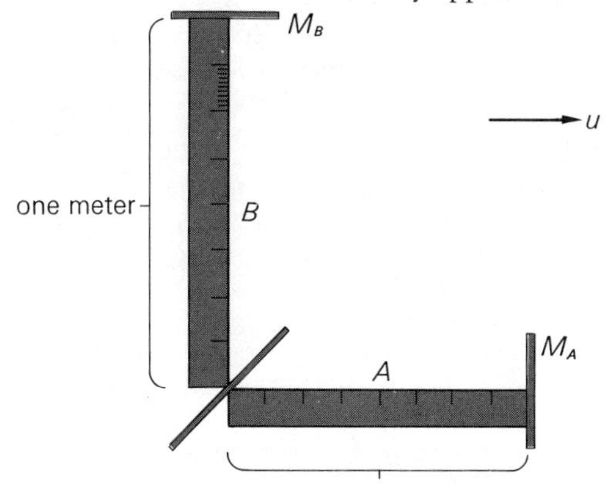

the ether. The meter stick along path B, on the other hand, does not shrink in length, so the length of path B is really one meter.

How much shrinking is necessary to produce the result $t_A = t_B$? Assuming that the distances from the half-silvered mirror to the mirrors M_A and M_B were each equal to L, we found earlier that

$$t_A = \frac{2L}{c} \frac{1}{(1 - u^2/c^2)} \quad \text{and} \quad t_B = \frac{2L}{c} \frac{1}{\sqrt{1 - u^2/c^2}}$$

Assuming instead that the apparatus shrinks in the direction of its motion relative to the ether so that the distance to M_A is not L, but rather $L\sqrt{1 - u^2/c^2}$, then the time required for the pulse to travel along path A would be:

$$t_A = \frac{2L\sqrt{1 - u^2/c^2}}{c} \frac{1}{(1 - u^2/c^2)} = \frac{2L}{c} \frac{1}{\sqrt{1 - u^2/c^2}}$$

which is the same as t_B. FitzGerald therefore assumed that material objects shrink in the direction of their motion through the ether in such a way that an object of length L contracts to a length $L\sqrt{1 - u^2/c^2}$. With this hypothesis he was able to explain the absence of a time difference in the Michelson-Morley experiment.

Furthermore, both FitzGerald and Lorentz, who arrived at the same suggestion independently, saw a possible cause for such a contraction. In 1892, Lorentz wrote:

> Now, some such change in the length of the arms in Michelson's experiment . . . is so far as I can see, not inconceivable. What determines the size and shape of a solid body? Evidently the intensity of the molecular forces;* any cause which would alter the latter would also influence the shape and dimensions. Nowadays we may safely assume that electric and magnetic forces act by means of the intervention of the ether. It is not far-fetched to suppose the same to be true of the molecular forces.[6]

* At that time, the nature of the "molecular" forces that bind the molecules of a solid body together was not well understood. As discussed in Chapter 10, it is now believed that these forces are simply electromagnetic.

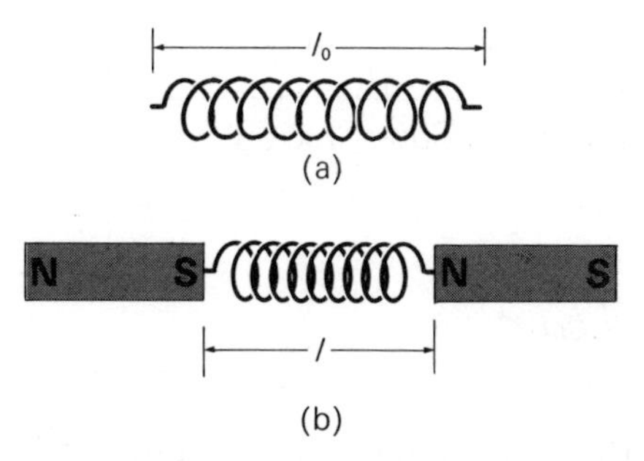

Figure 12–11. The distance between the magnets is determined by a balance between the attractive magnetic force and the repulsive spring force. (a) An uncompressed spring. (b) The spring compressed between two attracting magnets.

The way in which the "intensity of the molecular forces" determines the size of a solid body can be understood by analogy. Imagine a system composed of two magnets connected by a spring as shown in Figure 12–11(b). The two magnets, oriented as shown, attract each other. However, as the spring is compressed to less than its relaxed length l_0, it tends to push the magnets apart. The equilibrium distance l between the magnets is determined by the balance between the attractive magnetic force and the repulsive spring force. If the spring force is weakened, for example by heating the spring, the two magnets will come closer together and the system will contract.

In an analogous fashion, the size of a particular solid body is determined by a balance of the internal forces acting between its atoms. Suppose that these forces were electromagnetic. Their strength should then be affected by any motion of the solid through the ether. As we discussed earlier, charges at rest interact by purely electrostatic forces, while the same charges in motion experience additional magnetic forces. Thus the size of the solid at rest would be determined solely by a balance of the electrostatic forces between its atoms, while the size of the same solid in motion would be determined by a combination of these electrostatic forces and additional magnetic forces. Consequently the size of the solid in motion might be expected to differ from the size of the same solid at rest.

This established a plausible mechanism for the Lorentz-FitzGerald contraction. The details of the process remained obscure, however, because the molecular forces responsible for the cohesion of solids were not well understood.

The Lorentz Theory

By the turn of the twentieth century, there had been a number of unsuccessful attempts to measure the earth's motion with respect to the ether, and a comparable number of attempts to explain their negative results while maintaining the ether hypothesis. Noting the *ad hoc* nature of much of the discussion, the French mathematician and physicist Henri Poincaré in 1895 suggested a new point of view:

> Experiment has revealed a multitude of facts which can be summed up in the following statement: it is impossible to detect the absolute motion of matter, or rather the relative motion of ponderable matter with respect to the ether; all that one can exhibit is the motion of ponderable matter with respect to ponderable matter.[7]

That is, Poincaré proposed that the principle of Galilean relativity, previously restricted to mechanical phenomena, should be extended to electromagnetism and indeed to all physical phenomena. In fact, Poincaré was the first to focus explicitly on a "principle of relativity" and call it by this name. In enumerating the basic principles of physics, he listed:

> The principle of relativity, according to which the laws of physical phenomena should be the same, whether for an observer fixed, or for an observer carried along in a uniform motion of translation; so that we have not and could not have any means of discerning whether or not we are carried along in such a motion.[8]

One of the "laws of physical phenomena" in question was Maxwell's equations, which seemed to describe electromagnetic and optical phenomena with striking success. Poincaré's statement of the principle of relativity, then, requires that Maxwell's equations should take on the same form when written with respect to the ether or with respect to the earth moving through the ether. This is consistent with the Michelson-Morley experiment, which gave the result, $t_A = t_B$, that one would have expected had the apparatus been at rest with respect to the ether.

Lorentz responded to Poincaré's suggestion of a principle of relativity applicable to all physical phenomena by incorporating this principle in a new theory of the behavior of moving electric charges. He introduced the theory in 1904 with the following remarks:

> Poincaré has objected . . . that, in order to explain Michelson's negative result, the introduction of a new hypothesis [the Lorentz-FitzGerald contraction] has been required, and that the same necessity may occur each time new facts will be brought to light. Surely this course of inventing special hypotheses for each new experimental result is somewhat artificial. It would be more satisfactory if it were possible to show by means of certain fundamental assumptions . . . that many electromagnetic actions are entirely independent of the motion of the system.[9]

To do this, Lorentz noted first that if one begins with Maxwell's equations, written in terms of the coordinates x, y, z of the rest frame of the ether, and then uses the Galilean transformation,

$$x' = x - ut$$

$$y' = y$$

$$z' = z$$

to re-express Maxwell's equations in terms of the coordinates x', y', z' of the rest frame of the earth, the resulting expression of Maxwell's equations in the rest frame of the earth has a different form. This appears to violate Poincaré's principle of relativity.

Lorentz managed to find another transformation which, used in place of the Galilean transformation, did leave Maxwell's equations the same when expressed in the rest frame of the earth. This transformation was the following:

$$x' = \frac{1}{\sqrt{1 - u^2/c^2}}(x - ut)$$

$$y' = y$$

$$z' = z$$

$$t' = \frac{1}{\sqrt{1 - u^2/c^2}}\left(t - \frac{u}{c^2}x\right)$$

We shall meet this transformation again and explore its meaning more fully in Chapter 14. Here, let us briefly examine the interpretation given it by Lorentz and Poincaré.

Let us consider an event E, which has coordinates x, y, z, t in the rest frame of the ether and x', y', z', t' in the rest frame of the earth, which moves with speed u relative to the ether. The Galilean transformation, according to Lorentz, specifies how the true coordinates of the event are related in the two reference frames. In particular, we should have $t' = t$—time is time, whatever the reference frame. But the earth's motion through the ether causes strange effects to occur. In particular, measuring rods shrink, so that a coordinate measured with a shrunken meter stick in the earth frame is not the true coordinate, but a distorted one. Thus, Lorentz showed, the value of x' at which the event is measured to occur is not $x - ut$, but rather

$$\frac{1}{\sqrt{1 - u^2/c^2}}(x - ut)$$

Even more striking, it appeared from the above transformation that the operation of clocks at rest on the earth must somehow be affected by their motion through the ether so that they measure not true time $t' = t$, but a distorted time

$$t' = \frac{1}{\sqrt{1 - u^2/c^2}}\left(t - \frac{u}{c^2}x\right)$$

In summary, then, the Lorentz theory asserts that all meter sticks and clocks at rest on the earth undergo physical changes as a result of their motion through the ether. Consequently, our measurements of distance and time are not true measurements, but distorted ones.

Naturally, we would like to determine the speed of the earth relative to the ether to correct our distorted distance and time measurements. But any attempt to determine this speed, such as the Michelson-Morley experiment, involves distance and time measurements. In fact, according to the Lorentz theory, no matter what our speed relative to the ether really is, these physical changes in our measuring rods and clocks due to that motion are just such as to make our experiments yield the apparent result that our speed relative to the ether is zero. It seems almost as if a conspiracy of nature prevents our detection of the ether frame. This leaves the ether with a very peculiar status. Lorentz assumed its existence and granted it a central role in the propagation of light and electromagnetic effects. Yet any direct detection of its presence seemed destined to be forever beyond the reach of man.

Suggested Reading

Shankland, R. S. "The Michelson-Morley Experiment." *Scientific American*, **211** (November 1964): 107–114. A non-mathematical description of the experiment and its historical context.

Cooper, L. *An Introduction to the Meaning and Structure of Physics*. New York: Harper & Row, 1968. Chapter 29 analyzes the Michelson-Morley experiment from a point of view somewhat different from ours.

Williams, L. P., ed. *Relativity Theory: Its Origins and Impact on Modern Thought*. New York: Wiley, 1968. This reprint collection contains several articles relevant to this chapter, including ones by Lorentz, Poincaré, and the paper of Michelson and Morley on their experiment.

Questions

1. The ether is described on page 295 as "massless, invisible, and all-pervasive." What observational evidence can you cite to indicate that the ether, if it exists, has these properties?
2. The frame of reference in which the ether is motionless, its rest frame, is cited as a natural candidate for an absolute

space in physics. Why should one suppose that there is a reference frame in which the ether is at rest? Might it not swirl around so that if a bit of ether here is at rest another bit somewhere else is moving? Can you think of any observational evidence that bears on this question?

3. According to the argument on page 296, if the earth is moving relative to the ether with its orbital speed, $u = 3 \times 10^4$ m/sec, the speed of light measured in the direction of the earth's motion would not be c, but rather $c + u$. How precisely would the speed of light have to be measured to detect this effect directly?

4. Using Figure 12–5, explain clearly why Michelson and Morley originally intended to repeat their experiment at three-month intervals.

5. Explain clearly why it was expected that at some time of the year the earth should be moving relative to the ether with a speed of at least 3×10^4 m/sec. If the speed of the earth relative to the ether were 3×10^4 m/sec along the direction of path A of the Michelson-Morley apparatus, as indicated in Figure 12–6, what would be the difference in the times for the light to traverse the two paths? I.e., calculate $t_A - t_B$. Assume the distance L to be that of the original Michelson-Morley experiment, 11 m. [Hint: This calculation is not difficult if you use the approximation

$$\sqrt{1 - x} \approx 1 - \tfrac{1}{2}x$$

valid when x is very much smaller than 1. This approximation follows from $(1 - \frac{1}{2}x)(1 - \frac{1}{2}x) = 1 - x - x^2/4 \approx 1 - x$, since $x^2/4$ is much smaller than x if x is much less than 1.]

6. Each item below is an experimental observation that disagrees with theoretical prediction. Discuss the nature of the conflict in each case, and compare the various ways people attempted to resolve these conflicts.

   (a) The Michelson-Morley experiment vs. the ether theory.
   (b) A 100-lb ball reaching the ground two finger breadths ahead of a 1-lb ball vs. Galileo's theory.
   (c) No stellar parallax in the days of Copernicus, Kepler, and Newton vs. the heliocentric model.

(d) Tycho's data on the orbit of Mars vs. Kepler's application of the ideal of uniform circular motion.

7. In 1892 the British physicist Sir Oliver Lodge performed a variation of the Michelson-Morley experiment designed to test the ether-drag hypothesis. His experiment can be described in the following terms: A pulse of light from a source is split into two pulses by the half-silvered mirror shown in the figure. The two pulses reflect from the mirrors M_A, M_B, and M_C, traveling around the circuit in opposite senses. Any difference in transit time of the two pulses can be measured by the detector. The apparatus is sandwiched between two large steel discs that can be rotated rapidly about the axis shown.

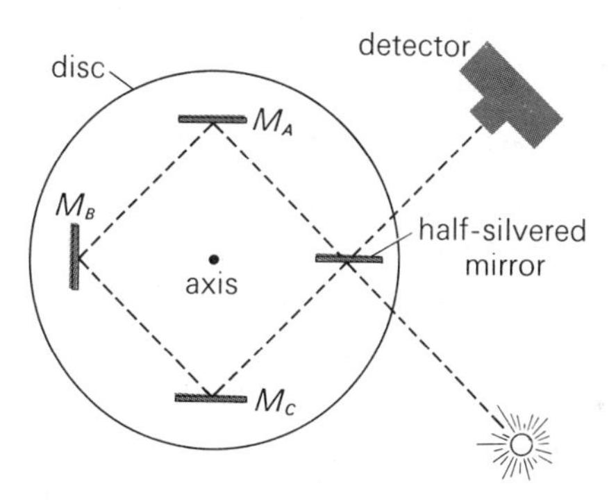

(a) If the discs are not rotating, what time difference would you expect?
(b) If the ether is dragged by the motion of material objects, what might be expected to happen to the ether between the plates when they are rotated?
(c) What effect would this be expected to have on the transit time of the two pulses? (Lodge found no such effect.)

8. Another way to analyze the Michelson-Morley experiment is from the point of view of the ether's rest frame. In this frame the ether is at rest and the apparatus is moving to the right with speed u, as shown in the figure. The light moves with speed c along both arms of the apparatus.

(a) Consider the light pulse that reflects from mirror M_A. Since the mirror is moving to the right with speed u, the light pulse traveling out toward the mirror must move through a distance greater than L before reaching the mirror. Let t_1 be the time it takes the pulse to reach M_A. During this time the mirror moves a distance ut_1 to the right so that the total distance the light pulse must travel is $L + ut_1$. Since the light moves with speed c, this distance is also equal to ct_1. The equation $ct_1 = L + ut_1$ can be solved for t_1. A similar analysis can be employed to find the time t_2 required for the pulse to return from M_A to the half-silvered mirror. In this case, since the half-silvered mirror moves to the right to meet the returning pulse, the light pulse moves through a distance less than L. Calculate t_2 and hence the total time $t_A = t_1 + t_2$ along arm A. Compare with the result derived in the text.

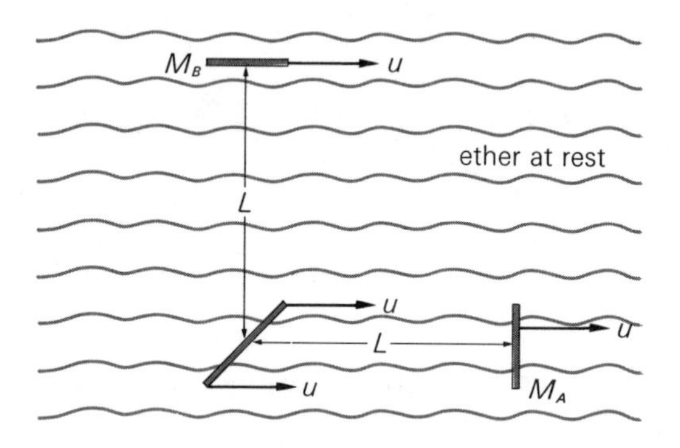

(b) Now consider the light pulse that reflects from mirror M_B. Since the apparatus is moving to the right, the pulse must follow the path shown in the figure. Use

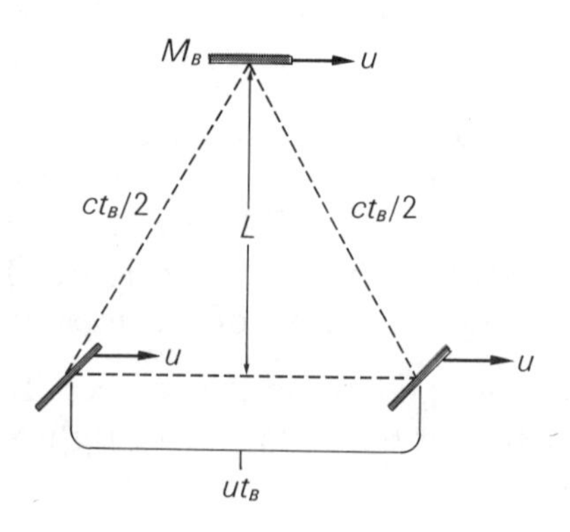

this figure and the Pythagorean theorem to determine t_B, the time required for the pulse to travel from the half-silvered mirror to M_B and back again. Again, compare with the result in the text.

9. Consider the apparatus shown in the figure: A light pulse leaves the source S, reflects from the mirror M, and returns to the source. The source and mirror are mounted at opposite ends of a metal rod. Connected to the source are a detector and a clock that measures the total transit time of the pulse.

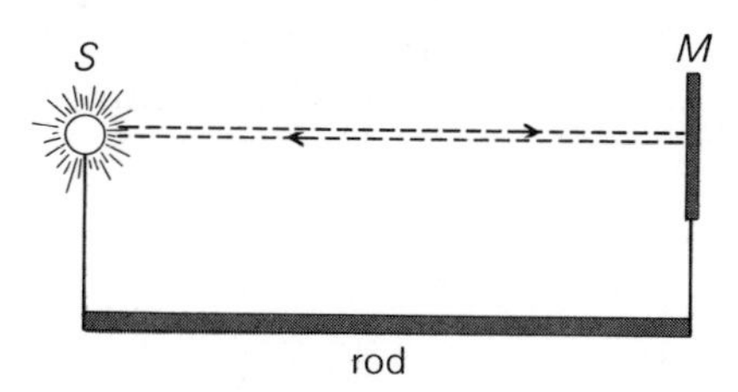

(a) Suppose the apparatus is at rest relative to the ether. If the distance from S to M (the length of the rod) is L, and if the pulse leaves the source at $t = 0$, find the time t at which the pulse returns to the source, in terms of L and c.

(b) Now suppose the same apparatus is set moving toward the right with speed u relative to the ether. If the length of the moving rod is L', and the pulse leaves the source at $t = 0$, find the time t' at which the pulse returns to the source, in terms of L', c, and u. (Hint: Recall the treatment of path A of the Michelson-Morley experiment.)

(c) According to the Lorentz theory, the length of the rod (and of a meter stick that measures this length) contracts due to its motion through the ether. Thus if an observer moving with the apparatus measures the length of the rod to be L, the actual length $L' = L\sqrt{1 - u^2/c^2}$. Rewrite your result from (b) in terms of the apparent length L.

(d) According to the Lorentz theory, the reading of the clock at the source is affected by its motion through the ether so that instead of the actual time t', it reads a time $t = t'\sqrt{1 - u^2/c^2}$. Using your result in (c), find an expression for the apparent transit time of the pulse t, in terms of L, the apparent length of the rod, and u and c. You should find precisely the same expression as in (a), illustrating how the Lorentz theory leads to the impossibility of detecting the motion of a reference frame relative to the ether.

10. If, as the Lorentz theory suggests, the ether can never be detected, can it be considered a meaningful concept? Can you think of other concepts that cannot be verified, yet are nevertheless widely accepted? (Don't limit your examples to physics.)

CHAPTER THIRTEEN

The Overthrow of Absolute Time

A radical new approach to light and electromagnetism, an approach that focused on the nature of time itself, was suggested in 1905 by an unknown young clerk in the Swiss patent office, Albert Einstein. Einstein was born in the small Bavarian town of Ulm in 1879, a son of the owner of a small electrochemical firm. His childhood and early adolescence were spent in Munich, where he attended secondary school. Because of the strict discipline and rote memory work emphasized in German schools of that day, Einstein disliked his schooling intensely, and his performance was mediocre, even quietly rebellious.

At an early age, however, he became interested in natural phenomena. In his autobiography, he recalled the time his father showed him a compass, whose needle moved in a determined way without any apparent material cause:

> I can still remember—or at least believe I can remember—that this experience made a deep and lasting impression upon me. Something deeply hidden had to be behind things.[1]

As a child, Einstein read widely in the popular scientific literature. At the same time he was introduced to mathematics by an uncle and studied algebra, geometry, and calculus largely on his own.

Scientific and mathematical subjects, however, were not emphasized in the schools, and at the age of fifteen, when his family moved to Italy following the failure of his father's business, Einstein left school a year short of receiving his diploma. After a year of wandering around Italy by himself, he decided to resume his education. However, he failed the entrance examination to the Zurich Polytechnic Institute because of deficiencies in language and descriptive biology. He was admitted a year later after completing his secondary education at another Swiss school.

Einstein had decided to study physics, but found the program at the Polytechnic Institute disappointing. The lectures were restricted to the older, established areas of physics, and offered little to excite someone attracted by the fundamental issues of the day. In particular, Maxwell's work on electromagnetism and light was not taught, and this was a subject Einstein was eager to learn. He studied on his own from what books he could obtain, and his unusual interests and probing questions soon caught the attention of the faculty.

After receiving his degree in 1901 at the age of twenty-two, Einstein expected to obtain a university position as assistant to an established faculty member. However, for reasons that were never made clear to him, such a position did not materialize. After intermittent employment as a tutor, he finally found a job in the Swiss patent office, examining, evaluating, and rewriting patent applications. During his tenure as a patent clerk Einstein began the serious work in physics that was to earn him the reputation as the most creative scientific mind since Newton.

As is so often the case with significant conceptual advances, Einstein's approach to electromagnetism and light was distinguished from that of his predecessors such as Lorentz by the different questions he asked. Lorentz had proposed his theory in 1904 at the age of fifty-one, after he had become recognized as one of the foremost physicists of his time. He had spent much of his life trying to reconcile his conception of the ether with the experimental evidence concerning light, and to him, as to most other physicists at the turn of the century, the puzzling question in this regard was, "Why does it not seem possible to detect the motion of the earth relative to the ether?"

Einstein, on the other hand, first approached the subject as a young man in his early twenties. He was then an unknown physicist, largely self-taught in this area of physics, and in his patent office job somewhat cut off from the discussions and ideas current in the physics community. He was apparently only vaguely aware of the ether experiments discussed in Chapter 12, and he did not know of the later papers of Lorentz and Poincaré. At least partly for these reasons, his view of the problem was significantly different. For him the fundamental question was, "Why should the laws of electromagnetism and light, alone among the laws of physics, allow the possibility of detecting the motion of an inertial reference frame?"

Early in his study of physics, Einstein had been impressed with the remarkable generality of Galilean relativity—that is, with the apparent impossibility of detecting the uniform motion of an inertial reference frame within the domain of the laws of mechanics. He later wrote:

> From the beginning it appeared to me intuitively clear that, judged from the standpoint of such an observer [moving relative to the earth], everything would have to happen according to the same laws as for an observer who, relative to the earth, was at rest.[2]

In all physics, the only exceptions to this general principle seemed to be light and electromagnetism. That such exceptions should exist deeply troubled Einstein. He preferred to believe that the principle should also encompass the laws of light and electromagnetism. In particular, even with light one should not be able to discern the absolute motion of an inertial reference frame—light should propagate in the same way relative to every inertial frame.

Simultaneity

Another aspect of light intrigued Einstein, and seemed to undermine the long-held notion of an absolute time with respect to which all events can be specified. He noted that, of all known natural phenomena, light (or, more generally, electromagnetic radiation of any type) travels most rapidly. Further, while the speed of light is almost incredibly large, it is not infinite. Consequently, the communication of information between two separate points in space must take some finite time.

Einstein then asked a question so basic that it had apparently never occurred to anyone before. What does it mean to say that two events that take place at spatially separated points are *simultaneous*? If information could be propagated with infinite speed the answer would be clear. An observer at the position of one event could have instantaneous knowledge of the other event and hence could determine if they occurred at the same time. As remarked above, however, as far as is known, the maximum speed of propagation of information is the speed of light. Thus, one apparently cannot have instantaneous knowledge of an event at a spatially distant point and consequently one must examine the concept of simultaneity with care.

Einstein suggested an *operational definition* of simultaneity, specifying a sequence of operations that can be used to determine whether two events are simultaneous. Let us consider two particular events E_A and E_B. (For example, E_A and E_B might be two explosions.) We shall assume that E_A and E_B occur at two spatially separated points A and B respectively, and that some other point M is equidistant from A and B, as shown in Figure 13–1. Finally, we suppose that when E_A occurs a flash of light is sent from A toward M and when E_B occurs a flash of light is sent from B toward M. Then Einstein's definition can be stated

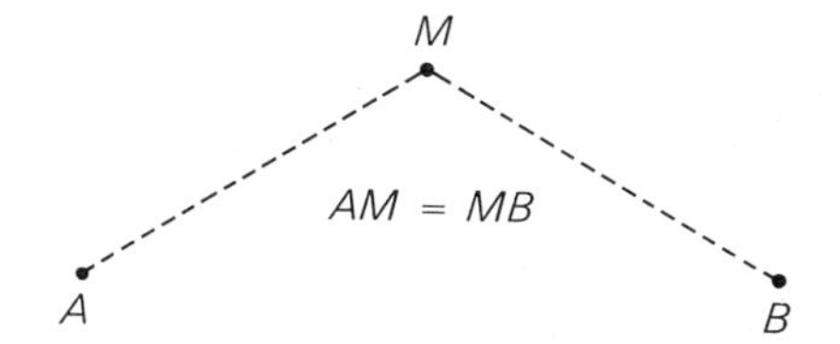

Figure 13–1. Point M is equidistant from A, the position of event E_A, and B, the position of event E_B. E_A and E_B are defined to be simultaneous if light signals sent out from A and B when E_A and E_B occur arrive at M at the same time.

as follows: *The events* $\mathrm{E_A}$ *and* $\mathrm{E_B}$ *are defined to be simultaneous if the light flash from* A *and the light flash from* B *reach* M *at the same time.**

This definition seems intuitively reasonable, although its appeal rests on the implicit assumption that the speed of light is independent of the direction of propagation. In the nineteenth-century view of light, this assumption is valid only in the rest frame of the ether. To Einstein, who felt that light should propagate in the same way in every inertial frame, the assumption should be valid in any such frame. However, it must be emphasized that, reasonable or not, Einstein adopted this as his *definition* of simultaneity—a statement of the logical equivalence of simultaneity and the operations specified. That is, in the theory to follow, when we say that two events at two separated points are simultaneous, we mean that if light signals were sent out when the events occurred they would reach an observer equidistant from the two points at the same time.

The Relativity of Simultaneity

Using Einstein's definition of simultaneity we immediately discover what at first seems a rather disturbing consequence: *Two events that are simultaneous when measured in one frame of reference*

* Notice that in this statement, the simultaneity of spatially separated events is defined in terms of the simultaneity of events that occur at the same point. The latter, sometimes called local simultaneity, would seem to present no difficulties.

are not in general simultaneous when measured in another frame of reference moving relative to the first.

To demonstrate this, Einstein conceived his famous train *gedanken* (thought) experiment. While one cannot perform this experiment in actual practice, so far as we know nothing is wrong with it in principle. Imagine a train traveling along a straight section of track with a speed u comparable to the speed of light c. We shall consider the two events shown in Figure 13–2: E_A is an explosion that leaves a record of where it occurred by burning marks at A' on the train and A on the embankment beside the tracks. E_B is another explosion that leaves burn marks at B' and B. Suppose there is an observer stationed at M', midway between A' and B' on the train and another observer stationed at M, midway between A and B on the embankment. Finally, we

Figure 13–2. Two explosions that are simultaneous with respect to the embankment are not simultaneous with respect to the train.

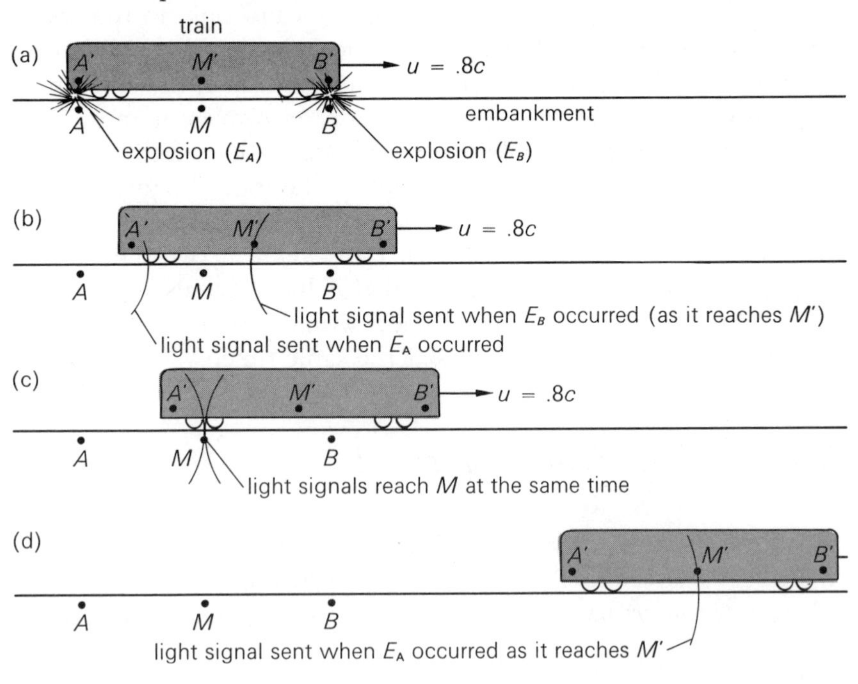

shall assume that the events E_A and E_B are simultaneous as measured in the S reference frame fixed to the embankment. This statement has a precise technical meaning given by Einstein's definition of simultaneity: a light signal emitted from A when E_A occurs and a light signal emitted from B when E_B occurs reach an observer at M, equidistant from A and B, at the same time.

Figure 13–2 illustrates the events as viewed from the S frame, corresponding to a train speed of $u = .8c$. In part (a) the explosions occur. Although not shown in the picture, light signals are sent out at this time toward point M. Part (c) shows that the light signals do indeed reach the point M, equidistant from A and B, at the same time. Hence, the events E_A and E_B are simultaneous as measured in the S frame.

Now let us transfer our attention to the S' frame fixed to the train. As seen in parts (b) and (d), the light signal emitted from B' when E_B occurred reaches the observer at M', equidistant from A' and B', *before* the light signal emitted from A' when E_A occurred. Hence, according to Einstein's definition, the events E_A and E_B are not simultaneous in the S' frame. In fact, we can say a bit more. Since the signal emitted when E_B occurred reaches M' before the signal emitted when E_A occurred, E_B occurred *before* E_A in the S' frame. Apparently two events that are simultaneous in one inertial reference frame are not simultaneous in another.

At this point, one might be tempted to say, "Of course the light signals didn't meet at M'; the train was moving to the right and M' simply moved forward with the train to meet the flash from B before the one from A caught up with it. That doesn't allow us to conclude that the two explosions weren't simultaneous." The trouble with this response is that it sees the simultaneity determination on the embankment as right because the embankment is at rest, while the simultaneity determination on the train is somehow wrong because the train is moving. But Einstein's definition of simultaneity applies to *any* inertial reference frame. Since light signals from the two events reach the point M, equidistant from the two events in the S frame, at the same time, the events are by definition simultaneous in the S frame. Since the light signals from the two events reach the point M', equidistant from the two events in the S' frame, at different times, the events are by definition not simultaneous in the S' frame.

Einstein realized that this result had profound implications for the notion of time itself. *The relativity of simultaneity directly precludes the existence of an absolute time independent of reference frame.* To see why this is true, let us return to the *gedanken* experiment. There we considered two explosions E_A and E_B that were simultaneous in the S frame (the rest frame of the embankment). That is, $t_A = t_B$. But according to the Galilean transformation, with its implicit notion of absolute time, the time of an event is independent of the reference frame in which the time is measured. Here, that means $t_A = t_A'$ and $t_B = t_B'$, which in turn implies that $t_A' = t_B'$—that is, the two explosions must also be simultaneous in the S' frame (the rest frame of the train). We know this to be false from the *gedanken* experiment. Hence the assertion of the Galilean transformation, $t' = t$, hitherto unquestioned—in fact usually unmentioned, as it was thought to be obvious—was called into question by Einstein. In fact the entire Galilean transformation, and hence our intuitive concepts of space and time, have to be modified to fit Einstein's view of the world.

Before discussing this modification, let us briefly reflect on why it took over two hundred years for this difficulty in Newtonian physics to become apparent. The primary reason is that prior to the twentieth century all the objects whose motion was studied had speeds far less than the speed of light. The effects Einstein uncovered are measurable only for objects moving with speeds comparable to c, such as the elementary particles encountered in present-day high-energy accelerator experiments. Indeed, Newtonian mechanics deals so successfully with such a wide range of phenomena that it is almost a necessary requirement of any new theory that it be consistent with the Newtonian results for velocities much less than c.

Of course, in our everyday environment, objects do not move with speeds approaching that of light, and our intuition is a product of our day-to-day experience. Hence, in dealing with objects moving with speeds near c, it is not obvious that the sequence of events can be visualized as a simple extrapolation of our experience. Our intuition may fail.

Postulates of the Theory

Einstein's analysis of the simultaneity of spatially separated events, and the resulting overthrow of absolute time, led to a new theory, now usually referred to as the *Special Theory of Relativity*.* The paper in which this theory was first presented, "On the Electrodynamics of Moving Bodies," appeared in 1905. In it, Einstein began by noting a simple example from electromagnetism. If a magnet is moved in the vicinity of a stationary electrical conductor, an electric current appears in the conductor. If, on the other hand, the magnet is stationary and the conductor is moved, exactly the same current appears. However, the explanations common at the time employed rather different concepts depending on whether magnet or conductor moved. Einstein continued:

> Examples of this sort, together with the unsuccessful attempts to discover any motion of the earth relative to the "light medium," suggest that the phenomena of electrodynamics as well as of mechanics possess no properties corresponding to the idea of absolute rest. They suggest rather that . . . *the same laws of electrodynamics and optics will be valid for all frames of reference for which the equations of mechanics hold good.* We will raise this conjecture (the purport of which will hereafter be called the "Principle of Relativity") to the status of a postulate, and also introduce another postulate, which is only apparently irreconcilable with the former, namely, that *light is always propagated in empty space with a definite velocity* c *which is independent of the state of motion of the emitting body*. These two postulates suffice for the attainment of a simple and consistent theory of the electrodynamics of moving bodies based on Maxwell's theory for stationary bodies. The introduction of a "luminiferous ether" will prove to be superfluous inasmuch as the view here to be developed will not require an "absolutely stationary space" provided with special properties . . .[3]

* As we shall see, Einstein later proposed a General Theory of Relativity.

Einstein's theory is particularly striking in its simplicity and generality. It flows in its entirety from the two postulates quoted above, which we may rephrase for future reference as follows:

I. *The Principle of Relativity: No physical measurement can distinguish one inertial reference frame from any other inertial reference frame.*

II. *The speed of light is the same in all inertial reference frames, independent of the speed of the light source.*

The first of these postulates is, of course, just that stated by Poincaré several years earlier. Its status in the minds of the two men, however, was quite different. To Poincaré, as we have seen, it was essentially an experimental conclusion. In the realm of terrestrial and celestial motion, it expressed the result that there seems to be no observation of motion that would allow one to decide which of two co-moving inertial reference frames is "really" moving and which is "really" at rest. In the realm of electromagnetism and light, it expressed the result that all attempts to locate the ether, a natural candidate for the inertial reference frame that was "really" at rest, had failed. Viewed in this light, the principle of relativity begs for explanation. Why is nature so perverse as to shield the ether from man's best efforts to find it?

Einstein, on the other hand, looked on the principle of relativity as a basic postulate about the world, a starting point for his theory. Like Aristotle's natural motion or Galileo's principle of inertia, it is so fundamental a principle that it need not or, perhaps, could not be explained.

In a sense, Poincaré and Lorentz had also anticipated the second postulate, the constancy of the speed of light. Here again, however, the postulate had a radically different interpretation in Einstein's theory. To determine the speed of light one could measure how long it takes light to travel a known distance. According to the Lorentz theory, any such measurement, performed in an inertial reference frame, would result in the same value c, which would be measured in the rest frame of the ether. That is, the speed of light will be *measured* to be the same in all inertial frames. The theory attributes this to distortions in meter sticks and clocks arising from their motion through the ether. The speed of light might be *measured* to be c in a reference frame moving relative to the ether, but it is *really* something different.

To Einstein, however, it seemed meaningless to discuss motion relative to the ether if that motion can never be detected. By the same token, it is also meaningless to say that the speed of light is really not c relative to a reference frame if one's measuring instruments always give the value c. On the basis of experience, man can apparently make no meaningful statement about the existence or nonexistence of the ether. Einstein therefore simply stated his second postulate without any discussion of an ether.

For Einstein, the second postulate simply asserted the general validity of Maxwell's theory, which predicted that light should travel in all directions with a speed c independent of the motion of the emitting object. If they are valid laws of physics, then according to the principle of relativity Maxwell's equations should have the same form in any inertial reference frame. Therefore, in any inertial reference frame light will have the same speed c, independent of the motion of the light source.

It should be noted that the failure of the various attempts to detect the ether is easily explained if Einstein's postulates are accepted. For example, the Michelson-Morley experiment was performed with an apparatus at rest in an inertial reference frame. The two arms were measured to have equal lengths. If the speed of light is c along both paths, as specified by the second postulate, then of course the two light pulses will arrive simultaneously at the detector. Again we have an illustration of the way one's theoretical framework determines which facts are significant. The Michelson-Morley result, which upset all of nineteenth-century physics, was natural to Einstein and thus hardly worthy of mention. As a matter of fact it was not mentioned at all in Einstein's 1905 paper on relativity.

Lengths Perpendicular to the Direction of Motion

From the two postulates, Einstein deduced a powerful and far-reaching theory. We begin our consideration of his results with a useful definition:

The *proper length* of an object is its length as measured in its rest frame—that is, in the reference frame in which the object is at rest.

We may use this definition to state the first consequence of the postulates: *The length of a moving object perpendicular to its direction of motion is equal to its proper length.* This is probably not a surprising statement since in our everyday experience the dimensions of objects do not seem to change when they move. However, we shall see that Einstein's theory does predict a change in the length of a moving object parallel to its direction of motion. It is therefore important to examine the other dimensions of the object.

Suppose we have two rods, 1 and 2, of the same proper length. One moves relative to the other with speed u, as shown in Figure 13–3. Let S and S' be the rest frames of the rods, and suppose one of the rods is outfitted at the top with a brush. As the two rods pass each other the brush will leave a dab of paint somewhere along the y-axis of the S frame. As a result of this procedure, observers at rest in the S frame now have a record of the length of the moving rod, 1, relative to their reference frame. In particular, if the dab of paint is above A, they can conclude that the length of the moving rod is greater than its proper length; conversely, if the paint is below A, the length of the moving rod is less than its proper length.

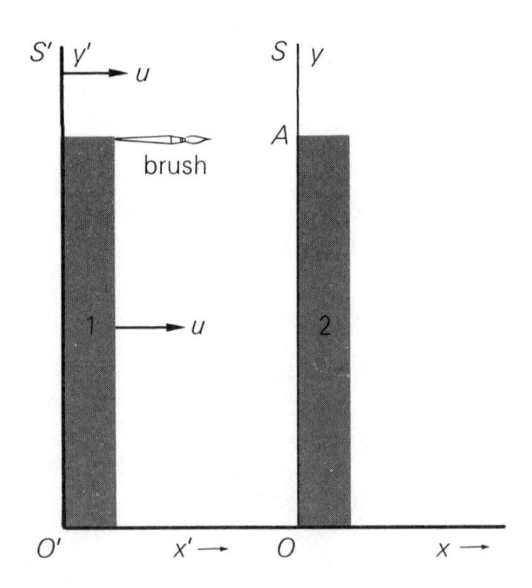

Figure 13–3. An experiment to determine the length of a moving rod perpendicular to its direction of motion.

It follows immediately from the principle of relativity, however, that the paint dab must coincide with A, implying that the length of the moving rod is equal to its proper length. Suppose this is not true—for example, suppose the paint mark is below A. An observer at rest in frame S, who sees rod 1 moving, could then formulate the physical law, "A moving rod shrinks in a direction perpendicular to its motion." According to the principle of relativity, however, this law must also be valid in the S' frame. But to an observer at rest in S', rod 1 is at rest and rod 2 is moving to the left with speed u. This observer, applying the same physical law, would reason that the moving rod 2 is shorter than rod 1 and hence the paint mark must be above A. But we began by supposing the mark to be below A. This assumption has led to a contradiction.

The assumption that the paint mark was above A would lead in like manner to a contradiction. Consequently we may conclude that if the first of Einstein's postulates, the principle of relativity, is valid, the mark must coincide with A and hence the length of a moving rod perpendicular to its motion is equal to its proper length.

Time Dilation

Another consequence of Einstein's postulates is time dilation, sometimes stated more dramatically as "moving clocks run slow." Again we begin with a definition:

> The *proper time* interval between two events is the time interval measured in the reference frame in which the two events occur at the same position. Time intervals between two events that occur at different positions are called *improper*.

The time-dilation result may then be stated formally as: The proper time interval between two events is less than any improper time interval between the same two events. In other words, relativity theory predicts that the time interval between two events will depend on the inertial reference frame in which the interval is measured and is least in that inertial frame in which the two events occur at the same place.

Figure 13–4. The mirror apparatus. A pulse of light reflects back and forth between the mirrors.

To help us prove that time dilation is a consequence of the postulates of relativity, we shall need the apparatus illustrated in Figure 13–4. It consists of two mirrors separated by a distance L. A light pulse is sent out from a point on one mirror, reflects from the second mirror, and arrives back at the original mirror, where it is detected. In the rest frame of the apparatus, the time it takes the pulse to traverse this path of total length $2L$ is clearly $2L/c$.

Now let us consider a *gedanken* experiment in which we measure the time between two events with respect to different inertial reference frames. For example, let the two events be two explosions that occur at different times at two locations on a laboratory table. (See Figure 13–5.)

To formalize our description of the events, consider a reference frame S, which is at rest relative to the laboratory and

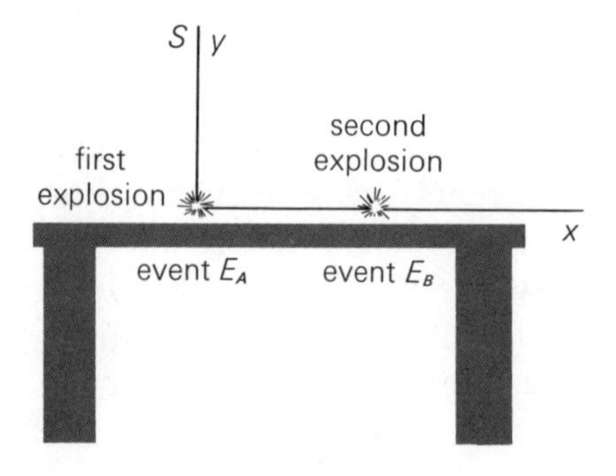

Figure 13–5. Two events, E_A and E_B. E_A occurs at the origin of the S frame at time $t = 0$, and E_B occurs at a later time with respect to the S frame.

whose origin, O, coincides with the location of the first explosion. We shall refer to the first explosion as event E_A, and the second explosion as event E_B.

Let us now imagine a second inertial frame S', which is moving relative to S in the $+x$ direction, as shown in Figure 13–6. By properly choosing the S' frame and its velocity u, we may arrange for the origin O' to coincide first with the location of event E_A when event E_A occurs, and then with the location of event E_B when event E_B occurs. In other words, both events occur at the same position (the origin O') in frame S'. Therefore, the time interval between the events, as measured in S', is a proper time interval.

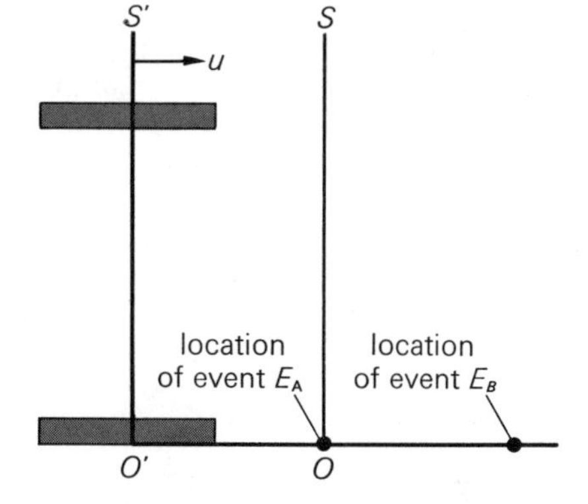

Figure 13–6. The S' frame moves with speed u in the $+x$ direction relative to the S frame of Figure 13–5.

Suppose that a mirror apparatus such as the one described above is at rest in S' with its base at O'. Thus the apparatus is moving relative to the laboratory in the $+x$ direction with speed u. Let us arrange to have a light pulse leave the lower mirror of the apparatus when the origins O and O' coincide (that is, when event E_A takes place). Finally, let us assume that the distance L between the mirrors has been adjusted so that the pulse returns to the base just as event E_B takes place. Figure 13–7 summarizes the situation.

In Figure 13–7(a) we see the sequence of events as observed in the S frame. Event E_A occurs as O and O' coincide. At this time the light pulse leaves the lower mirror of the apparatus. It takes a time Δt for the pulse to reflect off the upper mirror and return to the base. By this time the lower mirror has moved a distance $u\Delta t$ to the right and coincides with the position of event E_B when it occurs.

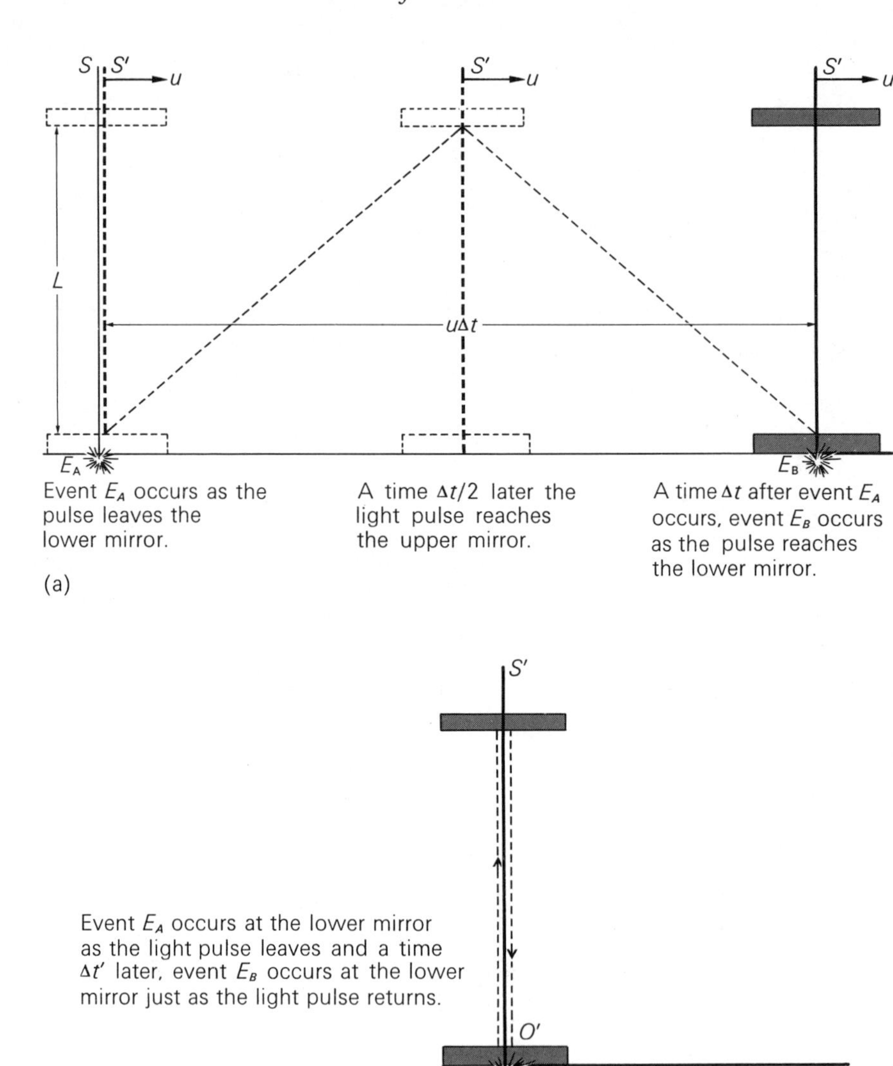

Figure 13–7. An experiment to compare the time interval between two events as measured in the S frame with the corresponding time interval as measured in the S' frame, in which the events occur at the same position. (a) The sequence of events as observed in the S frame. (b) The same sequence of events as observed in the S' frame.

Using this information together with Einstein's second postulate, we may calculate Δt, the time between events E_A and E_B, as measured in the S frame. Postulate II asserts that the

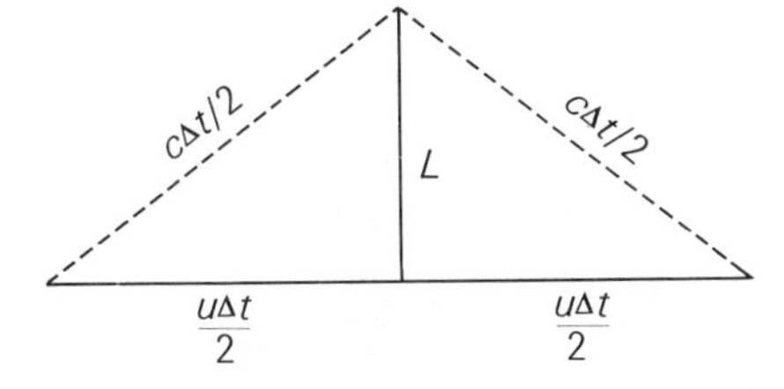

Figure 13–8. Some relevant distances from Figure 13–7(a).

speed of light in the S frame is c. Hence the total distance traveled by the light beam in Figure 13–7(a) is $c\Delta t$. Since the length of a moving object perpendicular to its direction of motion equals its proper length, the distance between the mirrors of the apparatus is still L in the S frame. Referring to Figure 13–7(a) and Figure 13–8, we may use the Pythagorean theorem to write:

$$\frac{c^2(\Delta t)^2}{4} = L^2 + \frac{u^2(\Delta t)^2}{4}$$

or, solving for Δt:

$$\Delta t = \frac{1}{\sqrt{1 - u^2/c^2}}\left(\frac{2L}{c}\right) \tag{13}$$

We now wish to compare this result, the time interval between events E_A and E_B as measured in the S frame, with the time interval $\Delta t'$ between the same events as measured in the S' frame. To determine $\Delta t'$, we must consider the sequence of events as observed in the S' frame, illustrated in Figure 13–7(b). In the S' frame the apparatus is at rest and events E_A and E_B both occur at the same place, the position of the lower mirror. Event E_A occurs as the light pulse leaves the lower mirror and event E_B occurs a time $\Delta t'$ later when it returns to this mirror. The light travels a distance $2L$ with speed c in time $\Delta t'$. Hence, $\Delta t' = 2L/c$. Comparing this result with equation (13), we see that:

$$\boxed{\Delta t' = \sqrt{1 - u^2/c^2}\,\Delta t} \tag{14}$$

Thus we conclude that *the time interval between the two events in the* S′ *frame, in which the two events occur at the same place, is shorter by a factor of* $\sqrt{1 - u^2/c^2}$ *than the time interval between the two events in the* S *frame, in which they occur at different places.* We may also note that in our derivation of equation (14), there was no reason why S had to be the laboratory frame. S could have been any inertial frame in which the events occur at different positions. We may therefore interpret equation (14) as relating the proper time interval between two events—the time interval as measured in the special inertial frame in which the events occur at the same place—to the time interval as measured in any other inertial reference frame, where u is the relative velocity of the two frames.

Experimental Tests of Time Dilation

Newtonian physics, with its implicit assumption of an absolute time, predicts no difference in the measurement of time intervals in different inertial frames. The time-dilation effect of equation (14) is thus a truly revolutionary prediction of relativity theory. The question immediately arises whether this effect can be observed.

To answer this question let us put in some numbers drawn from a "real-life" situation. Imagine a chicken farm of the not-too-distant future. Suppose that agricultural engineering has produced a desirable breed of hen which lays an egg that hatches in precisely ten seconds. It would be convenient to have the egg laid directly on a conveyor belt to be transported to the hatchery for hatching and care of the chicks. (See Figure 13–9.)

Figure 13–9. The egg is transported with speed u to the hatchery, where it hatches ten seconds after it was laid.

The engineers, never having studied relativity, do not realize that the egg, which hatches in a time $\Delta t = 10$ sec in its own rest frame, will remain unhatched longer in the rest frame of the farm. Need we fear for the success of their operation? Of course not. A typical conveyor belt speed might be 3 m/sec. The velocity of light is $c = 3 \times 10^8$ m/sec. Consequently, the time before the eggs hatch in the rest frame of the farm is

$$\Delta t' = \frac{10 \text{ sec}}{\sqrt{1 - \left(\dfrac{3}{3 \times 10^8}\right)^2}} = \frac{10 \text{ sec}}{\sqrt{1 - 10^{-16}}}$$

$$= 10.0000000000000005 \text{ sec}$$

which is negligibly different from 10 sec. One doesn't have to know relativity to run a chicken farm!

The point is that in this example, indeed in all everyday human experience, one deals with objects moving with relative velocities u so small compared with the speed of light c that $\sqrt{1 - u^2/c^2}$ is negligibly different from 1. Hence, we are not ordinarily aware of time-dilation effects.

Only in the study of elementary particles—electrons, protons, and other sub-microscopic constituents of atoms—does one encounter objects moving so fast that relativistic time dilation can be tested experimentally. Here the prediction of equation (14) has been corroborated to a high degree of accuracy.

A striking demonstration of time dilation concerns the lifetime of μ mesons in cosmic rays. Cosmic rays are the sub-atomic particles that continually rain down on the surface of the earth. These particles are apparently produced when high-energy protons of extra-terrestrial origin* crash into the outer portion of the atmosphere, some ten thousand meters above the surface of the earth. When such a proton collides with a molecule of the atmosphere, many particles are created, among them μ mesons.** These μ mesons have a very high energy and move with a speed nearly that of light.

* The source of the high-energy protons is uncertain, although there has been speculation that they are flung off from the recently discovered, rapidly rotating stars known as pulsars.

** The name meson comes from the Greek word for middle; the μ meson has a mass between that of the very light electron and the heavier proton.

A useful characteristic of the μ meson for the present discussion is its short lifetime. When a μ meson is created at rest, it lives for approximately 2×10^{-6} sec and then decays radioactively, transforming into other elementary particles. We might think of the creation of a μ meson as analogous to the laying of a 2×10^{-6} sec hatching-time egg in the chicken-farm example.

If a μ meson created at the top of the earth's atmosphere lives for only 2×10^{-6} sec and travels toward the earth with a speed of nearly $c = 3 \times 10^8$ m/sec, Newtonian physics would predict its range (the distance it travels before decaying) to be:

$$\text{range (Newtonian)} = 2 \times 10^{-6} \text{ sec} \times 3 \times 10^8 \text{ m/sec}$$
$$= 600 \text{ m}$$

Hence a Newtonian physicist would not expect the μ mesons, created at a height of 10,000 m, to reach the ground. In fact, however, a large number of μ mesons are observed at the surface of the earth. They shower down upon us and pass through our bodies, constituting one part of the cosmic-ray background radioactivity to which we are subjected all our lives.

That so many μ mesons do reach the earth's surface can be accounted for very simply as a time-dilation effect. While in its own rest frame the meson lives a time $\Delta t = 2 \times 10^{-6}$ sec, in the rest frame of the earth, in which the meson is moving rapidly, the time between its creation and decay will be longer, namely

$$\Delta t' = \frac{2 \times 10^{-6} \text{ sec}}{\sqrt{1 - u^2/c^2}}$$

where u is the speed of the meson relative to the earth. While the speeds u vary rather widely depending on the details of the reaction in which the meson was created, let us take a typical value, $u = .9992c$. In this case the lifetime of the meson in the earth's rest frame is:

$$\Delta t' = \frac{2 \times 10^{-6} \text{ sec}}{\sqrt{1 - (.9992)^2}} = \frac{2 \times 10^{-6} \text{ sec}}{\sqrt{1 - .9984}} = \frac{2 \times 10^{-6} \text{ sec}}{\sqrt{.0016}}$$
$$= 5 \times 10^{-5} \text{ sec}$$

Hence, according to the theory of relativity, the expected range of this μ meson is not 600 m, but rather:

$$\text{range (relativistic)} = 5 \times 10^{-5}\ \text{sec} \times 3 \times 10^{8}\ \text{m/sec}$$
$$= 15{,}000\ \text{m}$$

This accords with the observation of μ mesons at the earth's surface.

"Moving Clocks Run Slow"

Finally, let us remark on the sense in which time dilation implies that moving clocks run slow. Suppose we are given a clock that ticks every second in its rest frame. If this clock is set in motion with speed u relative to the laboratory, then the time between successive ticks as measured in the laboratory will be greater than one second, namely $1/\sqrt{1 - u^2/c^2}$ sec. This conclusion follows directly from the time-dilation result, equation (14).

Let us take S to be the rest frame of the laboratory and S' to be the rest frame of the clock. Suppose we focus on two successive ticks of the clock. Figure 13–10(a) depicts these two events as observed in the S frame while Figure 13–10(b) depicts

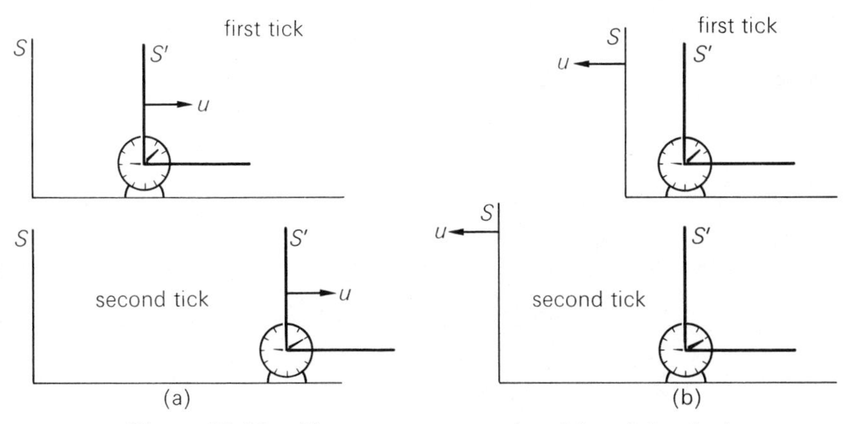

Figure 13–10. Two events, successive ticks of the clock, as observed in the laboratory and in the rest frame of the clock. (a) Successive ticks occur at different points in the S frame, the rest frame of the laboratory. (b) Successive ticks occur at the same point in the S' frame, the rest frame of the clock.

the same two events as observed in the S' frame. Clearly the two events take place at the same point in the S' frame and hence the time interval $\Delta t' = 1$ sec is a proper time interval. The two events occur at different positions in the S frame and the corresponding time interval Δt is improper. According to equation (14),

$$\Delta t = \frac{\Delta t'}{\sqrt{1 - u^2/c^2}} = \frac{1 \text{ sec}}{\sqrt{1 - u^2/c^2}}$$

which is greater than one second. Thus observers in the S frame note that in a time greater than one second, the moving clock shows only a one-second advance. That is, the moving clock runs slow.

To understand the origin of this effect, it may help to examine a particularly simple kind of clock—the mirror apparatus described previously. Let the first "tick" of the mirror clock be the event "pulse leaves lower mirror" and the second "tick" the event "pulse returns to lower mirror." Figure 13–11(a) illustrates these two ticks of the mirror clock as viewed in the S' frame, the rest frame of the clock. If the distance between the mirrors is L, the time between ticks in the S' frame is $\Delta t' = 2L/c$. Figure 13–11(b) shows the same two successive ticks as observed in the S frame, in which the clock is moving. The path the light beam traverses is clearly greater than $2L$, and according to the second postulate the speed of light is also c in the S frame. Following an

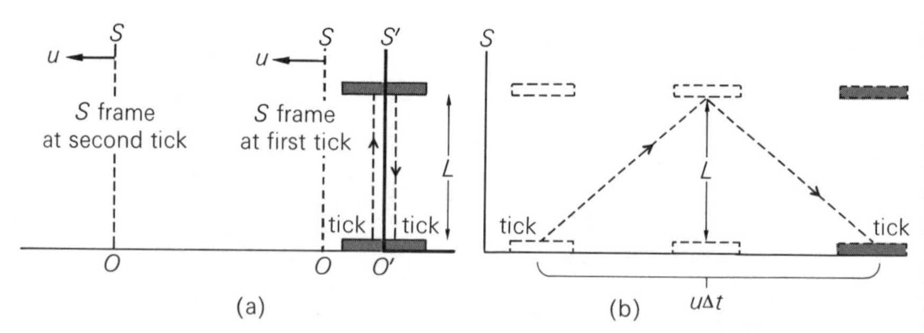

Figure 13–11. The mirror clock as viewed in the laboratory and in the rest frame of the clock. (a) Two successive ticks as viewed in the S' frame, the rest frame of the mirror clock. (b) The same two ticks as viewed in the S frame, the rest frame of the laboratory.

argument identical to that which led to equation (14), it is easy to show that the time between successive ticks as measured in the S frame is

$$\Delta t = \frac{2L}{c} \frac{1}{\sqrt{1 - u^2/c^2}} = \frac{\Delta t'}{\sqrt{1 - u^2/c^2}}$$

—that is, longer than $\Delta t'$.

It might be objected that we have considered only a very special kind of clock. How can we be sure that an ordinary clock would exhibit this effect? Suppose we build two identical ordinary clocks and two identical mirror clocks, and adjust them so that every time the ordinary clocks tick the mirror clocks tick. One ordinary clock and one mirror clock are left at rest in the laboratory while the other two clocks are transported at a high speed relative to the laboratory. The fact that the mirror clock and the ordinary clock remain synchronized in the laboratory implies that the other two clocks must also remain synchronized. If the other two clocks did not remain synchronized, this would provide a physical measurement that could distinguish between the two inertial reference frames—a violation of the principle of relativity.

Thus, relativity theory predicts that *any* moving clock runs slow. In fact this assertion is completely equivalent to the assertion of the time-dilation effect. We can say the moving meson lives longer because of time dilation or we can consider the meson to be a primitive clock that ticks only twice. The first "tick" occurs when it is created and the second "tick" occurs when it decays 2×10^{-6} sec later in its rest frame. In any other inertial frame the meson clock runs slow, in the sense that the time interval between the first tick (creation) and the second tick (decay) is longer than 2×10^{-6} sec.

Length Contraction—A Qualitative Analysis

Another important consequence of the postulates of special relativity is the phenomenon of *length contraction*. This result may be stated formally as follows: The length of an object parallel to its direction of motion is less than its proper length. In other words, Einstein's theory predicts that in general the length of an object will depend on the inertial frame in which the length is measured, and is greatest in the inertial frame in which the object is at rest.

A moment's reflection about how one would measure the length of a moving rod may make this conclusion less surprising. The most straightforward way would involve determining the positions of both ends of the rod *at the same time*. For example, suppose the rod is at rest in the S frame and has a proper length L as shown in Figure 13–12. Observers at rest in an S' frame, moving in the $+x$ direction with speed u, undertake to measure the length L' of the rod relative to S'. With respect to these observers, the rod is moving in the $-x'$ direction with speed u (Figure 13–13).

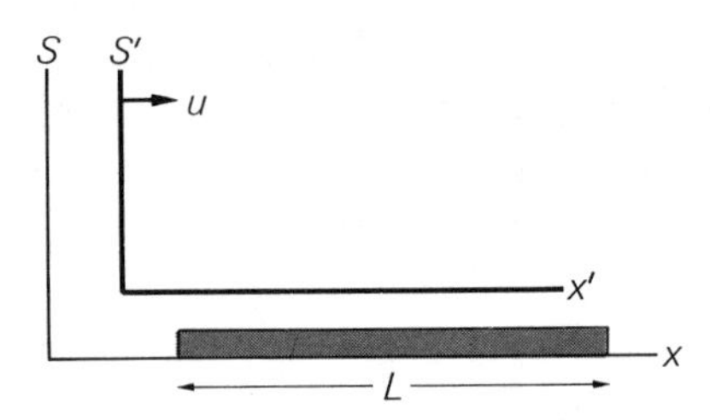

Figure 13–12. The rod is at rest in the S frame. Its length as measured in this frame is its proper length, L.

Figure 13–13. The rod is moving relative to the S' frame. Its length as measured in this frame is L': that is, the ends of the rod pass observers at A' and B', a distance L' apart, at the same time according to S' clocks.

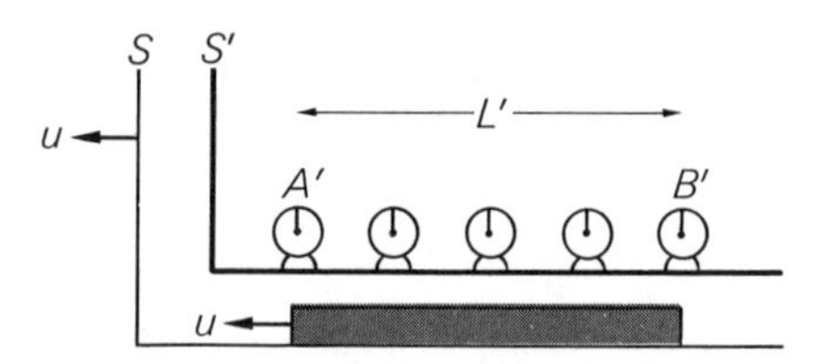

The S' observers distribute themselves along the x' axis, each seated expectantly with his own clock. Instructions are sent out: "At 12:00 noon, look to see if either end of the rod is passing by you. If it is, raise your hand." The rod comes flying along the x' axis, the clocks strike 12:00 noon, and two observers A' and

B' raise their hands. (Figure 13–13.) At their leisure, the other S' observers measure the distance between A' and B' and announce this to the expectant world as the length of the moving rod.

However, the success of this procedure requires that the S' clocks all be synchronized—that is, that they be set to strike 12:00 simultaneously in the S' frame. We have already seen that spatially separated events that are simultaneous in one inertial frame, such as the striking of noon by two of the S' clocks, are not simultaneous in another inertial frame. Thus the two clocks of observers A' and B', synchronized in S', will not appear to strike noon at the same time in the S frame. This leads to the length-contraction phenomenon, as we shall now see.

How might A' and B' have synchronized their clocks initially? A possible method would be just the inverse of the process described earlier to judge the simultaneity of two events: From the point M' halfway between the positions of the two clocks, a flash of light is set off. It travels out toward A' and B' with speed c relative to the S' frame. When the flash reaches each clock, it is started; the two clocks will then be synchronized in the S' frame [Figure 13–14(a)].

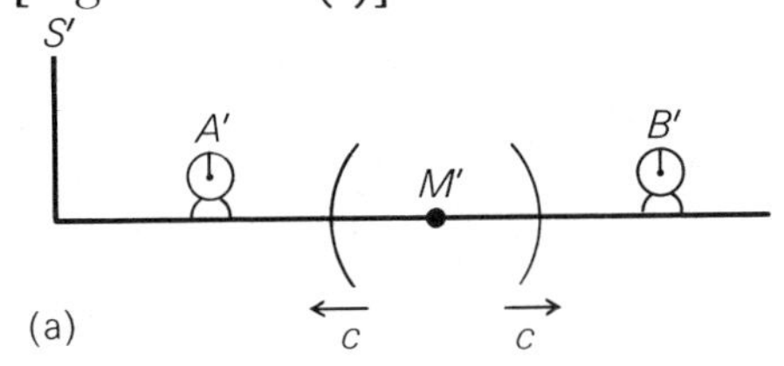

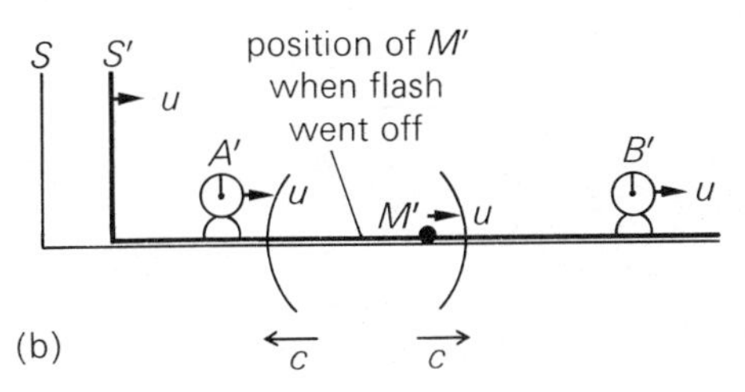

Figure 13–14. The synchronization of the S' clocks. Each clock, A' and B', is started when the light from the synchronizing flash reaches it. (a) According to S' observers, the clocks are started simultaneously. (b) According to S observers, clock A' is started before clock B'.

Let us consider the same synchronization from the point of view of observers at rest in the S frame [Figure 13–14(b)]. The synchronizing flash, according to Einstein's second postulate, also travels at speed c in the S frame. But clocks A' and B' move toward the right with speed u. A' moves toward the flash and hence is struck first; B' moves away from the flash and hence is struck later. Thus to observers at rest in S, clock A' was started before clock B'.

Now let us examine the above measurement of the length L' of the moving rod by the S' observers as it will appear to observers at rest in the S frame. Since clock A' was started before clock B', according to the S observers, the event "left end of rod coincides with clock A' when A' reads noon" occurs first, as shown in Figure 13–15(a). Clock B' does not yet read noon. Time passes and the S' frame with its clocks moves to the right [Figure 13–15(b)]. Then "noon" appears on clock B' as it passes the right end of the rod [part (c)]. But since the S' frame

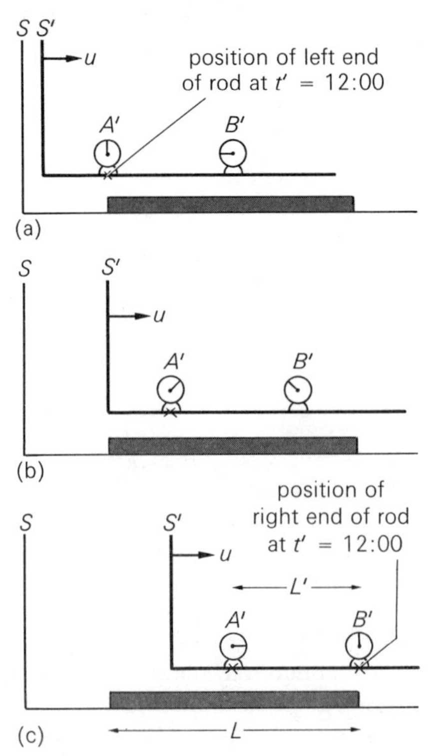

Figure 13–15. The measurement of the length of a rod by observers at rest in the S' frame, shown with respect to the S frame.

has moved to the right in the meantime, the distance between A' and B'—the length of the rod L' as measured in S'—is less than the rod's proper length L. Thus the relativity of simultaneity results in the measurement of a contracted length for a moving rod.

Length Contraction—A Quantitative Analysis

We could determine the exact amount of this contraction directly from the above argument; however, a less direct method will prove considerably simpler. Another way to measure the length of the moving rod in S' is to take a single clock at rest in S' and note the time $\Delta t'$ it takes the rod to go by. Knowing the rod's speed u, we can easily calculate its length. Figure 13–16 shows the situation with respect to both the S and S' frames. With respect to S' the clock is at rest and the rod is moving with speed u in the $-x'$ direction. With respect to S, the rod is at rest and it is the clock that is moving with speed u in the $+x$ direction.

The time $\Delta t'$ in the S' frame between the event "left end of rod passes clock" and the event "right end of rod passes clock" is measured [Figure 13–16(a)]. We can then calculate

$$L' = u\Delta t' \tag{15}$$

where L' is the length to be determined. Now let us consider the same two events as measured in the S frame [Figure 13–16 (b)]. Again, the time interval Δt between them is related to the length L of the rod by

$$L = u\Delta t \tag{16}$$

But our previous result concerning time dilation allows us to compare $\Delta t'$ (a proper time interval, since it refers to events occurring at the same position in S') with Δt:

$$\Delta t' = \Delta t\,\sqrt{1 - u^2/c^2}$$

Substituting this in equation (15), we find

$$L' = u\Delta t\,\sqrt{1 - u^2/c^2}$$

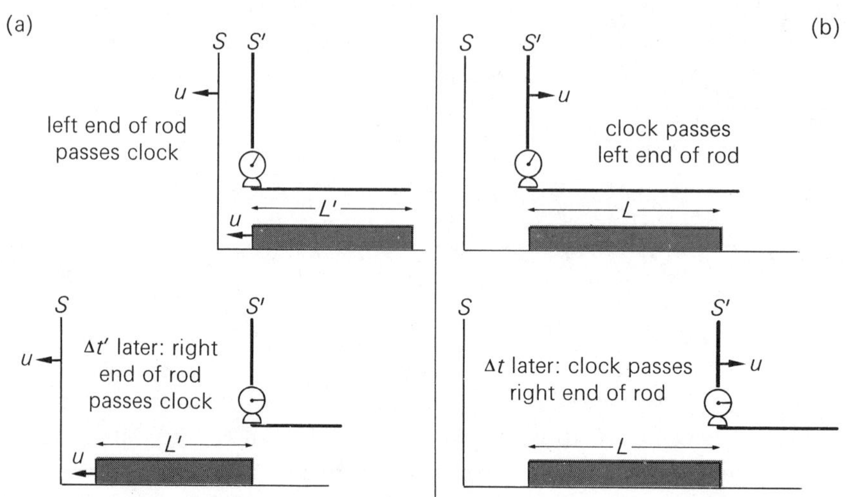

Figure 13–16. A measurement of length contraction. The length L' of a rod moving with speed u is measured by observing the time $\Delta t'$ it takes to pass a clock. If u and $\Delta t'$ are measured, the contracted length can be determined from $L' = u\,\Delta t'$. (a) Relative to the S' frame the clock is at rest and the rod, which has length L', is moving in the $-x'$ direction with speed u. It takes a time $\Delta t' = L'/u$ for the rod to pass the clock. (b) Relative to the S frame the rod, which has length L, is at rest and the clock is moving in the $+x$ direction with speed u. It takes a time $\Delta t = L/u$ for the clock to pass the rod.

But equation (16) says that $u\Delta t$ is just L. Therefore,

$$\boxed{L' = L\sqrt{1 - u^2/c^2}}$$

This is the length-contraction result. *The length of the rod as measured in the* S′ *frame where it is moving with speed* u *is less than its proper length by a factor of* $\sqrt{1 - u^2/c^2}$.

Notice that this is exactly the contraction assumed by Lorentz and FitzGerald to explain the null result of the Michelson-Morley experiment. They attributed it to changes in the internal forces in the rod brought about by motion through the ether. In Einstein's theory, only the interpretation has changed, but in a radical way. To Lorentz, the contraction was an effect like the contraction of a bar of iron when it is cooled. To Einstein, the effect was a consequence of the relativity of simultaneity.

Experimental Evidence for Length Contraction

Just as for time dilation, support for the length-contraction result is provided by observations of elementary particles in cosmic rays and those created artificially by high-energy accelerators. In fact, since the length-contraction phenomenon is so closely related to time dilation, the same experiments that support time dilation can also be interpreted as evidence for length contraction.

For example, let us consider the μ mesons discussed previously. Recall that a μ meson created 10,000 m above the earth and having a lifetime in its rest frame of 2×10^{-6} sec can reach the surface of the earth, although even with speed $c = 3 \times 10^8$ m/sec it will only travel 600 m in 2×10^{-6} sec. Let S' be the rest frame of the meson and S the rest frame of the earth, as shown in Figure 13–17. One way to explain this phenomenon

Figure 13–17. The arrival of μ mesons at the earth's surface can be explained as a length-contraction effect. (a) As viewed in the S frame fixed to the earth, a moving meson is created a distance 10,000 m above a stationary earth. (b) As viewed in the S' frame fixed to the meson, a stationary meson is created a length-contracted distance $\sqrt{1 - u^2/c^2} \times 10{,}000$ m above a moving earth.

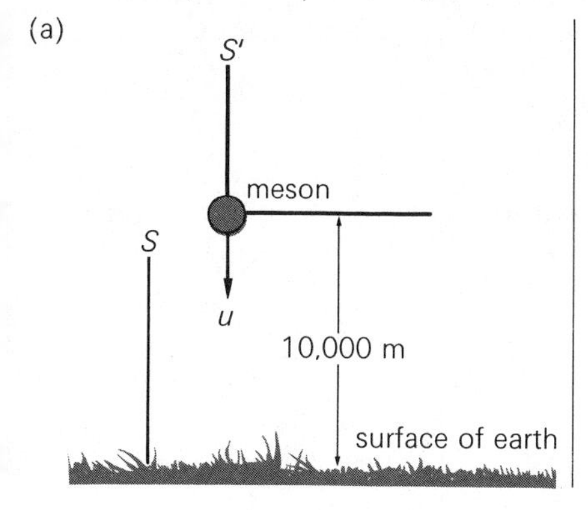

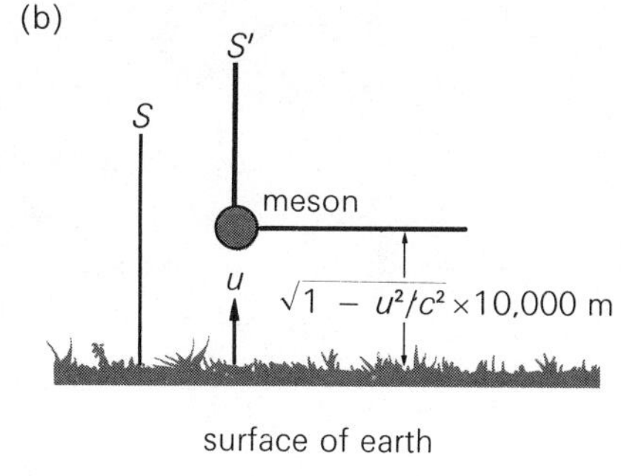

is to consider the situation from the S' frame—that is, from the point of view of the meson—as illustrated by Figure 13–17(b). In this frame the meson is stationary and the earth is moving toward it with velocity u. That the meson survives to contact the earth may be explained by the fact that in the S' frame the distance from meson to earth is not 10,000 m, but rather the length-contracted distance $\sqrt{1 - u^2/c^2} \times 10{,}000$ m. If u is sufficiently large, the earth will reach the meson before the meson decays.

Reception of the Theory

The relativity of simultaneity, the contraction of moving objects, and the slowing of moving clocks are important consequences of the theory of relativity that are incompatible with the Newtonian concepts of absolute space and time. These Newtonian concepts are so much a part of our intuitive perception of the world that they can be relinquished only with great difficulty. In fact, Einstein's theory by no means gained the immediate acceptance of the entire community of physicists. A debate over the theory of relativity and its peculiar consequences raged for years. The focus of the debate ranged from philosophical and aesthetic judgments of the theory to technical arguments concerning possible alternative hypotheses or conflicting interpretations of experimental evidence.

In 1911, for example, William Francis Magie, president of the prestigious American Physical Society, devoted a major portion of his presidential address to a critical evaluation of the philosophical implications of the relativity theory. He concluded:

> . . . I cannot see in the principle of relativity the ultimate solution of the problem of the universe. A solution to be really serviceable must be intelligible to everybody, to the common man as well as to the trained scholar. All previous physical theories have been thus intelligible. Can we venture to believe that the new space and time introduced by the principle of relativity are either thus intelligible now or will become so hereafter? A theory becomes intelligible when it is expressed in terms of the primary concepts of force, space,

> and time, as they are understood by the whole race of man. When a physical law is expressed in terms of those concepts we feel that we have a reason for it, we rest intellectually satisfied on the ultimate basis of immediate knowledge. Have we not a right to ask of those leaders of thought to whom we owe the development of the theory of relativity, that they recognize the limited and partial applicability of that theory and its inability to describe a universe in intelligible terms, and to exhort them to pursue their brilliant course until they succeed in explaining the principle of relativity by reducing it to a mode of action expressed in terms of the primary concepts of physics?[4]

Other physicists, however, were struck by the beauty and elegance of Einstein's theory, particularly in its later development by Einstein and his colleague Herman Minkowski, and embraced the theory on these grounds. For example, a leading German physicist, Wilhelm Wien, wrote of relativity in 1909:

> What speaks for it most of all, however, is the inner consistency which makes it possible to lay a foundation having no self-contradictions, one that applies to the totality of physical appearances, although thereby the customary conceptions experience a transformation.[5]

While such philosophical and aesthetic judgments as those of Magie and Wien cannot by themselves determine the fate of a theory, they are important in directing the course of science. Individual scientists often decide on the basis of such judgments whether to spend years of their lives trying to modify a theory or building elaborate equipment to compare a theory's predictions with experiment. Supposedly non-scientific considerations like these clearly must play a significant role in science performed by people.

The Emission Hypothesis

Many physicists also questioned Einstein's theory on technical grounds. A particular source of controversy was the second postulate—that the speed of light is the same in all inertial reference frames, independent of the speed of the source. After

1905, this postulate was frequently called into question by those who wished to modify the theory.

For example, in 1910 Richard C. Tolman, writing in the *Physical Review*, said of the second postulate:

> This is the assumption which has forced the theory of relativity to its strange conclusions . . .
>
> A simple example will make the extraordinary nature of the second postulate evident.
>
> S is a source of light and A and B two moving systems. A is moving towards the source S, and B away from it. Observers on the systems mark off equal distances aa' and bb' along the path of the light and determine the time taken for light to pass
>
> S * a A a' b B b'
>
> from a to a' and b to b' respectively. Contrary to what seem the simple conclusions of common sense, the second postulate requires that the time taken for the light to pass from a to a' shall measure the same as the time for the light to go from b to b'. Such a consideration makes the path obvious by which the theory of relativity has been led to strange conclusions as to the units of length and time in a moving system.[6]

Tolman and, independently, several other physicists discussed an alternative hypothesis that might replace the second postulate. Instead of assuming that the velocity of light is c independent of the velocity of its source, one might assume that the velocity of light is c plus the velocity of the source. Thus if a source moving with speed v emits light in the direction it is moving, the light would have speed $c + v$. If it emits light in the opposite direction, that light would have speed $c - v$. (Figure 13–18.)

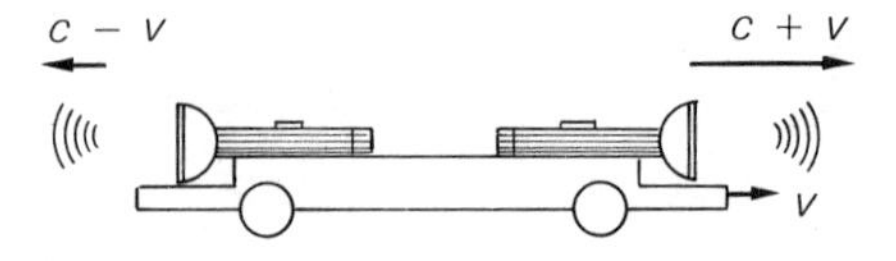

Figure 13–18. The emission hypothesis. A light pulse emitted in the direction a source is moving travels with speed $c + v$, while a light pulse emitted in the direction opposite to the motion of the source has speed $c - v$.

In fact, this *emission hypothesis* is just what one would expect if light consisted of particles emitted from a source like bullets from a gun. Let us imagine two cannons on a railroad car moving with speed v as shown in Figure 13–19. Suppose the cannonballs are fired with a speed c relative to the cannons. In other words, if the cannons were at rest the cannonballs would have speed c. However, due to the motion of the cannons, the cannonball shot in the direction of their motion will have speed $c + v$ relative to the ground and the cannonball shot in the opposite direction will have speed $c - v$ relative to the ground. This is just the behavior the emission hypothesis proposed for light.

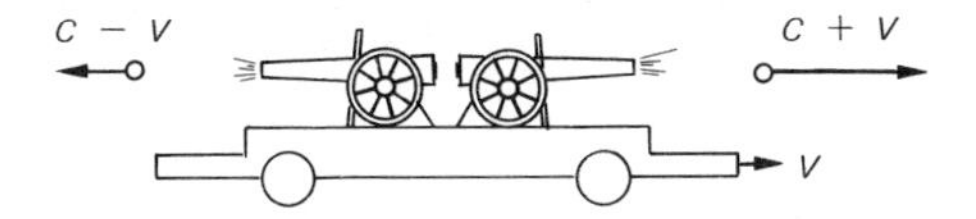

Figure 13–19. A cannonball shot in the direction the cannon is moving travels with speed $c + v$, while a cannonball shot in the direction opposite to the motion of the cannon has speed $c - v$.

Another statement of the emission hypothesis is that light travels with speed c in the rest frame of its source (rather than in all inertial reference frames, as claimed by the second postulate). An attractive feature of this hypothesis is that it leads to predictions in agreement with many of the experiments cited as support for the relativity theory, yet allows one to retain the ideas of absolute space and time. For example, consider the Michelson-Morley experiment. In the rest frame of the apparatus, the light source and the mirrors, which act like sources when the light reflects from them, are all at rest. Therefore, according to the emission hypothesis, the light pulses moving along both paths should move with the same speed c and no time difference is expected. This, of course, is just what Michelson and Morley found.

One objection to the proposed hypothesis was that it is inconsistent with Maxwell's theory of electromagnetic radiation. However, in 1908, Walther Ritz developed an emission theory of electromagnetic radiation, modifying Maxwell's theory and incorporating the emission hypothesis. In Ritz' theory light consists of particles whose velocity is additive with the velocity of the source.

However, some observations were difficult to reconcile with the emission theory. Perhaps the most important of these was pointed out by Willem de Sitter in 1913. He called attention to the motions of certain double-star systems in our galaxy, known as *eclipsing binaries*. We mentioned in Chapter 10 the discovery of double (binary) stars, in which one star is observed to move with respect to the other in apparent agreement with Kepler's (and thus Newton's) laws. A few of these double-star systems are oriented so that the plane of their orbits, when extended, contains the earth; that is, we see their orbits edge on. (See Figure 13–20.) Thus one star alternately eclipses and is eclipsed by the other.

The importance of such systems for the emission theory is that at the extremes of the orbit as seen from the earth, points A and B, the moving star is traveling directly toward (A) or away from (B) the earth. Thus, according to the emission theory, the light emitted at point A will travel faster toward the earth than the light emitted at point B. This would give rise to a rather strange apparent motion of the moving star as seen from the earth. In particular, if the earth is sufficiently far away, the light from A might arrive at the earth simultaneously with the slower light from B emitted a half-orbit earlier. The star would appear to be at both positions at once! Effects like this are not observed. The fact that the motion of the orbiting star appears to obey Newton's laws seems to indicate that we are getting a true picture, and that the motion of the star does not affect the speed of the light emitted by it.

De Sitter's argument was an important element in the rejection of the emission theory as an alternative to Einstein's second postulate. The absence of peculiar double-star effects was generally thought to disprove the emission theory. There is an ironic footnote to this account: recently, some fifty years later, the significance of the double-star observations has been questioned. There is now evidence that the stars in a binary system are surrounded by a gaseous envelope, which does not rotate

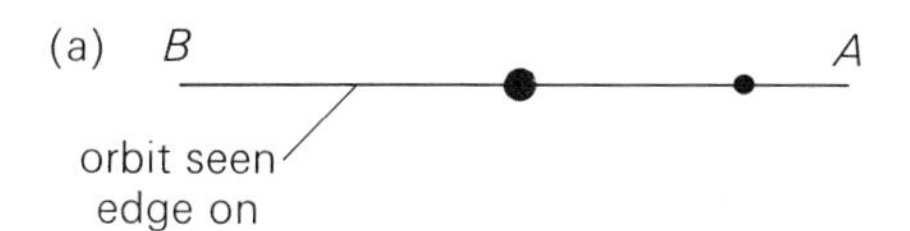

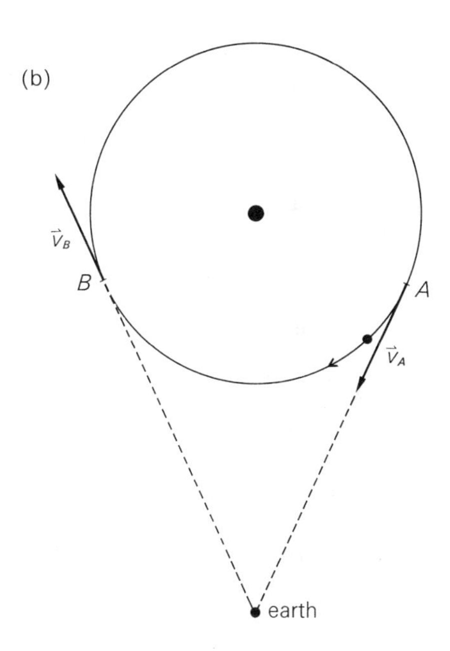

Figure 13–20. An eclipsing binary system. At point *A*, the small star moves directly toward the earth; at point *B*, it moves directly away. According to the emission hypothesis, the light emitted when the star is at *A* has a greater speed relative to the earth than the light emitted when the star is at *B*. (a) View from the earth. (b) View from above.

with the stars. It has been speculated that the light emitted by the stars is absorbed and then re-emitted by the surrounding gas. If so, the emission theory would predict that the light reaching the earth from each star would have a velocity c relative to the gas, not the star, and the effect predicted by de Sitter would not be seen even if the emission theory were correct. Perhaps the double-star observations did not disprove the emission theory after all.

Whether or not the emission theory was dismissed prematurely is really academic. As we shall see, there is now considerable experimental evidence in support of Einstein's theory. Moreover, relativity has held such a central role in physics for so long that it would be difficult to find a physicist today who felt it worthwhile devoting a significant effort to reviving the emission theory.

The Miller Experiments

In the years immediately following 1905, there was very little direct experimental evidence providing clear-cut support for Einstein's theory. In the minds of many physicists, the most convincing such evidence was the Michelson-Morley experiment. Regardless of its influence on Einstein when he formulated the theory, there is no doubt that the puzzling result of Michelson and Morley did establish a climate of opinion in which physicists were receptive to the theory of relativity. But this experiment was not decisive. We have mentioned, for example, that the emission theory of Ritz or the mechanical contraction of objects proposed by FitzGerald and Lorentz would also account for Michelson's and Morley's negative result.

There was also the lingering suspicion that there might be some flaw in the Michelson-Morley experiment that could lead to a less profound interpretation. Physicists continued to wonder whether there had been an error in experimental technique or whether something had been overlooked in the analysis. Such skepticism is natural when the alternative is a profound upheaval.

The repetitions of the Michelson-Morley experiment by Morley and Miller in 1902–1904, carried out at the urging of Lorentz, confirmed the original negative result. However, their experiment, like that of Michelson and Morley, was performed in a basement laboratory. Morley and Miller concluded the paper reporting their results with the statement:

> Some have thought that this experiment only proves that the ether in a certain basement room is carried along with it. We desire therefore to place the apparatus on a hill to see if an effect can be there detected.[7]

Soon afterward, Morley and Miller did move their apparatus from the basement laboratory to a hill on the outskirts of

Cleveland. It was placed in a building of very light construction with windows in the direction of the expected ether wind to preclude the possibility that the ether would be trapped in the room and carried along with the apparatus. Nevertheless, in experiments performed in 1905–1906, Morley's and Miller's results were consistent with a relative speed of earth and ether no larger than one-tenth that expected.

After Morley's retirement, Miller carried on. From 1921 to 1926, he performed a series of measurements on California's Mount Wilson, some six thousand feet above sea level. In 1925, he announced in his address as president of the American Physical Society that he had found a definite positive effect. He measured the speed of the earth relative to the ether to be 10 km/sec with an experimental uncertainty of .5 km/sec.

Miller's announcement triggered a new round of debate between the proponents of relativity, now in the majority, and the critics who yearned for a return to what Michelson referred to as "the beloved old ether (which is now abandoned, though I personally still cling a little to it)."[8] But the debate was short-lived. Others repeated the measurement with more sophisticated apparatus and found no such effect. It was suggested that despite the care with which Miller had performed his experiments, perhaps extraneous effects such as variations in temperature had influenced his results.*

Dependence of Mass on Velocity

One aspect of Einstein's theory was quickly subjected to an explicit experimental test. This was his prediction that the mass of a particle increases as its speed increases. Unfortunately, the derivation of the result is too complicated to discuss here. Suffice it to say that it follows from the two postulates of

* In 1955, R. S. Shankland and his colleagues re-analyzed Miller's data and claimed to be able to explain his results in terms of uncertainties in measurements and temperature variations that produced minute but significant expansions and contractions of his apparatus. [Shankland, R. S., *et al.*, *Reviews of Modern Physics*, **27** (1955): 167.]

relativity that the mass m of a particle moving with speed v is given by:

$$m = \frac{m_0}{\sqrt{1 - v^2/c^2}} \qquad (17)$$

where m_0 is the mass of the particle when it is at rest—the so-called rest mass.

Figure 13–21 shows what equation (17) says about the dependence of mass on velocity. For small velocities v, v^2/c^2 is much smaller than 1 and equation (17) indicates that m is approximately equal to m_0, the ordinary Newtonian value.

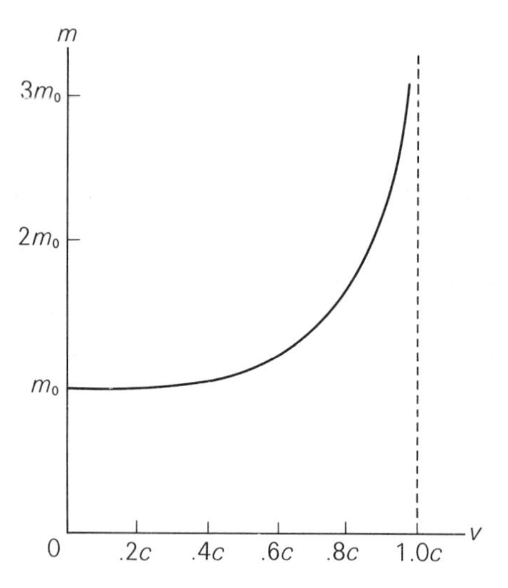

Figure 13–21. The dependence of mass on velocity, according to Einstein.

Only for velocities v very near the speed of light c does $\sqrt{1 - v^2/c^2}$ differ appreciably from 1 and the mass begin to increase significantly. Notice, in particular, that as v gets very close to c, $\sqrt{1 - v^2/c^2}$ approaches zero and m becomes infinitely large. As we have seen, m is a measure of the sluggishness of an object—its resistance to acceleration by a force. The significance of m approaching infinity as v approaches c is therefore evident. As the velocity of a particle approaches the velocity of light, its resistance to further acceleration (that is, to an increase in velocity) increases sharply. In fact, its speed can never reach c because this would require an infinitely large force.

Thus Einstein's theory predicts that the speed of a material particle must be less than the speed of light. Since the definition of simultaneity which lies at the heart of the theory assumes that light is the most rapid means of communication between two separate points, and hence that no particle can travel faster than light, this result means the theory is self-consistent.

Soon after Einstein published his 1905 paper, which included equation (17), this prediction was tested experimentally. First Walther Kaufmann in 1906 and later Alfred H. Bucherer in 1909 announced their measurements of the mass of electrons given off with varying velocities in the radioactive decay of radium. Once again the chain of events was more involved than the usual description of the scientific method would imply.

In his 1906 paper Kaufmann announced that his experiment disproved the relativity theory: "The measurement results are not compatible with the Lorentz-Einsteinian fundamental assumption."[9] He concluded from his data that the mass of electrons did increase with increasing velocity, but not in agreement with equation (17).

Not until 1909—four years after the relativity paper had been published—did Bucherer announce that he had measured the mass-velocity relation for electrons and that his results agreed with Einstein's prediction. With contradictory evidence like this, with alternative hypotheses to account for the failure to detect the ether, with the strong attachment many had for the ether theory of light, and with the peculiar predictions of relativity, it is not surprising that Einstein's theory did not gain immediate acceptance. Textbooks as late as 1915 described light as waves in the ether and, as we have mentioned, Miller's results touched off considerable debate in the late 1920s.

In the years since, however, Einstein's ideas have come not only to be accepted, but to play a prominent role throughout physics. Particularly in the realms of nuclear and elementary particle physics, they are absolutely central. To design a cyclotron, a synchrotron, or any other device to accelerate protons or electrons to speeds near that of light, relativity theory is essential. The longer lifetimes of rapidly moving unstable particles, the increase in particle mass with increasing velocity, and an upper limit on particle speeds are all facts of life in a contemporary high-energy physics laboratory. The esoteric theories proposed to explain the properties of elementary particles also make extensive use of the ideas of space and time Einstein developed.

Of course there are well-known practical consequences of relativity that affect all men. The theory is responsible for the notion that two hydrogen nuclei can be brought together to release an enormous amount of energy. Whether such energy will ultimately be used to provide abundant electrical power and other services to man, or whether it will be used to destroy him, remains to be seen.

Suggested Reading

Einstein, A. *Relativity: The Special and General Theory.* New York: Holt, 1921. The first sixty-five pages of this book present a clear, almost completely non-mathematical discussion of special relativity. The author is well qualified.

Barnett, L. *The Universe and Dr. Einstein.* 2d ed. New York: Sloane, 1957. A well-known, entirely non-mathematical presentation of Einstein's work.

Arons, A. *Development of Concepts of Physics.* Reading, Mass.: Addison-Wesley, 1965. Chapter 36 is a clear and careful exposition of Einstein's theory.

Kacser, C. *Introduction to the Special Theory of Relativity.* Englewood Cliffs, N.J.: Prentice-Hall, 1967. A more extensive discussion than either of the above, but requiring no more mathematics than Arons.

Mermin, N. D. *Space and Time in Special Relativity.* New York: McGraw-Hill, 1968. A slower, more discursive treatment than the others, in a refreshingly different style.

Frank, P. *Einstein, His Life and Times.* New York: Knopf, 1953. A biography by a colleague and friend of Einstein.

Holton, G. "Einstein, Michelson, and the 'Crucial' Experiment." *Isis,* **60** (1969): 133–197. A detailed discussion of the place of the Michelson-Morley experiment in the development and acceptance of the relativity theory.

Williams, L., ed. *Relativity Theory: Its Origins and Impact on Modern Thought.* New York: Wiley, 1968. A number of important articles are reprinted in this collection, including the non-mathematical part of Einstein's original relativity paper. The final third samples scientific and lay reaction to the theory.

Questions

1. Comment on Magie's criticism of the theory of relativity quoted on page 350. In particular, do you agree that "A theory becomes intelligible when it is expressed in terms of the primary concepts of force, space, and time, as they are understood by the whole race of man?" Was Newton's interpretation of these concepts that which is understood by the whole race of man?

2. Speculate on the consequences to the theory of relativity of the following hypothetical discoveries:

   (a) It is found that information can be propagated instantaneously from one place to another by extrasensory perception.
   (b) A new type of electron is discovered that moves at twice the speed of light.

3. Two observers, each at rest relative to the S frame, are stationed at spatially separated positions A and B. They wish to synchronize their clocks. What is wrong with the following methods? In each case, both clocks initially read 12:00.

   (a) The observer at A waves his hand as he starts his clock. The observer at B watches A through a telescope and starts his clock when A waves his hand.
   (b) Both clocks are brought to point A and started together. One clock is then carried to point B.
   (c) A long steel bar is connected between the starting levers of the two clocks located at A and B. When the lever of the clock at A is moved, the bar simultaneously moves the lever of the clock at B.

4. An explosion occurs on the earth. 800 sec later, earthbound astronomers see an explosion on Mars (1.5×10^{11} m away on that day). It is suspected that a flying saucer from outer space set off each explosion as it flew past the earth and then Mars.

   (a) How long did it take the light from the explosion on

Mars to reach the earth telescopes?

(b) What was the time interval between the two explosions, as measured in the rest frame of the earth? Could these events have been set off by a flying saucer?

5. Two events occur 1 sec apart at positions separated by 2.4×10^8 m, both relative to an S reference frame.

   (a) How fast must an S' reference frame travel so that both events occur at the same position relative to S'?

   (b) What is the proper time interval between the events?

6. An explosion occurs in New York. One second later (as measured on properly synchronized earth clocks), another explosion occurs in San Francisco, 4000 km away.

   (a) How fast must a rocket ship travel if it is to be present at both explosions?

   (b) What will be the time interval between the explosions, as measured by the rocket ship pilot?

7. (a) Is there always a proper time interval between any two events? That is, given any two events, is it always possible to find a reference frame in which the time interval between the events is a proper time interval?

   (b) If your answer to (a) is no, how would you characterize events for which there does exist a proper time interval?

   (c) Can you see any correlation between the possibility of a cause-effect relation between two events and the existence of a proper time interval between these events?

8. How fast must a train move so that observers at rest on the embankment measure its length to be one-half its proper length?

9. Time bombs are placed at points A and B, which are 100 m apart as measured in the rest frame of the earth. As a rocket ship, moving at speed $v = \sqrt{3}\,c/2$, passes A, the bomb there explodes. As it passes B, the bomb there explodes. According to a clock on the rocket ship, the second explosion occurs a time $\Delta t'$ after the first. Clocks synchronized in the rest frame of the earth measure a time interval Δt between the explosions.

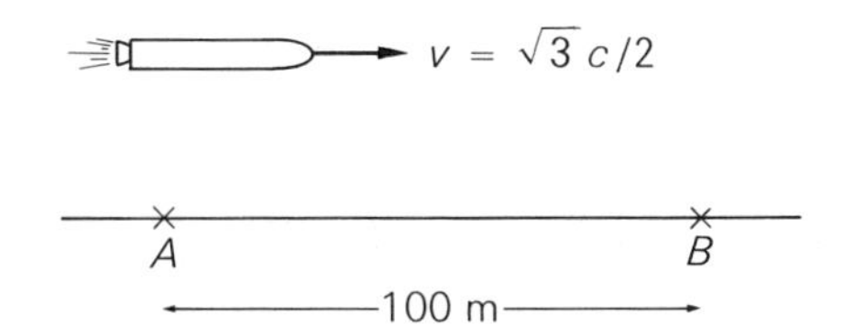

(a) Determine Δt.
(b) Determine $\Delta t'$ by an argument involving time dilation.
(c) Determine $\Delta t'$ by an argument involving length contraction. Compare with the result of (b).

10. Point out the error in the following argument: A meter stick, at rest with respect to the S frame, has a length L as measured in that frame. With respect to the S' frame, the meter stick is moving and is measured to have a shorter length L'. Therefore it is possible to distinguish between two inertial reference frames, and Einstein's principle of relativity is not valid.

11. What is the fastest moving vehicle you have ever seen? How long does it take to make a typical trip? According to the theory of relativity, what would a passenger aboard the vehicle measure the time of the trip to be? (You may find useful the approximation mentioned in Question 5, page 316.) Estimate how much the mass of the vehicle increases because of its motion.

12. How fast must an electron move relative to the laboratory so that its mass will be doubled?

13. Two guns, A and B, are mounted a distance 10 m apart on the embankment beside the railroad tracks. The barrels of the guns project out toward the tracks so that they almost brush the speeding express train ($v = \sqrt{15}\,c/4$) as shown in the figure below.

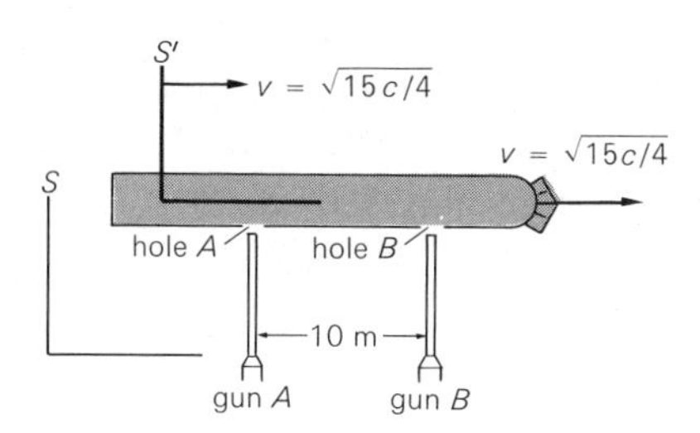

Suppose that the guns are fired simultaneously in S, the rest frame of the embankment, producing two bullet holes in the train a distance 10 m apart as measured in the S frame.

(a) If the bullet holes are 10 m apart as measured in the S frame (in which the train is moving), how far apart are they in the S' frame, the rest frame of the train?
(b) How far apart are the guns as measured in the S' frame?
(c) Since the two guns fire simultaneously in the S frame, you are aware that they do not fire simultaneously in the S' frame. Use the operational definition of simultaneity proposed by Einstein to argue which gun fired first as measured in the S' frame.
(d) Using the results of (b) and (c), discuss how observers at rest on the train account for the fact that the bullet holes are spaced as you calculated in (a). In this discussion deduce the time interval between the firing of the guns as measured in the S' frame.

14. Two hostile spaceships, S and S', approach each other on a near-collision course at relativistic velocities. Each ship has the same proper length. From the tail of S menacingly protrudes a space gun pointing perpendicular to the direction of relative motion of the ships. The noses of the two ships are labeled A and A', while their tails are labeled B and B', as shown in part (a) of the figure.

The pilot of S wishes to fire a warning shot across the bow of S'. He fires his gun (located at B) simultaneously with the coincidence of A and B' in his frame. Since in the frame of S the other ship is length-contracted, when A and B' coincide the pilot of S argues that his gun is ahead of the nose of S' as shown in part (b). Hence he expects the shot to miss S'.

On the other hand, when A and B' coincide, an observer at rest in S', for whom ship S is length-contracted, would claim the situation to be that shown in part (c). Will the bullet hit S' or not? Don't attempt a quantitative solution, but explain your answer in words.

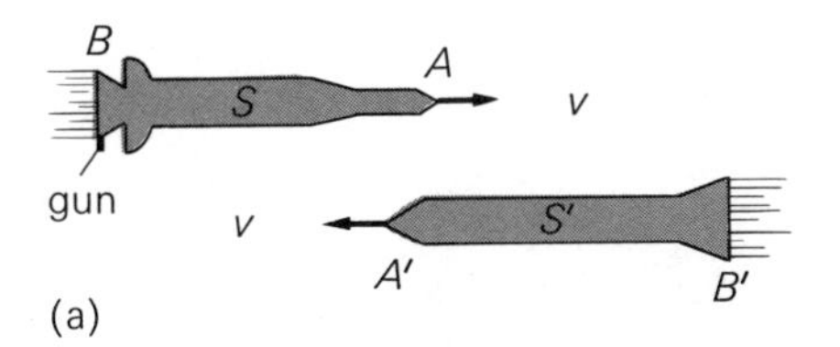
B
A
S
v
gun
v
S′
A′
B′
(a)

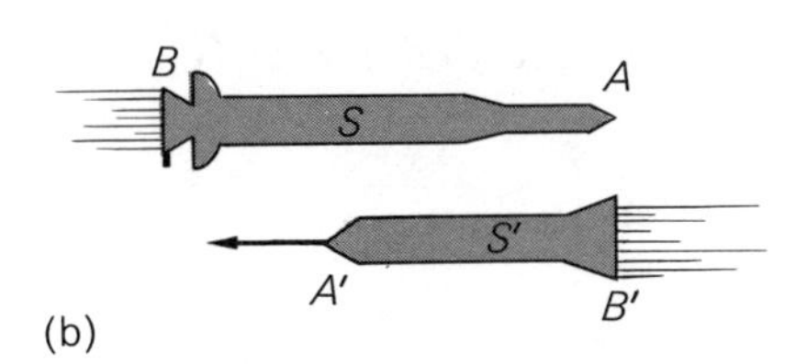
B
A
S
S′
A′
B′
(b)

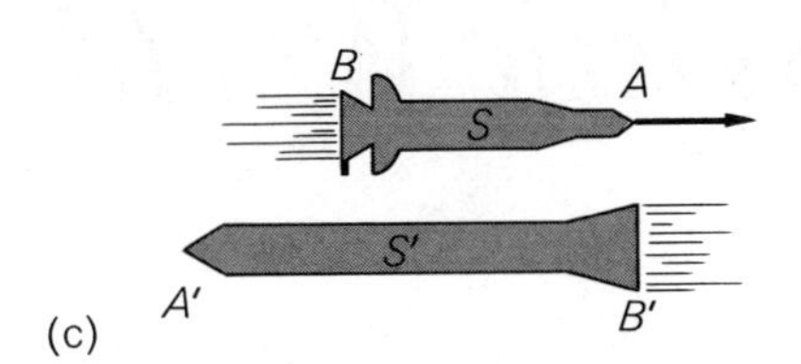
B
A
S
S′
A′
B′
(c)

CHAPTER FOURTEEN

The Logic of a Relativistic World

If the predictions of relativity seem strange or even nonsensical, perhaps it is because our intuition in these matters is a product of our everyday environment, where relative speeds are so small that we are not aware of time-dilation and length-contraction effects. In this chapter, we shall consider a number of *gedanken* experiments that indicate how different our world would be if speeds near that of light were commonplace. We shall expose some apparent paradoxes of relativity and examine how the theory deals with them.

Relativistic Space Travel

Let us consider briefly the implications of relativity for space exploration. Anyone who contemplates travel beyond our solar system is immediately impressed by the enormous distances involved. The distance from the sun to any of the other stars is so immense that interstellar distances are measured in light-years, the distance light travels in one year. As can easily be calculated, one light-year is 9.45×10^{15} m or six trillion miles. The nearest star (actually a double-star system), Alpha Centauri, is 4.3 light-years away. For manned exploration of other planetary systems to be feasible it would seem that spaceships would have to travel near the speed of light. Even so, a ship would take more than 4.3 years to reach Alpha Centauri.

Suppose a spaceship could achieve speeds near that of light. (Of course this is quite impractical today; to accelerate a macroscopic object to such a high speed would require an expenditure of energy far beyond the limits of present technology.) Let us imagine that such a spaceship is sent out from the earth to explore the planetary system of a star 100 light-years (or 100 light-years $\times$ 9.45×10^{15} m/light-year $= 9.45 \times 10^{17}$ m) from the earth (Figure 14–1). Suppose the spaceship travels with a speed $u = .995c$ relative to the earth. We can then calculate quite easily the time it takes the ship to make the journey, as measured by an observer on the earth. Between the event E_A, "ship leaves earth," and the event E_B, "ship arrives at star," the ship moves a distance of 9.45×10^{17} m at a speed of $.995c$, as shown in

Figure 14–1. A voyage to a distant star.

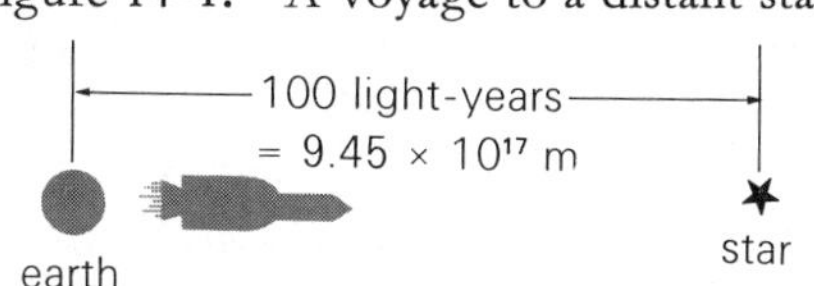

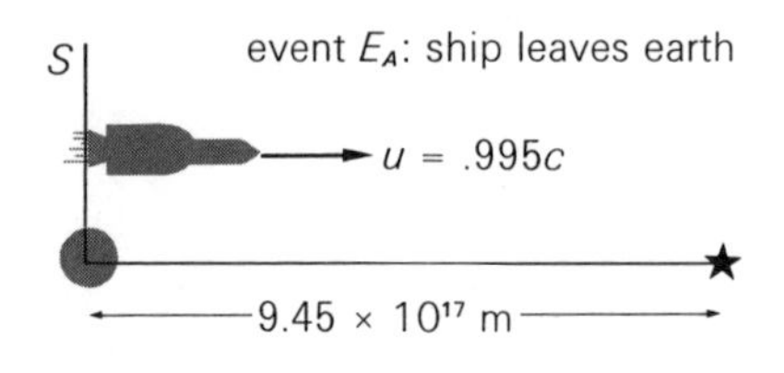

Figure 14–2. Events E_A and E_B as observed in the S frame, the rest frame of the earth.

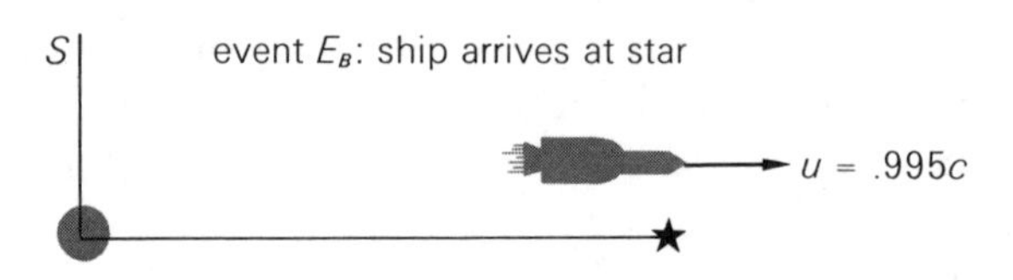

Figure 14–2. Hence the time interval between these two events as measured in the S frame, the rest frame of the earth, is simply:

$$\Delta t = \frac{9.45 \times 10^{17} \text{ m}}{.995 \times 9.45 \times 10^{15} \text{ m/year}} = 100.5 \text{ years}$$

Thus very few people who were alive when the ship departed from the earth would still be living when it reached the star. Certainly no one who was alive when the ship left the earth would ever receive any direct evidence that the mission was successful. Even if the pilot radioed back immediately upon reaching the star, it would take the radio signal, which travels with the speed of light, an additional 100 years to reach the earth, making a total time of 200.5 years from the beginning of the expedition until news of its success was received on earth.

What about the pilot of the spaceship? Would he survive to explore the planetary system of the star? Let us consider the trip from his point of view. Let S' be the rest frame of the ship. In this frame, or, equivalently, from the point of view of the spaceship's pilot, the earth and star are moving to the left with speed $u = .995c$, as shown in Figure 14–3. Note in particular that the two events E_A and E_B take place at the same point, the origin, in the S' frame. Consequently the time interval $\Delta t'$ between the two events, which is the time of the trip as measured by the pilot, is a proper time interval and is less than the time of the trip as measured from the earth by the time-dilation factor $\sqrt{1 - u^2/c^2}$.

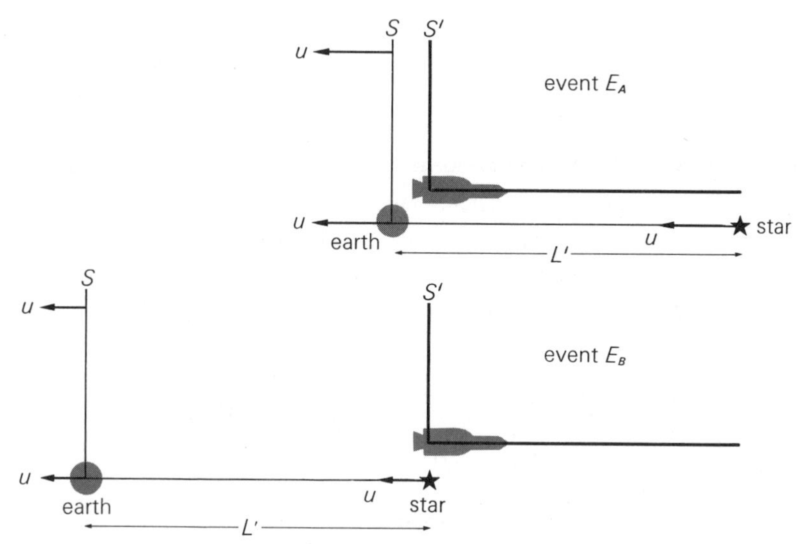

Figure 14–3. Events E_A and E_B as observed in the S' frame, the rest frame of the spaceship.

$$\Delta t' = \sqrt{1 - u^2/c^2}\,\Delta t = \sqrt{1 - (.995)^2} \times 100.5 \text{ years}$$

$$= \sqrt{1 - .99} \times 100.5 \text{ years} = \sqrt{.01} \times 100.5 \text{ years}$$

$$= 10.05 \text{ years}$$

Hence, according to the pilot, the trip takes only a little over ten years.

This result can also be obtained in another way. Suppose we consider the distance from the earth to the star as measured by the pilot. As shown in Figure 14–4, the earth-star distance is 100 light-years or 9.45×10^{17} m as measured in S, the rest frame of the earth and star. It may be convenient to think of a straight rod 9.45×10^{17} m long stretching from the earth to the star, as shown in Figure 14–4. From the point of view of the spaceship pilot (i.e., in the S' frame), the rod is moving to the left with speed $u = .995c$ and is consequently length-contracted. The length of the rod (earth-star distance) in the S' frame is:

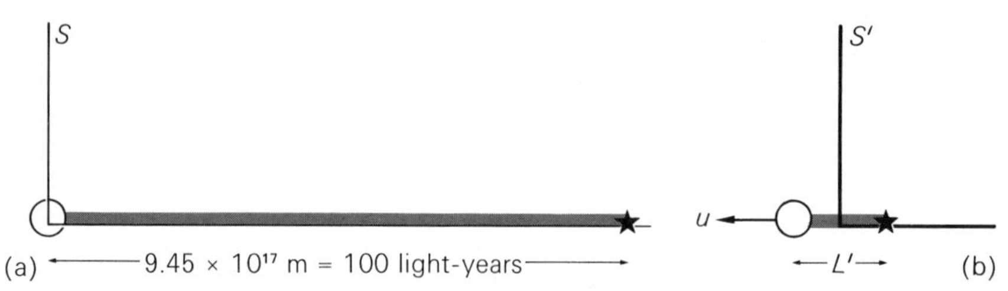

Figure 14–4. The earth-star distance as observed in the S and S' frames. (a) In the S frame, the distance is 100 light-years. (b) In the S' frame, the distance is length-contracted.

$$L' = \sqrt{1 - (.995)^2} \times 9.45 \times 10^{17} \text{ m}$$
$$= 9.45 \times 10^{16} \text{ m}$$
$$= 10 \text{ light-years}$$

Thus, according to the spaceship pilot the star is moving toward him with a speed of $.995c$, but it is initially only 10 light-years distant. Consequently the star takes only

$$\frac{9.45 \times 10^{16} \text{ m}}{.995 \times 9.45 \times 10^{15} \text{ m/year}} = 10.05 \text{ years}$$

to reach him. This is precisely the result obtained previously.

As the trip takes only 10.05 years in the S' frame, the pilot will age only 10.05 years. Hence, at the end of the trip he will have sufficient life ahead of him to explore the planetary system of the star.

Alternatively, he may turn the ship around and head back to earth with speed $u = .995c$. A detailed consideration of the return trip would be almost identical to our discussion of the outgoing trip. Now the rest frame of the spaceship is S'', which moves relative to S in the $-x$ direction with speed $.995c$, as shown in Figure 14–5. As measured from the earth, the return trip would take 100.5 years. From the point of view of the pilot, the earth-star separation is 10 light-years and the earth is moving toward him at $.995c$. Consequently the return trip takes 10.05 years. On his return to the earth the pilot, who has aged 20.1 years, would find a society that has advanced 201 years, with his generation long dead.

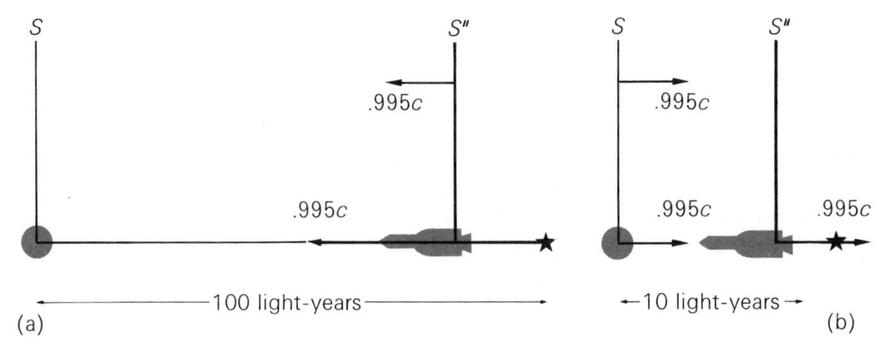

Figure 14–5. The return trip of the spaceship. (a) As observed in the S frame. (b) As observed in the S'' frame.

The validity of this account depends, of course, on the assumption that the biological processes of the pilot operate according to the passage of time measured on spaceship clocks. But life processes can be considered a crude sort of clock—and we argued earlier that, according to the principle of relativity, all clocks must be affected in the same way by the relative motion of an inertial frame. Thus Einstein's theory demands that the beats of the pilot's heart must be slowed, according to an observer on the earth, just like the ticks of any other spaceship clock.

The Twin Paradox

The result just discussed can be restated in a particularly vivid fashion, which leads to one of the most famous and controversial "paradoxes" of the special theory of relativity. Let us consider two twins, A and B. Suppose that B boards a spaceship, heads out to a nearby star at a speed near that of light, turns around and returns to earth. This sequence of events as observed by A (that is, in the rest frame of the earth and star) is illustrated in Figure 14–6(a). We have just argued that, on his return to earth, B will be younger than his twin A. But, one might protest, what if we consider the trip from B's point of view? According to B, he is at rest and the earth and twin A travel away at a speed near c and then return. The observations of B are shown in Figure

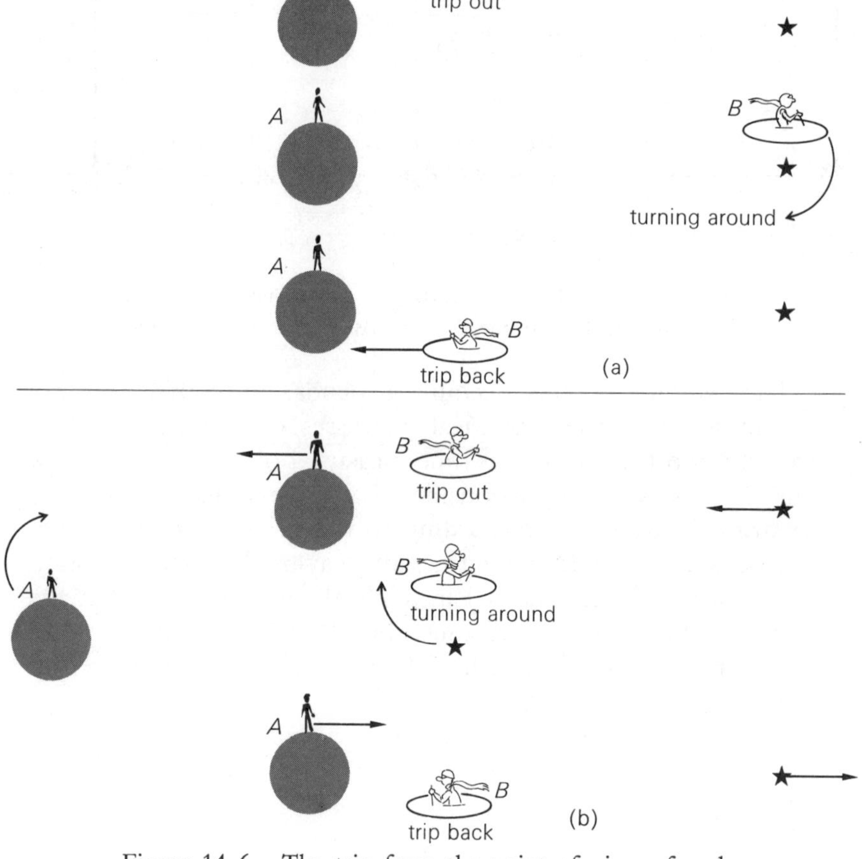

Figure 14–6. The trip from the point of view of each twin. (a) As described by *A*. (b) As described by *B*.

14–6(b). Thus, from *B*'s point of view, it is *A* who has been moving and consequently *A* who comes back younger. But it is not possible for *B* to be younger than *A* and *A* to be younger than *B*. Hence the apparent paradox.

On the surface this argument seems quite reasonable. For many years, the "twin paradox" was a controversial subject. A number of physicists argued that such a result casts doubt on the validity of Einstein's theory. A heated exchange of letters on the topic took place in the British journal *Nature* as recently as 1956–57. However, most physicists today accept the view that a careful analysis leads to a consistent conclusion: Both A and B agree that B is younger at the end of the trip.

Since the analysis is relatively long and involved, we shall not present a detailed treatment here. However, it is easy to appreciate the resolution of the paradox in general terms. The essence of the argument that led to the twin paradox was the apparent symmetry of the motions of A and B. From A's point of view, B travels away and then back with speed u. From B's point of view, A travels away and then back with speed u. A observes B to be younger, and B observes A to be younger.

The trouble with this argument is that it fails to take into account the central role of inertial reference frames in special relativity theory. The motions are not symmetrical. A remains at rest in an inertial frame throughout the trip. B does not. B is originally at rest in the inertial frame attached to the earth. After lift-off he is at rest in another inertial frame moving with speed u to the right relative to the earth's rest frame. After turning around, B is at rest in still another inertial frame, this one moving with speed u to the left relative to the earth's frame.

This asymmetry is not merely technical; it has obvious physical consequences. Consider, for example, what happens when B turns around in the vicinity of the star. This will involve a tremendous deceleration and then an acceleration of the spaceship back toward earth. One may say glibly: "The motion is symmetrical. From B's point of view A decelerates and then accelerates." But if the doubter were B on the spaceship and didn't have his seat belt fastened, he would rapidly be convinced of the asymmetry!

Naturally this experiment has not yet been performed with people. However, a comparable experiment with μ mesons has recently been carried out at CERN (the European Center for Nuclear Research) near Geneva, Switzerland. It uses the fact that μ mesons are electrically charged and that charged particles move in a circular path when under the influence of a magnetic field. As illustrated in Figure 14–7, high-speed μ mesons are injected into an evacuated circular ring. Large magnets surrounding the ring cause the μ mesons to travel in a circular path within the ring.

In this experiment the speed of the μ mesons is approximately $u = .995c$ so that $1/\sqrt{1 - u^2/c^2} = 10$. This means that if a single

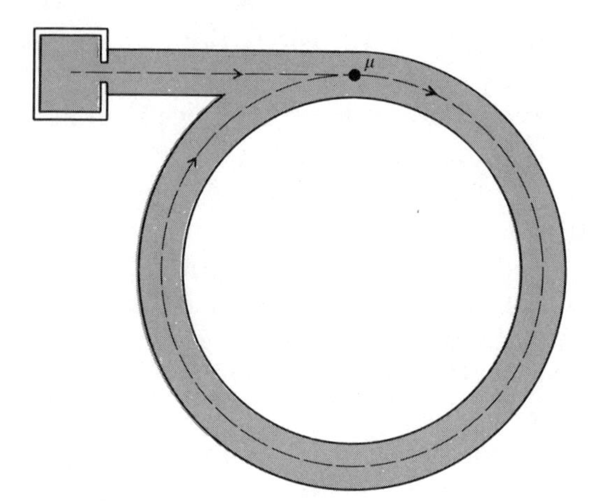

Figure 14–7. High-speed μ mesons move in a circular path, returning to their starting point.

μ meson is injected into the ring at this speed it should live about ten times longer than its "twin," a μ meson that remains at rest in the laboratory. The mesons in the ring are in fact observed to live about 2×10^{-5} sec while mesons at rest live 2×10^{-6} sec. Thus, a μ meson sent out around the ring in this experiment will return to find its twin long dead.

The Lorentz Transformation

We were able to discuss space travel and the twin paradox using only the basic relativistic results of length contraction and time dilation. However, in treating more complicated examples, it is often convenient to use a general relation between the position and time coordinates of an event in one inertial frame and its position and time coordinates in another inertial frame. Such a relation is easily derived with our length-contraction and time-dilation results.

We have previously derived the corresponding relation in Newtonian physics—the Galilean transformation. According to that transformation, an event described by the coordinates x, y, and t in one inertial frame has coordinates $x' = x - ut, y' = y$, and $t' = t$ when measured with respect to another inertial frame moving with relative speed u in the $+x$ direction.

The Galilean transformation was a consequence of Newtonian ideas of absolute space and time; we assumed that lengths and durations were unaffected by motion. Within the Einsteinean framework, this transformation must be changed. In his 1905 paper, Einstein deduced the appropriate relation, which proved identical to the transformation Lorentz had introduced in an *ad hoc* manner the previous year. In relativity theory, this transformation is an immediate consequence of time dilation and length contraction, as we shall now demonstrate.

Let us consider two inertial reference frames S and S' moving with relative speed u, as shown in Figure 14–8. As usual, we shall suppose the relative motion to be along the x direction, and we shall let the time at which the two origins O and O' coincide be $t = t' = 0$. Let us focus on an event E which, as measured in the S frame, takes place at time t at a position whose coordinates are x and y. A perpendicular line drawn from this position toward the x- and x'-axes crosses them at a point we shall designate as A.

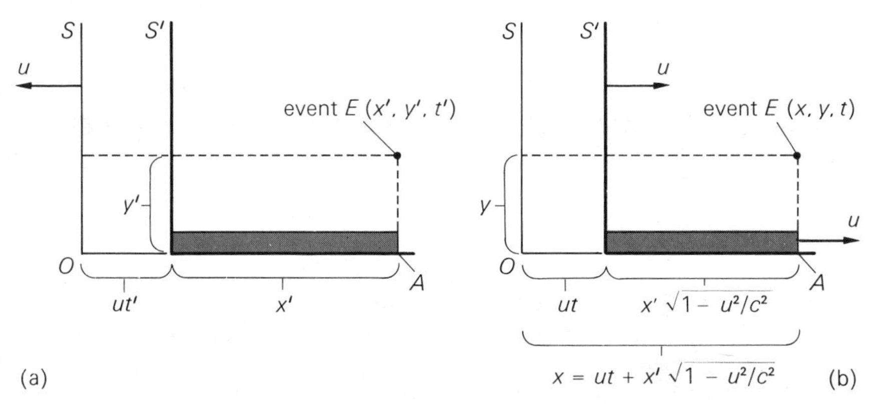

Figure 14–8. The coordinates of an event as measured in two inertial reference frames. The rod, at rest in the S' frame, is length-contracted as measured in the S frame.

Since lengths perpendicular to the direction of motion are not contracted, the distance EA is the same in both reference frames. This means that the y- and y'-coordinates of the event are the same:

$$y' = y \tag{18}$$

The relation between x and x' is somewhat more complicated. From Figure 14–8(a) it is clear that if we were to lay a rod at rest in the S' frame from O' to A, its length would be x'. This rod would be moving when viewed from the S frame and hence would be length-contracted so that its length in S would be $x' \sqrt{1 - u^2/c^2}$. We can then read directly from Figure 14–8(b): $x = ut + x' \sqrt{1 - u^2/c^2}$. This equation can be solved for x' with the result:

$$x' = \frac{1}{\sqrt{1 - u^2/c^2}}(x - ut) \tag{19}$$

Alternatively we could consider a rod at rest in the S frame with ends at O and A [Figure 14–9(a)]. It would have length x in the S frame and a contracted length $x \sqrt{1 - u^2/c^2}$ in the S'

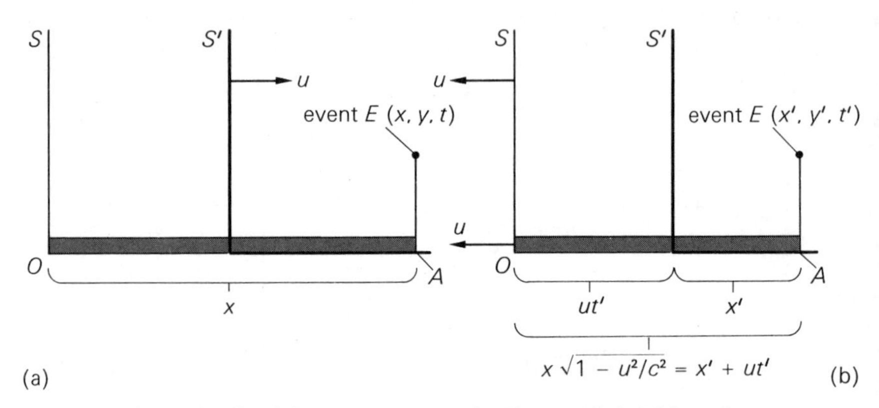

Figure 14–9. The same event as in Figure 14–8. This rod, at rest in the S frame, is length-contracted as measured in the S' frame.

frame. From Figure 14–9(b) we see that $x' + ut' = x \sqrt{1 - u^2/c^2}$ or, solving for x,

$$x = \frac{1}{\sqrt{1 - u^2/c^2}}(x' + ut') \tag{20}$$

If equation (19) for x' is substituted in equation (20), we find:

$$x = \frac{1}{\sqrt{1 - u^2/c^2}}\left[\frac{x}{\sqrt{1 - u^2/c^2}} - \frac{ut}{\sqrt{1 - u^2/c^2}} + ut'\right]$$

Solving for t', we find:

$$t' = \frac{1}{\sqrt{1 - u^2/c^2}}\left(t - \frac{u}{c^2}x\right) \tag{21}$$

Equations (18), (19), and (21) relate the coordinates x, y, and t of an event in S to the coordinates x', y', and t' of the same event in S'. These equations constitute the Lorentz transformation. For convenience, we collect these results, along with the equivalent equations in which x, y, and t are given in terms of x', y', and t':

Lorentz transformation $S \mid S' \mid \rightarrow u$

$$x' = \gamma\,(x - ut) \qquad x = \gamma\,(x' + ut')$$

$$y' = y \qquad y = y'$$

$$t' = \gamma\left(t - \frac{u}{c^2}x\right) \qquad t = \gamma\left(t' + \frac{u}{c^2}x'\right)$$

$$\text{where } \gamma = \frac{1}{\sqrt{1 - u^2/c^2}}$$

We have adopted a special symbol γ (the Greek gamma) for the frequently used factor $1/\sqrt{1 - u^2/c^2}$. Since $\sqrt{1 - u^2/c^2}$ is less than 1 provided u is less than c, it follows that γ is greater than 1. γ is, of course, just the factor that appeared earlier in the length-contraction and time-dilation results.

As we pointed out above, this transformation, which results directly from Einstein's concepts of space and time, is identical to that obtained by Lorentz. It is interesting that the same equations result from two theories so radically different in conception. Lorentz obtained the transformation in an *ad hoc* manner—with this transformation Maxwell's equations of electromagnetism have the same form in a moving reference frame as in the rest frame of the ether. For Lorentz, the transformed coordinates x' and t' were not real positions and times, but only the apparent coordinates measured by clocks and rods that are distorted by their motion through the ether. For Einstein, the ether need not be discussed, and x' and t' are just as real as x and t.

A final property of the Lorentz transformation should be noted. If the relative speed u of the reference frames is small compared with the speed of light c, the factor γ is approximately 1, and the Lorentz transformation closely approximates the Galilean transformation of Newtonian physics. Since Newtonian physics has demonstrated its success in dealing with the motion of objects at speeds small compared with that of light, this is a reassuring—almost necessary—result. Only as technological advances allowed experiments in a new domain, that of high speeds, did the need for changes in the Newtonian point of view become apparent.

Example: Time Dilation Again

As a simple example of the way the Lorentz transformation can be used, let us again consider time dilation. We will hardly be surprised to find that the Lorentz transformation predicts time dilation, since we used the time-dilation result to derive it. Nevertheless, this will be an instructive prelude to our later discussion.

Suppose that a clock at rest at position X' in the S' frame records a time interval $\Delta T' = T_2' - T_1'$ [Figure 14–10(a)]. What is the corresponding time interval as measured in the S frame, in which the clock is moving with speed u [Figure 14–10(b)]? Let us consider the two events E_1 and E_2 corresponding to the clock's striking T_1' and T_2' respectively. We first ask, "What do we know about these events?" In this case, we know the position and time of each event in the S' frame. It is convenient to make a table of the known information:

E_1, "clock strikes T_1'":

$$x_1' = X' \qquad x_1 = ?$$

$$t_1' = T_1' \qquad t_1 = ?$$

E_2, "clock strikes T_2'":

$$x_2' = X' \qquad x_2 = ?$$

$$t_2' = T_2' \qquad t_2 = ?$$

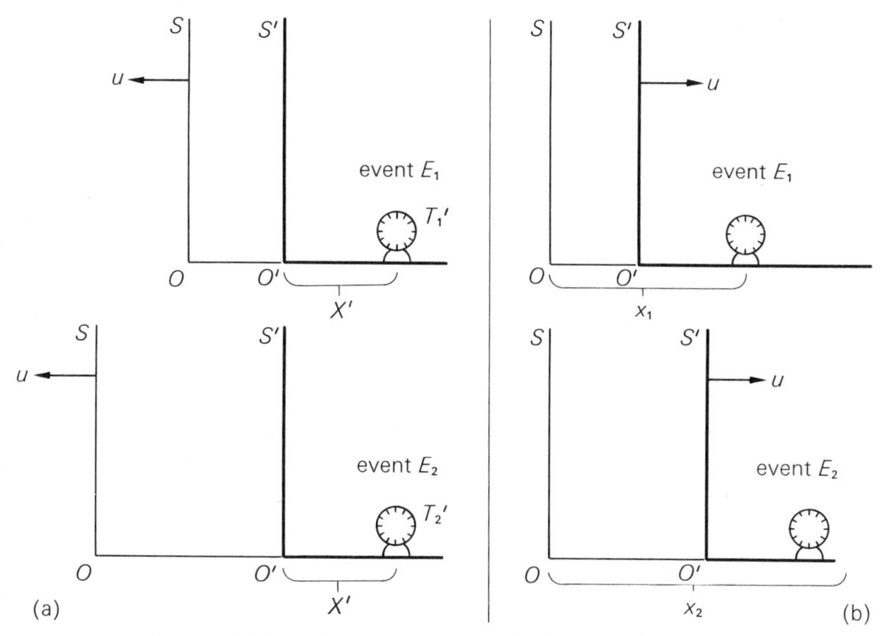

Figure 14–10. Two events: E_1, "clock strikes T_1'", and E_2, "clock strikes T_2'". Both events occur at the same position in the S' frame.

We wish to find the time interval $\Delta T = t_2 - t_1$ between E_1 and E_2 as measured in the S frame. According to the Lorentz transformation:

$$t_2 = \gamma\,(t_2' + ux_2'/c^2) = \gamma\,(T_2' + uX'/c^2)$$

$$t_1 = \gamma\,(t_1' + ux_1'/c^2) = \gamma\,(T_1' + uX'/c^2)$$

Hence,

$$\begin{aligned}\Delta T &= t_2 - t_1 \\ &= \left(\gamma\,T_2' + \frac{\gamma u X'}{c^2}\right) - \left(\gamma\,T_1' + \frac{\gamma u X'}{c^2}\right) \\ &= \gamma\,(T_2' - T_1') \\ &= \gamma\,\Delta T'\end{aligned}$$

Suppose that T_1' and T_2' are the times of successive ticks of the clock. Since γ is greater than 1, we have once again shown that the time interval ΔT between successive ticks of a clock is longer in the S frame, in which the clock is moving, than the time interval $\Delta T'$ in the S' frame, in which the clock is at rest.

The Pole Vaulter and the Barn

The Lorentz transformation is a powerful tool for analyzing even quite complex situations in a straightforward manner. We shall now use it to examine in detail how an apparent paradox of relativity, the case of the pole vaulter and the barn, can be resolved.

Let us imagine a pole vaulter, carrying a pole of proper length 20 m, dashing toward a barn with a speed $u = \sqrt{15}\, c/4$, as shown in Figure 14–11(a). (This peculiar speed was chosen

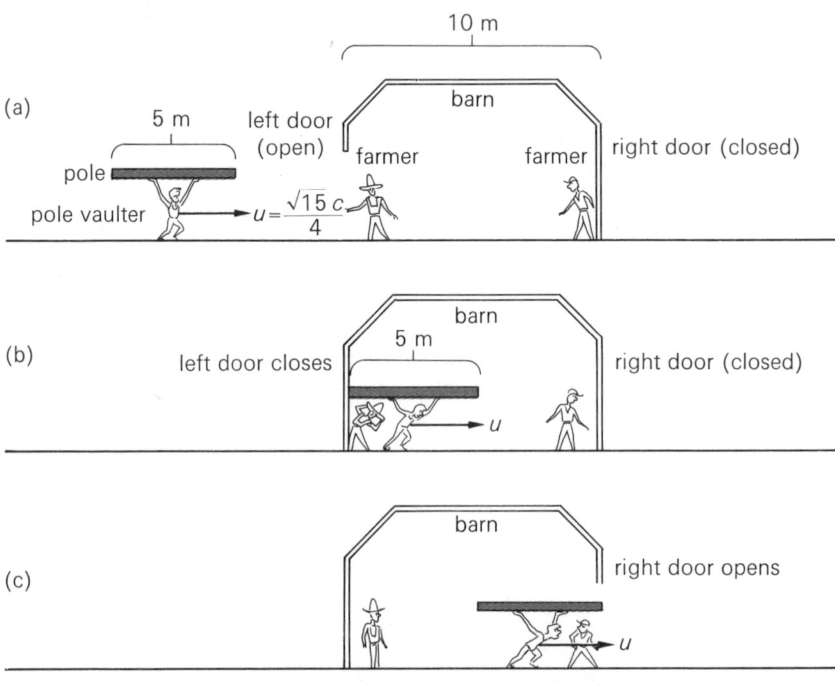

Figure 14–11. The pole passing through the barn, as described by the farmers. The left door closes before the right door opens. (a) Before the pole reaches the barn, the left door is open and the right door is closed. (b) When the pole is entirely in the barn, the left door is closed. (c) The right door is opened, allowing the pole to emerge.

because it corresponds to a simple value of γ, $\gamma = 4$.) The barn has a proper length of 10 m. To farmers stationed at the left and right doors of the barn, the pole is length-contracted and hence measures only 5 m long, while the barn is 10 m long.

The farmers relate the following sequence of events. [See parts (b) and (c) of Figure 14–11.] Originally the left door of the barn was open and the right door closed. As soon as the pole was entirely in the barn, the left door was closed. When the pole reached the right door, that door was opened and the pole passed through.

However, the pole vaulter has his own story. According to him, the pole is 20 m long and the length-contracted barn only 2.5 m long. (See Figure 14–12.) Hence the pole could not possibly fit in the barn. At first this might seem paradoxical, since the farmers report that for a while the pole is in the barn with both doors closed. However, according to the theory of relativity there really is no paradox. Both the farmers' *and* the pole vaulter's descriptions are correct.

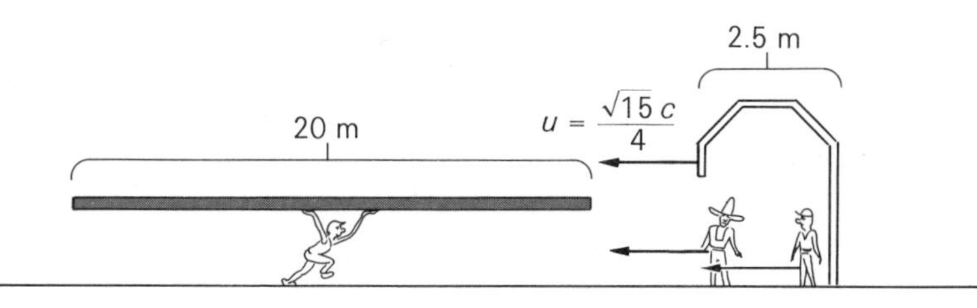

Figure 14–12. The situation before the barn reaches the pole, as described by the pole vaulter.

To analyze the situation, let us consider two inertial reference frames. S is the barn's rest frame, with its origin at the left door of the barn. S' is the pole's rest frame, with its origin at the left end of the pole. As usual, we employ the convention that $t = t' = 0$ when the origins O and O' coincide. A somewhat more abstract version of Figure 14–11 is given in Figure 14–13, showing the sequence of events in the S frame, while an abstract version of Figure 14–12 is shown in Figure 14–14.

Let us focus on two events, E_1, "left door closes," and E_2, "right door opens." It should be clear from Figure 14–13 that at $t_1 = 0$, when the left door closes, the right end of the pole is located at $x = 5$ m. The pole is moving to the right with speed

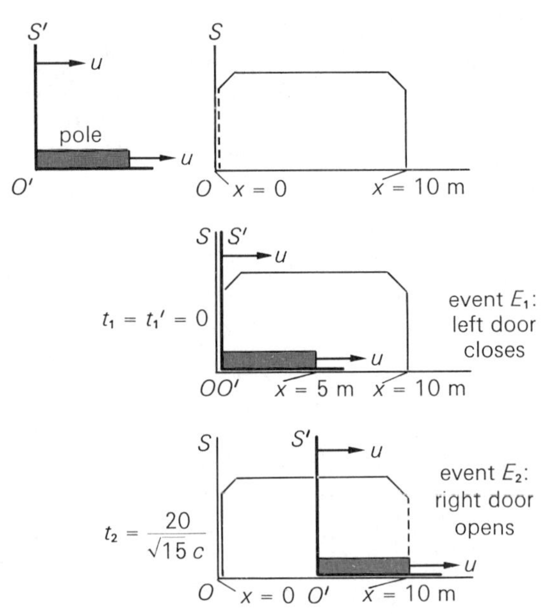

Figure 14–13. The sequence of events relative to the S frame. This is an abstract version of Figure 14–11.

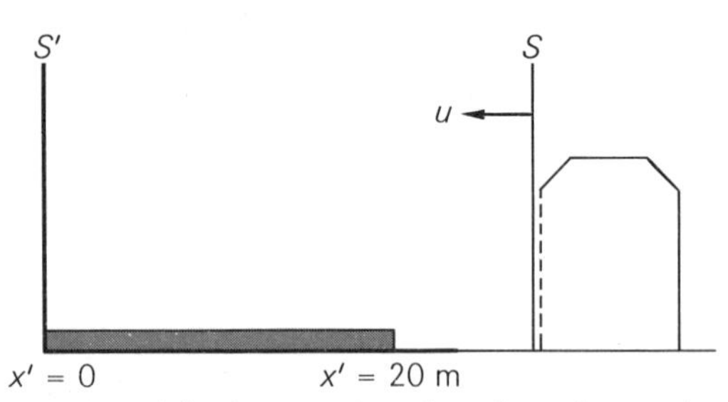

Figure 14–14. The barn approaches the pole, as observed in the S' frame. This is an abstract version of Figure 14–12.

$u = \sqrt{15}\,c/4$ and the right door opens when the right end of the pole reaches the door. As the pole moves a distance of 5 m with speed $\sqrt{15}\,c/4$, the time of event E_2, "right door opens," is therefore

$$t_2 = \frac{5\text{ m}}{(\sqrt{15}/4)c} = \frac{20}{\sqrt{15}\,c}$$

This calculation and Figures 14–13 and 14–14 then imply the following information about events E_1 and E_2:

$$
\begin{array}{llll}
E_1, \text{“left door closes”:} & x_1 = 0 & & x_1' = 0 \\
& t_1 = 0 & & t_1' = 0 \\
E_2, \text{“right door opens”:} & x_2 = 10\text{ m} & & x_2' = 20\text{ m} \\
& t_2 = \dfrac{20\text{ m}}{\sqrt{15}\,c} & & t_2' = ?
\end{array}
$$

We may use the Lorentz transformation to determine t_2', the time at which the right door opens in the S' frame:

$$
\begin{aligned}
t_2' = \gamma\left(t_2 - \frac{u}{c^2}x_2\right) &= 4\left(\frac{20\text{ m}}{\sqrt{15}\,c} - \frac{\sqrt{15}\,c}{4c^2} \times 10\text{ m}\right) \\
&= \frac{4}{c}\left(\frac{20\text{ m}}{\sqrt{15}} - \frac{\sqrt{15} \times 10\text{ m}}{4}\right) \\
&= \frac{4}{3 \times 10^8\text{ m/sec}}(5.16\text{ m} - 9.69\text{ m}) \\
&= -6.0 \times 10^{-8}\text{ sec}
\end{aligned}
$$

The important thing about this result is that t_2' is negative. Since $t_1' = 0$, this means that in the S' frame event E_2 occurs before event E_1. (See Figure 14–15.) In other words, while in

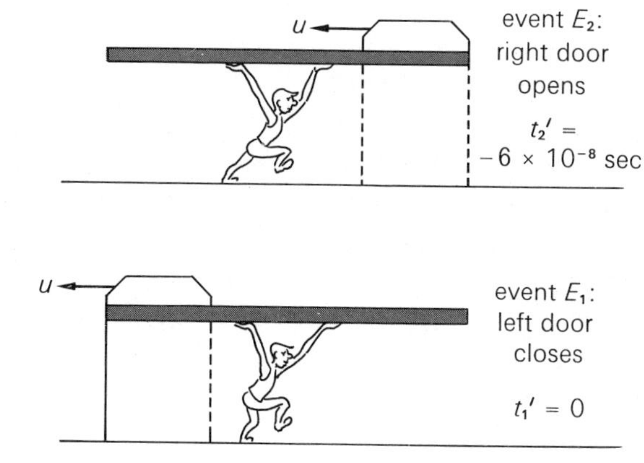

Figure 14–15. The sequence of events as observed by the pole vaulter. The left door closes after the right door opens.

the farmers' rest frame the left door closes before the right door opens and hence the pole spends some time within the closed barn, according to the pole vaulter the right door opens first and then, later, the left door closes. In the framework of the theory of relativity, these descriptions are compatible and the apparent paradox is resolved.

Causality

In this example, we considered two events whose time order depended on the particular inertial frame in which the events were observed. In the S frame, the left door closed before the right door opened, while in the S' frame the left door closed after the right door opened. That the time order of two events may differ when they are viewed from different reference frames suggests a possible difficulty when dealing with causally related events. We naturally expect a cause to precede its effect in time. Are there inertial reference frames in which the time order of cause and effect is reversed?

For example, one might imagine the following sequence of events relative to some inertial frame S: First a gun fires, then the target shatters. If, relative to some other inertial frame S', the sequence occurred in the reverse order—first the target shattered, then the gun fired—the theory of relativity would be in grave difficulty. Someone could disarm the gun after the target shattered, but before the gun fired!

In fact, there is no such difficulty in relativity theory. The Lorentz transformation allows an ambiguity in the time order of events only when the events are so close together in time and far apart in space that no cause-effect relationship could possibly exist between them.

To see that this is indeed the case, let us first discuss the space and time relationship between two events that are cause and effect. Since no signal can travel faster than the speed of light, one event cannot cause another unless a light signal sent out from the position of the first event (cause) when it occurs would arrive at the position of the second event (effect) at or before the time it occurs. Figure 14–16 is a *space-time diagram* illustrating

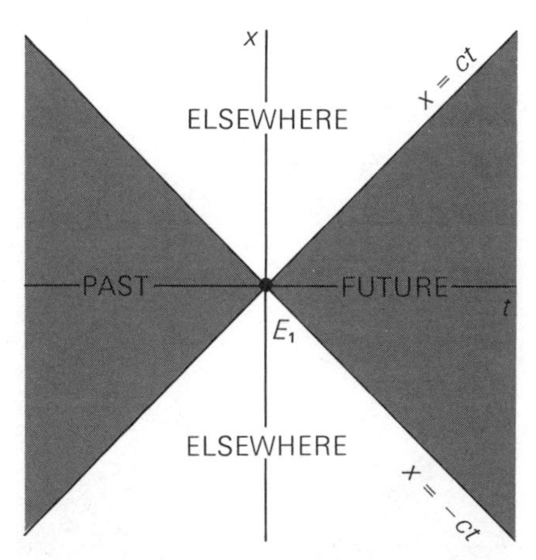

Figure 14–16. A space-time diagram. Events in the region labeled FUTURE may be caused by the event E_1. Events in the region labeled PAST may be the cause of E_1. Events in the region labeled ELSEWHERE cannot be causally related to E_1.

this relationship. Suppose the event E_1 occurs at time $t = 0$ at position $x = 0$ in the S frame. The region labeled FUTURE indicates the position and time coordinates of all the events on the x-axis that might be caused by E_1. The region labeled PAST indicates the position and time coordinates of all events on the x-axis that might be the cause of E_1. The region labeled ELSEWHERE indicates the position and time coordinates of all events that could not be causally related to E_1 because a signal traveling with a speed greater than c would be required to connect them.

For example, E_1 might be the event "gun fires," which we assume to occur at the origin of the S frame at $t = 0$ as illustrated in Figure 14–17. The flight of the bullet as it travels toward the target with constant speed v is represented on the space-time diagram by a straight line with slope v. The event E_2, "bullet strikes target," is also indicated on the diagram. It is an effect of E_1, the firing of the gun. Since v must be less than c, the path representing the flight of the bullet in the space-time diagram always remains in the region labeled FUTURE. Thus, any effect of the firing of the gun, such as E_2, must lie in that region. This is a direct consequence of our assumption that no signal can propagate faster than the speed of light.

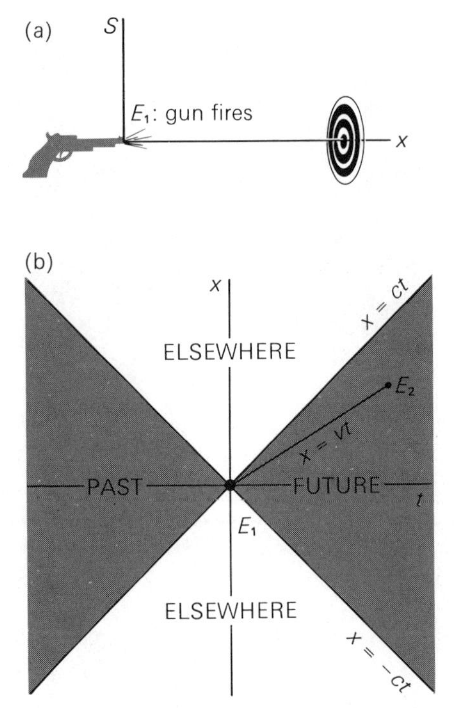

Figure 14–17. A gun fires a bullet that strikes a target. Event E_1, "gun fires," is the cause of event E_2, "bullet strikes target." (a) Event E_1, "gun fires." (b) The space-time path of the bullet. Note that the event E_2, "bullet strikes target," lies in the region labeled FUTURE.

What is more, if the position and time coordinates of the causally related events E_1 and E_2 are measured in any other inertial frame, the cause E_1 always occurs before its effect E_2. This can be seen by considering another inertial frame S' moving with speed u in the $+x$ direction so that the origin of S' coincides with the origin of S at the instant the gun fires. As usual, we shall choose this time to be $t = t' = 0$.

If, as measured in the S frame, the target is a distance L from the gun and the bullet speed is v, the bullet will strike the target at time $t = L/v$. (Figure 14–18.) The following table then summarizes what we know about the position and time coordinates of events E_1 and E_2:

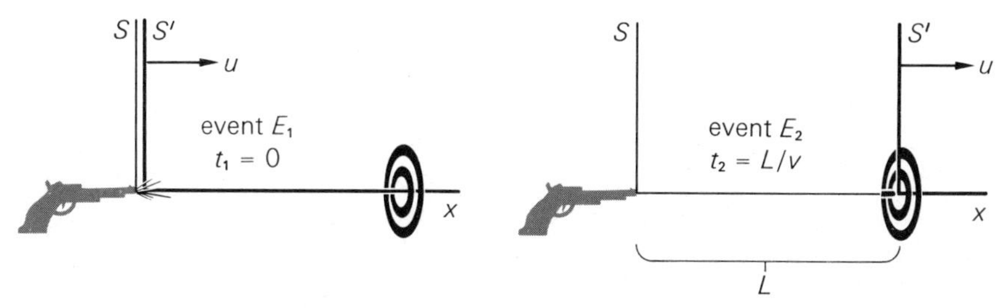

Figure 14–18. The two causally related events, E_1 and E_2, as observed in the S frame.

$$
\begin{array}{llll}
E_1: & x_1 = 0 & & x_1' = 0 \\
 & t_1 = 0 & & t_1' = 0 \\
E_2: & x_2 = L & & x_2' = ? \\
 & t_2 = L/v & & t_2' = ?
\end{array}
$$

Let us focus on t_2', the time of the effect E_2 as measured in the S' frame. According to the Lorentz transformation, t_2' is given by:

$$
\begin{aligned}
t_2' &= \gamma\,(t_2 - ux_2/c^2) \\
&= \gamma\left(\frac{L}{v} - \frac{uL}{c^2}\right) \\
&= \frac{\gamma L}{v}\left(1 - \frac{uv}{c^2}\right)
\end{aligned}
$$

If u and v are both less than c, $1 - uv/c^2$ is a positive number and hence t_2', the time of event E_2 as measured in S', is greater than zero. But $t_1' = 0$. Hence, t_2' is greater than t_1', i.e., as measured in the S' frame, the effect, E_2, occurs *after* the cause, E_1.

This conclusion can easily be generalized to all pairs of events that are cause and effect. If the speeds of signals v emanating from the cause must be less than c and if the relative speeds of reference frames are also less than c, then the Lorentz transformation indicates that the cause will precede the effect in all inertial frames. Since events in the region labeled PAST could be the cause of event E_1, the same argument indicates that these events will precede E_1 in all inertial frames. Only events that cannot have a causal relationship to E_1—those in the region labeled ELSEWHERE in the space-time diagram—either precede or follow E_1, depending on the inertial frame in which the times of the events are measured. While such an ambiguity may seem peculiar, it must be stressed that this does not lead to a logical inconsistency such as a cause occurring after its effect.

The Addition of Velocities

Another potential source of difficulty in the theory of relativity relates to the assumption that there is no means of transmitting information with a speed greater than that of light. Let us imagine, for example, a spaceship traveling with velocity u relative to the earth. Suppose a gun inside the ship shoots a bullet with a velocity v' relative to the gun (see Figure 14–19). According to Newtonian physics, the velocity of such a bullet relative to the earth, v, is equal to the velocity of the bullet relative to the spaceship, v', plus the velocity of the spaceship relative to the earth, u—that is, $v = v' + u$. If u and v' are both close to the speed of light, it would seem easily possible for the speed of the bullet relative to the earth to be greater than c, in contradiction to the initial assumption of relativity.

However, the relation $v = v' + u$ is based on the Newtonian ideas of absolute space and time, and, like the closely related

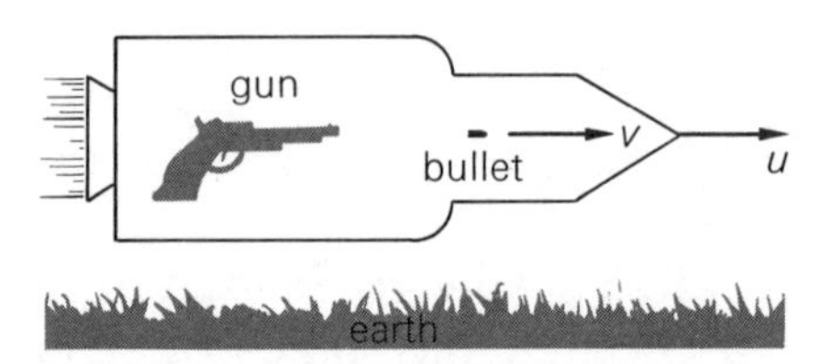

Figure 14–19. A gun at rest in a spaceship fires a bullet. The velocity of the bullet relative to the ship is v', and u is the velocity of the ship relative to the earth. It might appear that if v' and u are both near the speed of light, the speed v of the bullet relative to the earth would be greater than the speed of light.

Galilean transformation, it must be modified within the theory of relativity. This modification can be found with the Lorentz transformation.

Let S' be the rest frame of the ship and the gun, and let S be the rest frame of the earth. The origin of S' is chosen to coincide with the end of the barrel of the gun. For convenience, let us suppose that the bullet leaves the gun at $t = t' = 0$ when the origins of S and S' coincide. To measure the velocity of the bullet, we might measure the time T' it takes the bullet to reach a target a distance L' away from the gun in the spaceship.

Let us focus on two events: E_1, "bullet leaves gun," and E_2, "bullet strikes target." Figure 14–20 illustrates these events as observed in the S' frame. As usual, let us make a table of what we know of these two events:

E_1: "bullet leaves gun":	$x_1' = 0$	$x_1 = 0$
	$t_1' = 0$	$t_1 = 0$
E_2: "bullet strikes target":	$x_2' = L'$	$x_2 = ?$
	$t_2' = T'$	$t_2 = ?$

Of course, we are really interested in comparing v' and v, the velocities of the bullet as measured in the spaceship and on earth respectively. But v' and v can be simply calculated from the coordinates of the above events.

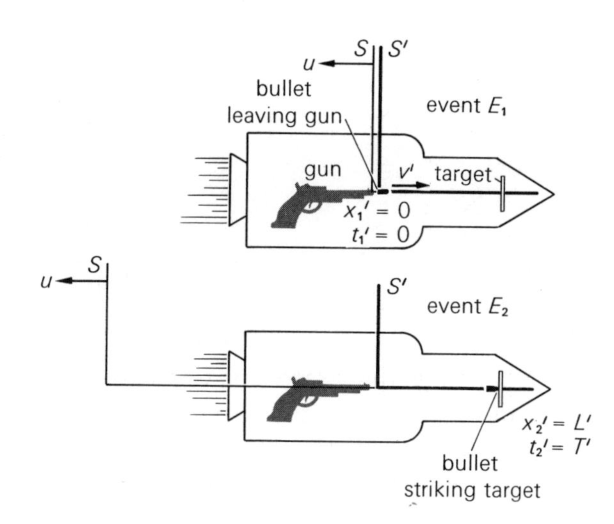

Figure 14–20. Events E_1 and E_2 as observed in the S' frame, the rest frame of the spaceship.

We find v' by dividing the distance traveled by the bullet by the travel time as measured in the S' frame:

$$v' = \frac{\Delta x'}{\Delta t'} = \frac{x_2' - x_1'}{t_2' - t_1'} = \frac{L'}{T'}$$

or

$$L' = v'T' \tag{22}$$

Similarly, v is the distance traveled by the bullet divided by the travel time as measured in the S frame:

$$v = \frac{\Delta x}{\Delta t} = \frac{x_2 - x_1}{t_2 - t_1} = \frac{x_2}{t_2}$$

The Lorentz transformation can then be used to calculate x_2 and t_2 (and hence v) since x_2' and t_2' are known:

$$x_2 = \gamma\,(x_2' + ut_2') = \gamma\,(L' + uT')$$

$$t_2 = \gamma\left(t_2' + \frac{u}{c^2}x_2'\right) = \gamma\left(T' + \frac{u}{c^2}L'\right)$$

Hence,

$$v = \frac{x_2}{t_2} = \frac{\gamma\,(L' + uT')}{\gamma\left(T' + \frac{u}{c^2}L'\right)}$$

We may cancel the γ's, and use equation (22), $L' = v'T'$, to rewrite this expression:

$$v = \frac{v'T' + uT'}{T' + \frac{u}{c^2}v'T'}$$

or

$$\boxed{v = \frac{v' + u}{1 + \frac{uv'}{c^2}}} \tag{23}$$

This is the relativistic expression that relates the velocity v of an object measured in one inertial frame to its velocity v' measured in another inertial frame moving in the same direction as the object. Notice that if u and v' are much less than c, then

uv'/c^2 is very small and the relation is approximately $v = v' + u$, the Newtonian result.

Let us examine how equation (23) resolves the difficulty suggested in Figure 14–19. Suppose the spaceship is moving with a speed $3c/4$ relative to the earth, and the bullet travels with a speed $3c/4$ relative to the spaceship. Then equation (23) implies that the speed of the bullet relative to the earth is not $3c/4 + 3c/4 = 3c/2$, but rather:

$$v = \frac{3c/4 + 3c/4}{1 + (3c/4)(3c/4)(1/c^2)} = \frac{\frac{3}{2}c}{1 + \frac{9}{16}} = \frac{24}{25}c$$

In fact, equation (23) shows that an object moving with a speed less than that of light in one inertial frame also has a speed less than that of light in any other inertial frame.

Finally, implicit in equation (23) is the second postulate of the theory of relativity, which states that light has the same velocity c in any inertial reference frame. Let us replace the gun in the spaceship with a flashlight. A pulse of light sent out by the flashlight has velocity $v' = c$ relative to the spaceship. If the spaceship is moving with speed u relative to the earth, what is the speed v of the light pulse relative to the earth?

$$v = \frac{c + u}{1 + \dfrac{uc}{c^2}} = \frac{c + u}{1 + \dfrac{u}{c}} = \left(\frac{c + u}{c + u}\right) c = c$$

Thus the light pulse has the same velocity c relative to the earth and the spaceship.

From these considerations, we see once again that Einstein's theory is internally consistent. In each of the cases we have examined in this chapter, we anticipated a possible logical contradiction that might arise from an application of the theory. Each time, however, the apparent contradiction was seen to result from an unjustified application of our intuitive, Newtonian ideas of space and time. Einstein's theory, while it conflicts with our intuition, nevertheless provides a consistent description of a possible world. The success of the theory in dealing with the phenomena of light and rapidly moving elementary particles leads one to assume that the Einsteinean conceptions of space and time are applicable to our world, and that the peculiar effects of relativity would be evident in our everyday lives if we moved with speeds near that of light.

This consistency, the way the entire theory flows so simply without contradiction from the two basic postulates, accounts for much of the aesthetic attraction of Einstein's theory. While the Lorentz theory, resulting as it did in the same equations, presumably could have led to the same agreement with experimental observations without the sacrifice of the Newtonian concepts of absolute space and time, the simplicity and consistency of the special theory of relativity have led to its universal acceptance among physicists today.

Suggested Reading

Arons, A. *Development of Concepts of Physics*. Reading, Mass.: Addison-Wesley, 1965. Chapter 36 gives a different way to derive the Lorentz transformation.

Kacser, C. *Introduction to the Special Theory of Relativity*. Englewood Cliffs, N.J.: Prentice-Hall, 1967. Chapter 3 has another derivation of the Lorentz transformation; Chapter 4 treats the twin paradox, velocity addition, and causality.

Mermin, N. D. *Space and Time in Special Relativity*. New York: McGraw-Hill, 1968. The Lorentz transformation is discussed in Chapter 13, while Chapter 16 is a particularly good analysis of the twin paradox.

Questions

1. Nature and Nature's laws lay hid in night:
God said, "Let Newton be!" and all was light.

—Alexander Pope

It did not last: the Devil howling, "Ho,
Let Einstein be!" restored the status quo.

—J. C. Squire

To what extent did Einstein's theory of relativity represent a revolutionary departure from the Newtonian world view? What elements of Newton's physics were changed? What elements remained intact?

2. Comment on the possible use of a high-speed spaceship as a "time machine." What major limitation would it have?
3. An enchanted prince, aged twenty years, has been turned into a frog and can only turn back into a prince if kissed by a beautiful princess on his twenty-first birthday. The only beautiful princess around is fifteen years old, and is not allowed to look upon a man until she is twenty-one. How far must the frog-prince travel, on how fast a spaceship, to return to the earth on the day that both he and the princess turn twenty-one?
4. The first manned trip to the moon took about three days (approximately 3×10^5 sec) each way. The distance from the earth to the moon is roughly 4×10^8 m. When they returned, how much younger were the astronauts than their twin brothers who remained on earth? (Assume the trip to be made at constant speed. You may find useful the approximation described in Question 5, page 316.)
5. A spaceship travels out from the earth at constant speed to explore the planetary system of a star 100 light-years from the earth, as measured in the rest frame of the spaceship. Suppose the trip takes 125 years as measured on clocks aboard the spaceship.
   (a) What is the speed of the spaceship?
   (b) How long does the trip take, as measured in the rest frame of the earth?
   (c) How far apart are the earth and star, as measured in the rest frame of the earth?
6. A μ meson is injected with speed $u = .8c$ into an evacuated circular ring surrounded by magnets, as shown in Figure 14–7. If the radius of the ring is 10 m and the μ meson lives 2×10^{-6} sec in its rest frame, how many times will it circle the ring before it decays?
7. Consider the train *gedanken* experiment illustrated in Figure 13–2. Let event E_A occur at $x = 0$ and $t = 0$, relative to the S reference frame fixed to the embankment. Let event E_B, which takes place simultaneously with E_A in the S frame, occur at $x = L$ relative to S. Use the Lorentz transformation to find the difference in the times at which E_A and E_B occur in S', the rest frame of the train.

8. Use the Lorentz transformation to derive the length-contraction result. The following approach may be simplest: Consider a rod of proper length L' at rest in the S' frame. The rod and S' frame move relative to the S frame with speed u in the $+x$ direction, as shown in the figure. Suppose at time T, relative to the S frame, the ends of the rod are located at x_1 and x_2. That is, the length of the moving rod in S is $L = x_2 - x_1$. Focus on two events: E_1, "left end of rod passes x_1," and E_2, "right end of rod passes x_2." Use the Lorentz transformation to express x_1' in terms of x_1 and $t_1 = T$, and x_2' in terms of x_2 and $t_2 = T$. This will enable you to find $x_2' - x_1' = L'$, the rod's proper length, in terms of L, u, and c.

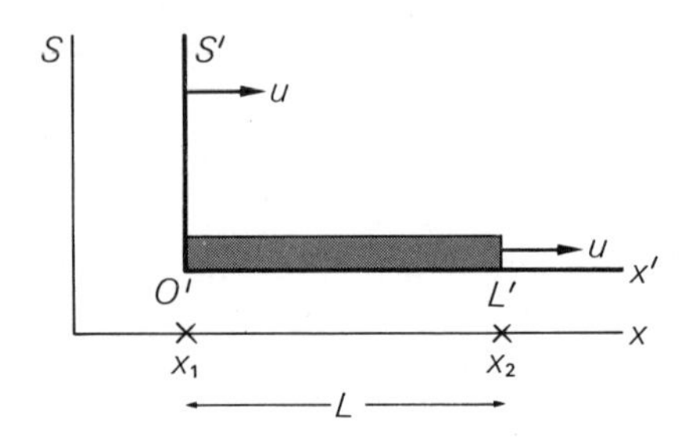

9. Two spaceships, A and B, travel in opposite directions relative to the earth, each with a speed $.9c$ relative to the earth. According to the pilot of A, how fast is B moving, assuming the applicability of

   (a) Newtonian physics
   (b) Relativity

   (Hint: Let S be the rest frame of the earth, and S' be the rest frame of ship A. Identify the velocities v', v, and u as used in the text. Be careful about signs.)

10. Two spaceships, A and B, are moving relative to the earth with speeds of $4c/5$ in opposite directions as shown.

    (a) What is the speed of spaceship A according to an observer on B?
    (b) According to an observer on B, how much does the separation of the spaceships increase in a time of one second, as measured on his clock?

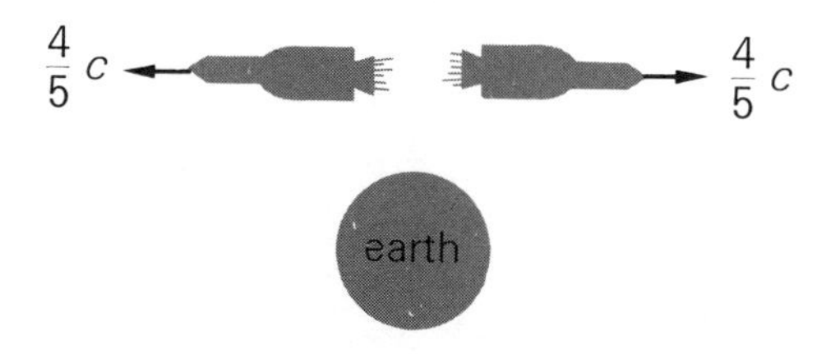

(c) According to an observer on the earth, how much does the separation increase in a time of one second as measured on his clock?

11. Consider again the train *gedanken* experiment illustrated in Figure 13–2. Let $x = 0$ be the position of event E_A with respect to the S reference frame fixed to the embankment, and let $t = 0$ be the time of E_A in this frame.

    (a) Draw a space-time diagram analogous to that of Figure 14–16, indicating events E_A and E_B relative to the S frame. In which of the three regions, PAST, FUTURE, or ELSEWHERE, does E_B lie?

    (b) What can you therefore conclude about the time order of E_A and E_B as measured in other inertial reference frames? Can you indicate an inertial frame in which E_A occurs before E_B? after E_B? (Hint: You might first consider the rest frame of the train.)

CHAPTER FIFTEEN

Does the Earth Really Move?

The development of physics described in this book shows a definite trend away from absolutism and toward relativity. In the geocentric universe of the Greeks, both position and motion were absolute. The center of the earth was the center of the universe, and all else took on meaning with respect to that special position. By the time of Newton, position was no longer absolute. An object dropped from above the earth was seen to obey the same laws and move in the same way as another object dropped from above another "earth" hundreds of light-years away.

But before Einstein, motion was seen as absolute. The ether was absolutely at rest and in principle all motion ought to be referred to the special reference frame fixed to the ether. Then came the special theory of relativity, which rejected in large measure the idea of absolute motion. According to Einstein's principle of relativity, no physical measurement will disclose which of two inertial reference frames moving relative to one another is really moving.

One last vestige of absolutism remained, however—the special role of the inertial reference frames. Einstein's special theory did not go so far as to say that *all* motion is relative. There are measurable effects that may be used to test whether an object is *accelerating* relative to the inertial frames. For example, there are observations which indicate that the frame of reference fixed to the earth is not quite inertial—i.e., that the earth accelerates relative to the inertial frames. These observations were long thought to provide direct evidence that the earth really moves in an absolute sense. However, we shall see that the so-called general theory of relativity, which Einstein later proposed, leads to a different interpretation of these observations by carrying the trend toward relativity one step further.

Detecting Inertial Frames

In principle, it is easy to determine whether a given reference frame is inertial. One simply observes the motions of objects relative to that frame and inquires whether these motions can be accounted for by the action of gravitational and electromagnetic forces through the equation $\vec{F} = m\vec{a}$. If they can, the frame is inertial; if not, the frame is not inertial.

The simplest case to consider is the motion of an isolated object. As discussed in Chapter 11, this means an object far from the other matter of the universe. Since both gravitational and electromagnetic forces decrease rapidly with distance, one can conclude that the force $\vec{F}$ on such an object is negligibly small. $\vec{F} = m\vec{a}$ therefore implies that the object's acceleration is zero, which means that it will move with a constant velocity relative to an inertial reference frame. For example, suppose the S frame of Figure 15–1(a) is inertial. An isolated object placed at rest at the origin of S will remain at this position.

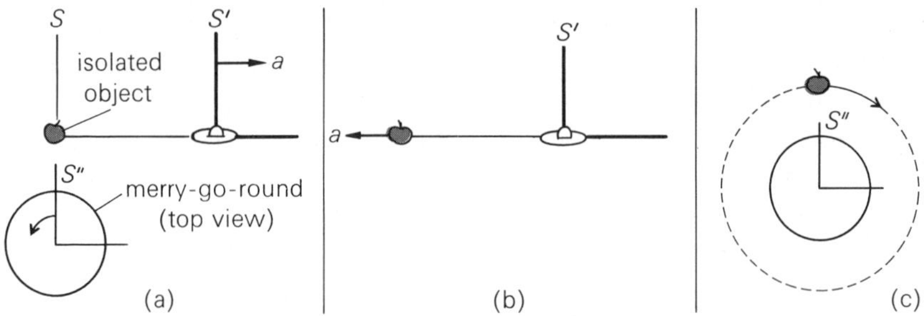

Figure 15–1. The motion of an isolated object depends on the reference frame in which its motion is measured. (a) The isolated object remains at the origin of the inertial frame S. (b) The object accelerates to the left relative to the non-inertial frame S', the rest frame of the spaceship. (c) The object moves in a circle relative to the non-inertial frame S'', the rest frame of the merry-go-round.

Now consider how the same object moves relative to non-inertial reference frames. Two such frames are shown in Figure 15–1(a): S'—the rest frame of a spaceship moving relative to S with acceleration a in the $+x$ direction, and S''—the rest frame of a merry-go-round rotating uniformly relative to S. Let us examine the motion of the isolated object relative to each of these reference frames.

As illustrated in Figure 15–1(b), the object will accelerate toward the left with respect to the rest frame of the spaceship; part (c) shows the uniform circular motion of the object relative to the rest frame of the merry-go-round. Thus an isolated object does not move with a constant velocity relative to either of these frames. Consequently, by observing the (accelerated) motion of an isolated object, a person at rest in either frame could conclude that he was "really" moving (relative to the inertial frames).

The Foucault Pendulum

It is often claimed that there is direct evidence that the earth is really moving—that there are observations of motion at the surface of the earth that do not quite agree with the predictions of $\vec{F} = m\vec{a}$, and therefore indicate that the rest frame of the earth is accelerating relative to the inertial frames. The observation most frequently cited is the motion of a *Foucault pendulum*, which is said to prove that the earth is rotating.

The Foucault pendulum consists of a heavy sphere attached to the end of a long string. The sphere is pulled to one side and released from rest. Many museums have such a display, and the result is striking. Instead of retracing the same path over and over again, the pendulum precesses—its plane of swing changes, slowly but perceptibly, eventually rotating through a full 360° and returning to its original orientation.

This observation would be difficult to explain if the earth were fixed in an inertial reference frame. The only forces experienced by the sphere, the gravitational attraction of the earth and the pull of the string, are in a vertical plane. (See Figure 15–2.) There are no lateral forces that would tend to change the direction of swing of the pendulum. According to $\vec{F} = m\vec{a}$, the pendulum would be expected to swing back and forth in a vertical plane. It would therefore not precess relative to an inertial reference frame. Suppose, however, that the pendulum were viewed from a merry-go-round rotating slowly beneath it. To an observer on the merry-go-round, the plane of swing of the pendulum would appear to rotate.

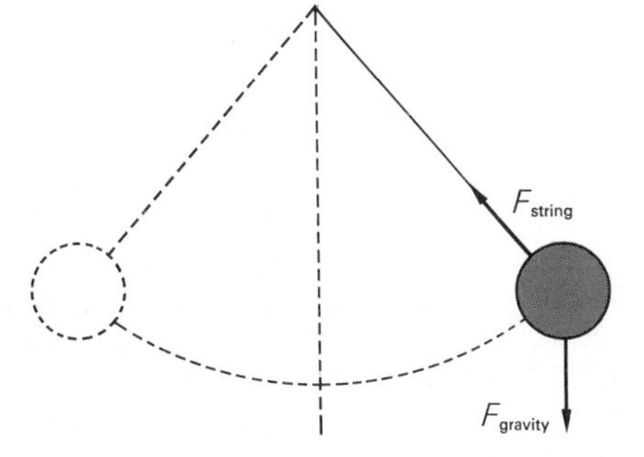

Figure 15–2. The forces on the Foucault pendulum are in a vertical plane, leading to the prediction that the pendulum will swing back and forth in this plane.

This suggests a simple explanation of the precession of a Foucault pendulum. Imagine the pendulum to be located at the earth's North Pole. Relative to the "fixed stars," presumed by Newton to be an inertial reference frame, the direction of swing of the pendulum will remain fixed. But if the earth acts like a gigantic merry-go-round, rotating beneath the pendulum, the direction of swing will seem to precess relative to the earth, completing a full rotation once each day. Though the analysis is somewhat more complicated if the pendulum is not located at one of the poles of the earth, a precession will still result from the earth's rotation. Such an effect, first demonstrated by the French physicist Jean Foucault in 1851, seems to indicate that the earth rotates relative to the inertial frames.

Fictitious Forces

It is often convenient to use Newton's laws to analyze data obtained in an accelerated reference frame. For example, if we made detailed measurements of the motion of the Foucault pendulum, we would surely make them with respect to a reference frame at rest on the earth. To compare these measurements with the predictions of Newton's laws, it would be simplest to make these predictions with respect to this non-inertial frame.

From the discussion in Chapter 11, it should be clear that such a program must be followed with caution. Newton's laws, with the two basic forces—gravity and electromagnetism—can be applied directly only in inertial frames. However, it is not difficult to show that satisfactory predictions of motion in a non-inertial frame can be obtained from Newton's laws in the usual fashion, provided we postulate additional *fictitious forces* that depend on the acceleration of the reference frame.

There are many familiar examples of fictitious forces. When a car's brakes are suddenly applied, its occupants are thrown forward—they accelerate relative to the car. There are two ways to look at this, as illustrated in Figure 15–3. In the rest frame of the earth, the car moves with acceleration $a = -A$ as a result of the frictional force between its wheels and the ground [Figure 15–3(a)]. The occupant experiences no such force and hence continues to move with the constant velocity he had before the brakes were applied—until he crashes into the windshield of the car, which is no longer moving with him.

Alternatively, one could consider the motion from the point of view of the non-inertial frame in which the car is at rest. As

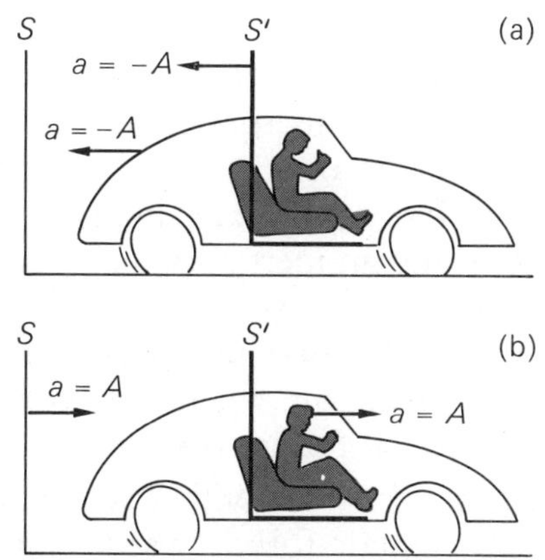

Figure 15–3. The motion of the occupant of a stopping car, as viewed in two reference frames. (a) The car, with brakes applied, has an acceleration toward the rear. The occupant continues to move with constant velocity relative to the earth. (b) Relative to the car, the occupant accelerates forward.

shown in Figure 15–3(b), the occupant accelerates forward with acceleration A relative to this frame. To attempt to explain this motion using $\vec{F} = m\vec{a}$, one must postulate that in this non-inertial frame the man experiences a fictitious force F_f, with magnitude equal to the mass of the man times the acceleration of the reference frame (i.e., $F_f = mA$), and direction opposite to the direction of the acceleration of the frame. Applying $\vec{F} = m\vec{a}$ in the accelerated frame, we then find:

$$mA = ma$$

or

$$a = A$$

indicating that the man will accelerate forward in the (non-inertial) rest frame of the car.

Another familiar example of a fictitious force is the outward thrust one experiences on a rapidly rotating merry-go-round. While this force and the one in the car example are called fictitious, they certainly seem real to the child on the merry-go-round or the occupants of the car. These new forces are called fictitious because in each case they can be attributed to the acceleration of the reference frame relative to the inertial frames.

The greater the acceleration, the greater the magnitude of the fictitious force that must be assumed. When the motion is described with respect to an inertial frame, these forces do not exist. We postulate them in order to apply Newton's laws in an accelerating reference frame; they are simply a correction for the fact that we are not using an inertial frame as Newton's first law says we should.

The notion of a fictitious force provides an alternative way to describe the test for deciding whether a particular reference frame is inertial. In inertial frames the motion of all objects can be accounted for using $\vec{F} = m\vec{a}$ with $\vec{F}$ restricted to the gravitational and electromagnetic force laws; in non-inertial frames, fictitious forces must be invoked as well.

Such at least was the view of Newtonian physics retained in the special theory of relativity. However, in the late nineteenth century the German physicist Ernst Mach called into question this distinction between real and fictitious forces.

Are Fictitious Forces Real?

Mach was a latter-day follower of the German philosopher and mathematician Leibniz, a contemporary of Newton, who had disputed Newton's assumption of an absolute space existing independent of the matter in it. Leibniz advocated instead a complete relativism. He insisted that all one could know about the position and motion of an object was its position and motion relative to other objects, and that therefore relative motion (including acceleration) was all that could enter into any physical theory. In answer to these objections Newton put forth his famous water-bucket experiment, illustrated in Figure 15–4:

> A bucket, suspended by a long cord, is turned about so that the cord is strongly twisted, then is filled with water, and held at rest together with the water [Figure 15–4(a)]; thereupon, by the sudden action of another force, it is whirled about the contrary way, and while the cord is untwisting itself, the bucket continues for some time in this motion; the surface of

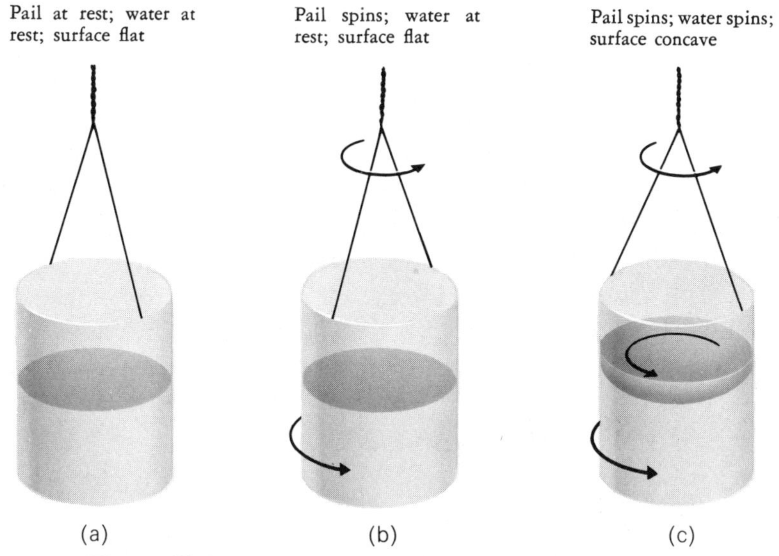

Figure 15–4. Newton's water-bucket experiment.

> the water will at first be level, just as before the bucket began to move [part (b)]; but after that, the bucket, by gradually communicating its motion to the water, will make it begin sensibly to rotate, and recede little by little from the middle, and ascend to the sides of the bucket, forming itself into a concave figure (as I have experienced), and the swifter the motion becomes, the higher will the water rise, till at last, performing its revolutions in the same times with the bucket, it becomes relatively at rest in it [part (c)].[1]

There is a straightforward Newtonian explanation for the concave shape of the spinning water surface. Consider a bit of water, as shown in Figure 15–5. At some instant it has a velocity $\vec{v}$ as shown, and by Newton's first law, it would continue moving in the absence of forces with unchanged velocity—in a straight

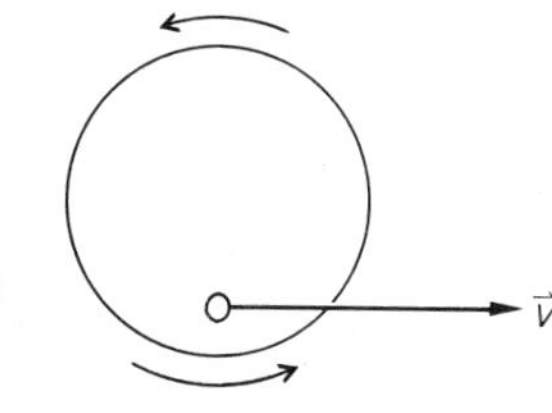

Figure 15–5. A bit of water in the spinning bucket. (Top view.)

line. But it cannot; the side of the bucket prevents it. Consequently it, and the rest of the water, tend to pile up around the inside surface of the bucket, giving a concave shape to the top of the water. Another way to look at this effect is from the point of view of the reference frame rotating with the water. In this non-inertial frame, each bit of water, like a child on a merry-go-round, experiences a fictitious force that throws it out toward the sides of the bucket.

Newton argued that the experiment refutes Leibniz' assertion, showing that there are physical effects that depend on more than simply the relative motion of objects. In Figure 15–4(a), when there was no relative motion of water and bucket, the water surface was flat. Then, in (b),

> when the relative motion of the water in the bucket was greatest, it produced no endeavor to recede from the axis; the water . . . remained of a level surface, and therefore its true circular motion had not yet begun. But afterwards, when the relative motion of the water had decreased, the ascent thereof towards the sides of the bucket proved its endeavor to recede from the axis; and this endeavor showed the *real circular motion* of the water continually increasing, till it had acquired its greatest quantity, when the water rested relatively in the vessel [as in (c)].[2]

In parts (a) and (c), there is the same relative motion of the bucket and water; a bug floating in the water in each case would see the bucket at rest relative to himself. But by observing the shape of the water's surface, the bug could tell whether he was really moving. The "real circular motion" of the water can thus be determined.

Mach, however, was reluctant to accept any such absolutes. As he wrote:

> No one is competent to predicate things about absolute space and absolute motion; they are pure things of thought, pure mental constructs, that cannot be produced in experience.[3]

He proposed a different interpretation of the water-bucket experiment, which maintained the principle that only relative motions are important:

> Newton's experiment with the rotating vessel of water simply informs us that [the curvature of the water's surface] is produced by its *relative* rotation with respect to the mass of the earth and the other celestial bodies.[4]

Thus, according to Mach, there is a significant difference, which Newton did not recognize, between parts (a) and (c) of Figure 15–4. The bulk of the matter of the universe, which Newton called the "fixed stars," must not be neglected. In (a) the stars are at rest with respect to the water; in (c) the stars are rotating (accelerating) with respect to the water. Might it be possible to attribute the curved surface in (c), not to an absolute rotation of the water, but rather to the rotation of the water relative to the stars?

Suppose, Mach suggested, that when one bit of matter accelerates relative to another bit of matter, it experiences a force. This force is different from the gravitational and electromagnetic forces we have discussed, but just as fundamental. If such a force exists, Newton's water-bucket experiment can be explained without introducing the idea of absolute rotation. In Figure 15–4(a), the stars are at rest relative to the water, so the water experiences no force. In part (c), however, the stars are rotating around the water and hence exert a force on it. According to Mach, it is this force, arising from the relative acceleration of the stars and the water, that accounts for the distortion of the water's surface.

The existence of this force would also explain the unique role of inertial reference frames in Newtonian physics and special relativity. In Mach's view, what singles out the inertial reference frames is that they are the frames moving with a constant velocity relative to the bulk of the matter in the universe. In these frames, the fixed stars do not accelerate, and therefore in them Mach's new force is zero. In any other frame the force is nonzero. Hence, according to Mach, there is nothing absolute about the inertial frames. If Mach's force exists, *any* reference frame should be valid for the application of Newton's laws, provided that this new accelerated matter force is included among the basic forces of nature.

At first encounter, Mach's assumption may seem a bit far-fetched. Is it not difficult to believe that the stars, so remote from the earth, could have such a large effect on earthly objects? In addition, if we can attribute the curved surface of the water in Figure 15–4(c) to the rotation of the stars relative to the water, why is Mach's force not observed when earthbound objects accelerate relative to one another?

In reply to these objections, one can argue that although the stars may be distant, there are quite a few of them, and conceivably their vast number could dominate over their great distance. Furthermore, production of an accelerating matter effect of measurable size in the laboratory might require the movement of masses so huge that the force has thus far escaped notice.

In 1894 an experiment to search for such an effect was attempted at Mach's urging. A detector of accelerating matter forces, which we represent schematically in Figure 15–6 as a bucket of water, was placed near the center of a massive flywheel. However, the force generated by the rotation of the flywheel, if any, was too small to be detected. Very delicate experiments of a rather different nature, but designed to measure such an effect, are currently being planned. They will be performed in an earth satellite to take advantage of the cold and vacuum of outer space.

Even in the absence of explicit experimental support, however, Mach's point of view appeals to many physicists. Let us em-

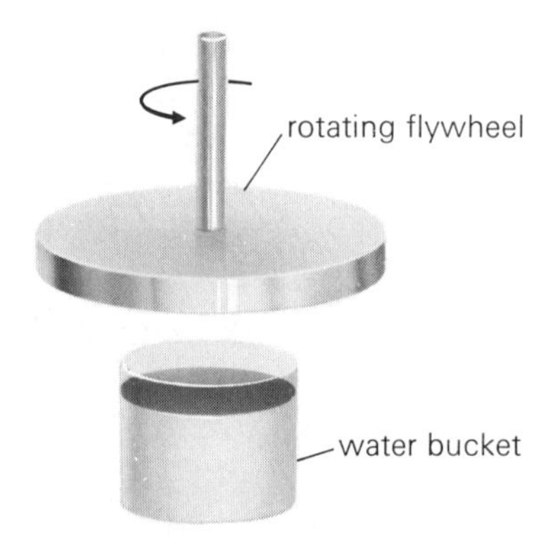

Figure 15–6. An experiment to detect Mach's new force. The acceleration of the matter in the flywheel relative to the bucket was expected to cause a curvature in the water surface.

phasize again the difference between Newton's and Mach's ideas of motion. To Newton, inertial reference frames had an absolute quality in some way related to the basic properties of space, and independent of any matter occupying that space. Even in a universe devoid of all matter, inertial and non-inertial reference frames would be meaningful concepts. In an otherwise empty universe, the water in a rotating bucket would attain a curved surface because of its absolute acceleration.

To Mach, all reference frames were taken to be equivalent, in the sense that the laws of physics could be applied equally well in any of them. The price of this generality was the assumption of a new force to add to the basic two. In an otherwise empty universe, the water in a rotating bucket would remain level because there would be no other matter to exert a force on it.

Mach's outlook also leads to a different perspective on the once-critical issue of whether the earth is really moving. To a Newtonian observer standing on the earth, the precession of the Foucault pendulum must be attributed to a fictitious force, indicating that the earth is not fixed in an inertial reference frame—thus, that the earth "really" moves. But from Mach's point of view, the precession can be attributed to forces arising from the rotation of the distant stars relative to the earth. We may equally well imagine the universe to be rotating relative to the earth, or the earth rotating relative to the universe. To Mach, it makes no sense to accept either point of view as more real than the other.

The Principle of Equivalence

Einstein was intrigued with Mach's thinking, and after completing his work on the special theory, he sought an extension of the principle of relativity that would apply in *any* reference frame. The result, now known as the general theory of relativity, was an ingenious reversal of Mach's thought: Instead of considering fictitious forces as real, Einstein tried to view real forces as fictitious—that is, as due simply to the acceleration of one's reference frame.

The theory was actually developed only for the gravitational force. The principal clue to the approach came from the apparent connection between inertial mass and gravitational force.

The dependence of the gravitational force on mass was not seen as noteworthy until the end of the nineteenth century. After all, Newton himself had not claimed to explain gravity, and in the two hundred years after his death, physicists were so preoccupied with the application of the Newtonian formalism to all aspects of the physical world that they felt no strong inclination to venture where Newton had failed. Newton's approach had proved so remarkably successful that there was little interest in probing more deeply into the origin of the gravitational force.

The first serious reconsideration of this issue came late in the nineteenth century. It was noted that mass is an inertial property, a measure of an object's resistance to acceleration. Is it not curious that this property is related to the strength of the gravitational interaction? In 1889, the Hungarian physicist Baron Roland von Eötvös performed an experiment to test whether the gravitational force was precisely proportional to inertial mass. To an accuracy of one part in 10^9, he found that two objects with the same mass interact with the same gravitational force independent of the material of which they are composed.

One might view the dependence of gravity on mass as simply the way things are and leave it at that. To Einstein, however, this dependence suggested an intriguing possibility. Perhaps gravity and inertia can be viewed as different aspects of the same phenomenon. Consider, by way of illustration, the two laboratories shown in Figure 15–7. In part (a) the laboratory is at rest on the surface of a massive object (e.g., the earth). If we ignore the small effects arising from the leisurely motion of the earth, we can say that the reference frame S', at rest with respect to the laboratory, is also at rest with respect to the fixed stars and is thus an inertial frame. Thus the laboratory is labeled $a = 0$. All objects released inside the laboratory accelerate downward relative to S' with the same acceleration $a' = 980 \text{ cm/sec}^2$. Note that Newton's second law predicts this result; because the object's mass m appears both in $F' = GM_e m/r_e^2$ and in ma', when we

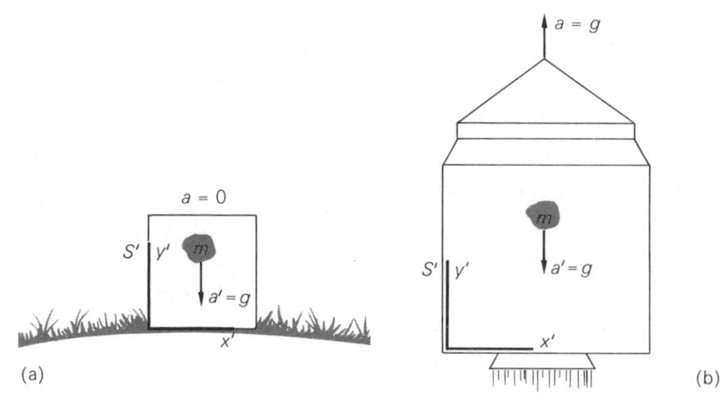

Figure 15–7. In each case, the object has an acceleration of magnitude $g = 980 \text{ cm/sec}^2$ in the $-y$ direction, relative to the S' reference frame. (a) An object released above the surface of the earth, as viewed from an inertial frame S' at rest on the earth. (b) An isolated object in outer space, as viewed from an S' frame with an acceleration g relative to the fixed stars.

apply $F' = ma'$ we find that the mass of the falling object cancels out:

$$\frac{GM_e m}{r_e^2} = ma'$$

or

$$a' = \frac{GM_e}{r_e^2} = 980 \text{ cm/sec}^2$$

In Figure 15–7(b), the laboratory is in a spaceship somewhere in outer space. Suppose that the spaceship is accelerating with respect to the fixed stars, with acceleration $a = 980 \text{ cm/sec}^2$. A reference frame S' at rest in this laboratory is not an inertial frame. Imagine a mass m released inside the laboratory. Observed from the fixed stars, there are no forces on the mass, and consequently it does not accelerate. However, relative to the laboratory (frame S'), the mass accelerates downward with acceleration $a' = 980 \text{ cm/sec}^2$. This clearly would be true for any mass not fixed to the laboratory.

Notice that in each case, (a) and (b), the motion of objects relative to the laboratory is the same, a constant acceleration of 980 cm/sec^2 toward the floor of the laboratory. The effects of gravity can be duplicated by accelerating the laboratory in outer space, far from any gravitating bodies. Einstein's insight was thus that gravitational effects can be obtained without the presence of an attracting mass by choosing an appropriate (uniformly accelerated) reference frame. A person in an enclosed room on the spaceship could not determine, by watching the free fall of objects in his room, whether the spaceship was accelerating or parked on a massive planet.

By the same token, the effects of gravity near a large body such as the earth can be eliminated by the appropriate choice of reference frame. Consider a laboratory in an elevator, as shown in Figure 15–8. Suppose the elevator cable is severed and the laboratory falls toward the earth. Everything in such a freely falling laboratory will accelerate toward the earth with the same acceleration, 980 cm/sec^2. Thus, relative to the laboratory, the objects within will not accelerate. An apple released from rest will remain suspended where it was released. A person enclosed in the elevator would be unable to determine by watching the free motion of objects whether someone had cut the elevator cable or had snatched the earth out from under him. Of course, he might guess which was more likely . . . and he might soon find out.

Thus, Einstein viewed the presence of the gravitational force in a laboratory at rest on the surface of the earth merely as a consequence of the choice of an inconvenient reference frame.

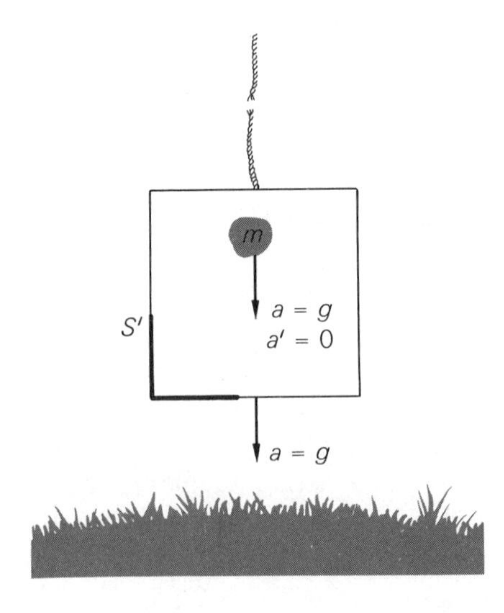

Figure 15–8. An object released above the surface of the earth does not move relative to a freely falling elevator.

With respect to a laboratory falling freely toward the earth, an isolated object moves with a constant velocity. Viewed from the earth, the same object is seen to be accelerated by gravity. Inertia in one reference frame is seen as gravity in another. In this fashion, Einstein attempted the first explanation of gravity since the seventeenth century.

Generalizing this view, Einstein postulated the complete indistinguishability of gravity and acceleration in his *principle of equivalence*: No physical measurement can distinguish a reference frame accelerating relative to the inertial frames from an inertial frame in a constant gravitational field. Upon this foundation, he erected his general theory of relativity, which provides insight into the origin of the gravitational force by linking it to the properties of space and time. Unfortunately, the details of the general theory involve abstract mathematics beyond that assumed of readers of this book. What is important for our purposes, however, is to recognize that the theory, by treating all reference frames as equivalent for the description of nature, climaxes a trend from absolutism to relativity in physics.

The Relativity of Physics

The general theory of relativity is yet another in a series of world views, each resulting from a revolution in the way man has looked upon his universe. Each of the men whose ideas we have studied contemplated the same universe, and each saw something different in it. Each asked different questions, saw different facts as significant, and accepted a different kind of explanation for what he saw.

Let us contrast once more the common view of science with the physics we have examined. Consider again a question often thought sufficiently objective and concrete to allow a "scientific" answer: Does the earth move around the sun, or the sun around the earth? It should be clear by now that there is no definitive answer to this question, and that it is not the business of science to supply one. The most one can say with certainty is that the relative motion of the earth and the sun in the heliocentric and geocentric models is the same. What one says beyond this depends on which aspects of the world appear most significant to him.

To the early Greek, concerned with reconciling the motions of the many stars in the heavens with his intuitive feelings about the symmetry of nature, the stability of the earth, and the central role of man, the geocentric view had the greatest appeal. To Newton, seeking to establish a connection between terrestrial and celestial motions, the heliocentric system provided a far more useful viewpoint. To Mach, who believed that only relative motion can enter into physical theories, and to Einstein, who sought to understand the relation between gravity and inertia, both models had to be seen as equivalent, neither more real than the other. To each man, and, we suggest, to all of physics, the goal is to make sense of the world, not to determine absolute truths.

Suggested Reading

Einstein, A. *Relativity: The Special and General Theory*. New York: Holt, 1921. The second half of this non-mathematical book deals with the general theory.

Mach, E. *The Science of Mechanics: A Critical and Historical Account of its Development*. Translated by T. J. McCormack. 3rd ed. Chicago: Open Court, 1907. Part 6 of Chapter 2, "Newton's Views of Time, Space, and Motion," sets forth Mach's ideas and philosophical position.

Cooper, L. *An Introduction to the Meaning and Structure of Physics*. New York: Harper & Row, 1968. Chapter 33 gives a brief account of the general theory of relativity and its experimental tests.

Dicke, R. H. "The Eötvös Experiment." *Scientific American*, December 1961, pp. 84–94. A non-technical account of this experiment.

Questions

1. A skeptical student says, "There is one good way to tell if the earth really moves or not. Get on a spaceship and travel outside the solar system. Then look back. It will be obvious what is rotating about what." Comment on this statement.

2. To what extent is Mach's point of view revolutionary? How much of Newtonian physics must one give up to adopt it?

3. The flywheel experiment described on page 406 is another instance of an experiment disagreeing with a theoretical prediction. What resolution of the disagreement is implied in the text? Compare this discrepancy with those listed in Question 6, page 316. Which does it resemble most closely, insofar as the nature of the resolution is concerned?

4. Did the Eötvös experiment *prove* that the gravitational force is the same on two objects of the same mass but different materials? When a similar experiment was performed recently with still more precise apparatus, the same result was obtained to a precision of 1 part in 10^{11}. Why do you suppose people went to the trouble of improving the precision of this test? Under what circumstances would it seem important to improve the precision to 1 part in 10^{13}?

5. Suggest examples other than the merry-go-round and accelerating car in which people experience fictitious forces.

6. Discuss how Newton and Mach would each explain the observation that objects float in a freely falling elevator.

7. Astronauts orbiting the earth experience the phenomenon of weightlessness (no net force in the rest frame of the satellite). How would Newton and Mach each explain this effect?

8. In each of the non-inertial reference frames S' and S'' of Figure 15–1, what is the magnitude and direction of the fictitious force that must be assumed to account for the motion of the isolated object? (Assume S'' rotates relative to the inertial frame S once every T seconds, and its origin is located a distance R from the origin of S.)

9. Imagine an object that is free to slide without friction on the floor of a merry-go-round. The merry-go-round rotates uniformly in a counterclockwise sense relative to the ground. Suppose the object is initially located at the center of the merry-go-round, and is then given an outward push.

(a) Describe in words the object's path relative to the ground.
(b) Describe in words the object's path relative to the merry-go-round.
(c) In the rest frame of the merry-go-round, describe in words what kind of fictitious force must be assumed to account for the motion of the object.

10. The earth exhibits nearly uniform circular motion relative to an inertial reference frame S whose origin is at the sun.

    (a) Describe the motion of the earth relative to a reference frame S' whose origin coincides with that of S, but which rotates relative to S once each year in the same direction as the earth.
    (b) What fictitious force must be assumed in S' to account for the observed behavior of the earth in this frame?

11. It is feared that an astronaut's physical condition will deteriorate under conditions of prolonged weightlessness. Consequently, it is likely that spaceships for future long-term space exploration will rotate. Use the principle of equivalence to explain how such rotation will produce an artificial "gravity." Sketch such a spaceship, indicating what dimensions and rotation rate would produce a "gravity" comparable to that on the surface of the earth.

12. The equivalence principle predicts that light rays experience a gravitational attraction and are thereby bent when in the vicinity of a massive object. According to this principle, a laboratory in a spaceship accelerating upward relative to the inertial reference frames is equivalent to a laboratory in which a gravitational force is acting downward. Suppose that a light beam propagating in a straight line relative to an inertial frame passes through window W into the laboratory, as shown in the figure. In this inertial frame the laboratory is accelerating upward and the light continues to move in a straight line.

    (a) Indicate the position of the window on the other side of the laboratory where the light exits.

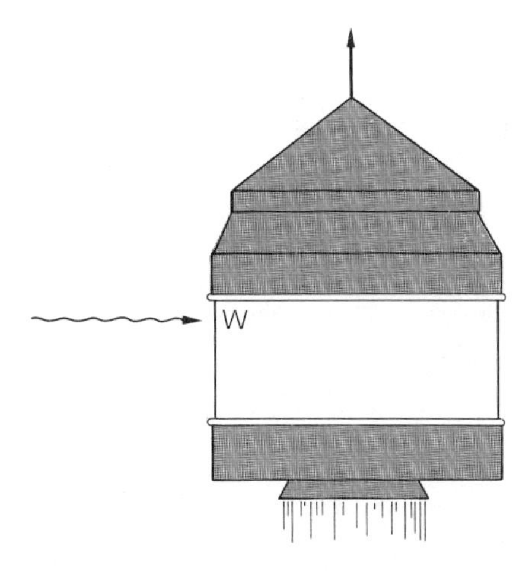

(b) Sketch the path of the light ray through the laboratory as seen by an observer in the laboratory (i.e., in the rest frame of the accelerating spaceship). Hint: Use the result of part (a).

(c) Explain how this result implies that light rays will bend in the presence of a gravitational force if the equivalence principle is valid.

(d) Suggest a way in which light from distant stars, passing near the sun on its way to the earth, could be used to detect such a bending of light by massive objects. (Successful observations of this sort were first made in 1919, when two British expeditions traveled to Brazil and Africa to observe the stars during a total eclipse of the sun. Their results, in agreement with Einstein's predictions, made headlines in newspapers around the globe, thrusting relativity into the public spotlight for the first time.)

EPILOGUE

What is happening has happened before in physics: the old way of looking at things, which was adequate for perceiving order in a limited number of observations, finally proved cumbersome and inadequate when the accuracy and range of observations increased . . . A similar situation may exist today in particle physics. The great unifying invention . . . is still not clearly in sight, but the experimental data are beginning to fall into striking and partly predictable patterns.[1]

—Geoffrey Chew, Murray Gell-Mann and Arthur Rosenfeld

Is the physics we have discussed in the preceding fifteen chapters relevant to physics as it is practiced today?

CHAPTER SIXTEEN

Elementary Particles: A Revolution in Progress?

In Newton's time, physics and astronomy were the primary concern of only a very few scholars, mainly in England and on the European continent. By 1905, when Einstein published his first paper on relativity, the physicists in the world numbered only a few thousand. By the 1970s, however, physics has become a popular profession; in the United States alone there are more than fifty thousand physicists employed by universities, industry, and the government. Significant contributions to the discipline are being made in almost every nation of the world. Of all the physicists who have ever lived, over eighty percent are active today.

Accompanying this growth in numbers has been a dramatic increase in the complexity and cost of the apparatus used in experimental physics, and an equally great change in the life style of the physicists who use it. For his data, Newton relied mainly upon John Flamsteed, the Astronomer Royal, who worked in relative isolation at the Greenwich Observatory. The apparatus of Michelson and Morley, while elaborate for its time, could not have cost more than a few thousand dollars, and the experiments were carried out by the two men alone in a basement room. But today a significant fraction of the research in physics requires budgets of hundreds of thousands and even millions of dollars. Radio telescopes hundreds of feet across, particle accelerators a mile in diameter, and nuclear reactors occupying whole buildings are expensive and complex, requiring scores of technicians for their operation. Experiments are often performed by teams of physicists, and research papers with five or more authors are common.

This vast expansion of effort in the twentieth century has brought many changes to physics. There has been one major conceptual revolution—the theory of *quantum mechanics*. In the early decades of this century, there was increasing evidence that Newton's approach was inadequate to describe atomic phenomena. Quantum mechanics, a highly abstract mathematical theory, was developed in the late 1920s to deal with atoms. While retaining some elements of the Newtonian scheme, such as a focus on interactions, quantum mechanics was truly a radical conceptual departure. It provided a world view consistent with data concerning the submicroscopic constituents of matter and with the Newtonian description of macroscopic phenomena as well.

It is probably fair to say that despite the intensive efforts of vast numbers of physicists, in the past half century there has been no further conceptual revolution in physics. Nevertheless, there has been enormous activity during this period. Operating within the conceptual framework provided by Newtonian mechanics, relativity, and quantum mechanics, physicists have intensively investigated phenomena ranging from atoms to galaxies. They have sought to explain the observed properties of solids, liquids, and gases for temperatures ranging from that of the sun to near absolute zero. They have probed the structure of the atom and in the process established the new discipline of nuclear physics. They have sought to understand why the stars shine and how galaxies evolve. The success of these efforts, and many others like them, seems to indicate that the theories of Newtonian mechanics, relativity, and quantum mechanics provide a highly satisfactory framework for viewing the world.

In light of this achievement, one is tempted to ask whether the succession of revolutions in physics has ended. Have we finally arrived at an ultimate description of the physical world? Or are there signs of a crisis that may herald yet another fundamental change in world view?

It is of course difficult to say whether a revolution is in progress today. One simply cannot be certain whether present concepts will be adequate to deal with currently unexplained phenomena. We can, however, point to areas that appear to be frontiers of research, where it seems most likely that the presently accepted theoretical framework will have to be modified or extended to provide satisfactory explanations. One such area is elementary particle physics.

Elementary particle physicists are concerned with experiments of the sort shown schematically in Figure 16–1. A charged nuclear particle is accelerated to a speed near that of light and then allowed to strike a target, such as a thin metal foil. In the interaction between the particle and the nuclei of the target atoms, new particles are created. Many properties of these particles, such as their mass and charge, can be measured. During the past two decades, thousands of such experiments have been performed, leading to the identification of over a hundred different so-called elementary particles.

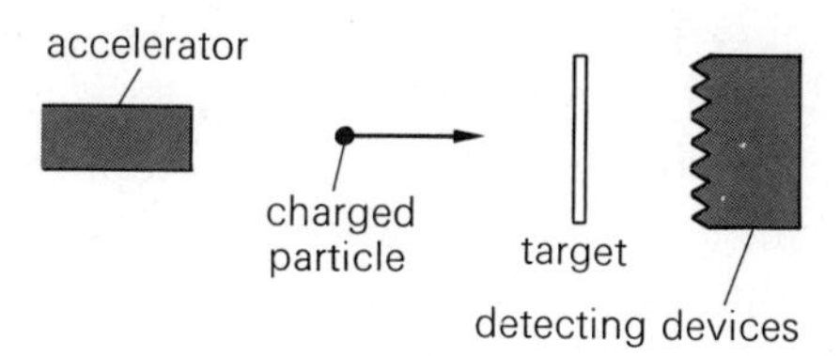

Figure 16–1. A typical elementary particle physics experiment. A charged nuclear particle emerges from an accelerator with a high speed and strikes a target, thereby creating new particles whose properties are analyzed by detecting devices.

We shall focus in this chapter on elementary particle physics as a possible example of a revolution in progress. So much observational evidence about elementary particles has been accumulated in recent years that many physicists now believe the time is ripe for theoretical understanding. But what form should a theory of elementary particles take? What questions should it answer? Which of the data now at hand are significant? How can they be displayed to bring out regularities that might otherwise lie hidden? These questions are being asked by physicists today, and the accompanying debate is reminiscent of that which preceded the conceptual revolutions of Newton and Einstein.

The fact that there is as yet no generally accepted solution to the elementary particle problem distinguishes it from the other topics we have examined. When we know the outcome of a revolution in physics, it is tempting to blame faulty logic or stubborn conservatism for the false starts and misread clues that preceded it. However, as physicists today struggle to understand elementary particles it is easy to appreciate the depth of the ambiguity that always accompanies the development of a physical theory.

The Search for the Ultimate Constituents of Matter

To understand the motivation behind elementary particle physics today, let us consider briefly its historical roots. Attempts to explain the basic structure of matter can be traced as far back as the early Greeks. This quest is a major theme in the development of physics, rivaling in importance the problem of motion.

We have mentioned one early Greek attempt to explain matter in terms of a single basic substance—Thales' Water. A more sophisticated proposal along the same lines was apparently made by Empedocles in the fifth century B.C. His suggestion that all matter is composed of four basic elements, Earth, Water, Air, and Fire, was widely accepted for almost two thousand years.

The related issue of whether matter is infinitely divisible or composed of discrete, irreducible chunks has been debated since Greek times. The dominant view until quite recently was that matter is a continuum—that it is indefinitely divisible and can be mixed in any proportions. The competing atomic picture was first put forth by Leucippus and Democritus in the fourth century B.C. It was adopted by others—notably Newton—in later times, but failed to gain widespread acceptance.

Developments in chemistry in the eighteenth and nineteenth centuries, however, led to the concept of a *chemical element* as a substance that could not be broken down into simpler constituents. Scores of chemical elements were identified and their properties catalogued. At the same time, studies of the way these chemical elements combine to form compounds revealed a number of regularities which a British schoolteacher, John Dalton, recognized could most easily be explained by assuming an atomic model. In this model, each chemical element is composed of vast numbers of identical, submicroscopic building blocks, the atoms of that element, and the chemical combination of two or more elements into compounds is brought about by a union of their atoms. Dalton proposed his theory in 1805, and by the end of the nineteenth century the view that matter is made up of over ninety different kinds of atoms was universally accepted.

The Constituents of Atoms

At the time it seemed a great simplification to be able to account for an enormous range of chemical properties in terms of just a few score of atoms. Viewed from another perspective, however, the atomic theory attributed an unattractive complexity to nature. It seemed to imply that matter is built up of ninety different kinds of elementary building blocks. Compared to the four elements of the Greeks, for example, this is a rather large number of basic constituents. Can nature be that complicated?

A key discovery in this regard was announced in 1869 by the Russian chemist Dmitri Mendeleev. He noted first that the elements can be grouped into families, the members of which have remarkably similar chemical properties:

> With respect to some groups of elements there are no doubts that they form a whole and represent a natural series of similar manifestations of matter. Such groups are: the halogens, the alkaline earth metals, the nitrogen group . . .[1]

He then pointed out a way of displaying the chemical elements that calls attention to these groupings:

> In the assumed system, the atomic weight of the element, unique to it, serves as a basis for determining the place of the element. Comparison of the groups of elements known up to now according to the weights of their atoms leads to the conclusion that the distribution of the elements according to their atomic weights does not disturb the natural similarities which exist between the elements but, on the contrary, shows them directly.[2]

Figure 16–2 shows Mendeleev's first attempt at such a "system," or *periodic table*, of the elements then known. Its rows correspond to the family groupings of atoms with similar chemical properties. For example, the halogens—fluorine (F), chlorine (Cl), bromine (Br), and iodine (J)—form one row, as does the nitrogen group—nitrogen (N), phosphorus (P), arsenic (As), antimony (Sb), and bismuth (Bi).

The existence of families of elements whose members occur periodically in a tabulation of the elements according to weight was an important clue in the search to understand the structure of matter. This regularity suggested that the scores of different atoms might not be the basic building blocks of matter. Instead, the atoms might themselves be combinations of more basic constituents.

A successful model of the atom based on this idea was gradually developed in the first third of the twentieth century, using the theory of quantum mechanics. In this model, atoms are composed of three different types of elementary particles—electrons, protons, and neutrons. The electron has a mass, m_e, of 9.1×10^{-28} gm and a particular negative electric charge which we shall symbolize by $-e$. The neutron and proton are nearly identical particles, each with a mass about 1800 times that of the electron. The neutron has no charge, while the proton has a positive charge, $+e$, equal in magnitude to the electron's

			Ti = 50	Zr = 90	? = 180
			V = 51	Nb = 94	Ta = 182
			Cr = 52	Mo = 96	W = 186
			Mn = 55	Rh = 104.4	Pt = 197.4
			Fe = 56	Ru = 104.4	Ir = 198
			Ni = Co = 59	Pd = 106.6	Os = 199
H = 1			Cu = 63.4	Ag = 108	Hg = 200
	Be = 9.4	Mg = 24	Zn = 65.2	Cd = 112	
	B = 11	Al = 27.4	? = 68	Ur = 116	Au = 197?
	C = 12	Si = 28	? = 70	Sn = 118	
	N = 14	P = 31	As = 75	Sb = 122	Bi = 210?
	O = 16	S = 32	Se = 79.4	Te = 128?	
	F = 19	Cl = 35.5	Br = 80	J = 127	
Li = 7	Na = 23	K = 39	Rb = 85.4	Cs = 133	Tl = 204
		Ca = 40	Sr = 87.6	Ba = 137	Pb = 207
		? = 45	Ce = 92		
		?Er = 56	La = 94		
		?Yr = 60	Di = 95		
		?In = 75.6	Th = 118?		

Figure 16–2. Mendeleev's 1869 periodic table of the elements. Each element is designated by its chemical symbol, followed by the approximate weight of its atom (with the weight of the lightest atom, hydrogen, arbitrarily chosen as 1). The question marks indicate elements suggested by the table but not then known (e.g., ? = 45), atomic weights suspected by Mendeleev to have been incorrectly determined (e.g., Te = 128?), and elements whose identification was uncertain (e.g., ?Er = 56).

charge. Each atom has a nucleus composed of a combination of neutrons and protons, with a diameter of roughly 10^{-15} m. Surrounding the nucleus is a "cloud" of electrons* with a diameter of roughly 10^{-10} m. The number of electrons in the

* The term electron "cloud" is a picturesque but somewhat misleading attempt to capture the quantum mechanical description of an electron. Quantum mechanics does not give a definite answer to the question, "Where is the electron?" It says that if the position of the electron in an atom were measured there is a rather good chance that it would be found in any one of a large number of different places around the nucleus. Thus the term "cloud" to describe its position.

atom is equal to the number of protons in its nucleus, leaving the atom with no net electric charge.

For example, hydrogen, the lightest atom, has a nucleus consisting of a single proton, surrounded by a single electron. Helium, the next lightest atom, has two protons and two neutrons in its nucleus, and two electrons outside the nucleus. As we proceed along the periodic table, for each new atom we add one electron and one proton. The number of neutrons is less regular, but roughly equals the number of protons.

This model accounts well for the masses and the nuclear charges of the atoms. In addition, the theory of quantum mechanics explains the chemical regularities displayed in Mendeleev's table in terms of the properties of the electron clouds surrounding the atomic nuclei. Thus, in the world view of 1930 physicists, all matter and its chemical properties could be understood in terms of three fundamental building blocks, the electron, the proton, and the neutron. This simple and appealing picture is the background to the story of elementary particle physics today.

The Nuclear Force

The acceptance of a model in which the atom has a nucleus of protons and neutrons raises an important question. What holds the nucleus together? It cannot be an electrostatic force. The neutrons are uncharged and hence do not interact electrically; the protons, being of like charge, tend to be repelled, not attracted, by their electrostatic interaction. The attractive gravitational force between the particles is far too weak to hold a nucleus together. There must be another kind of interaction between nucleons (as the constituents of nuclei—protons and neutrons—are sometimes called) that binds them together in a nucleus.

How can one learn about this interaction? To study gravity it was necessary to observe how an object accelerates when it is acted on by the gravitational force of another object. The accelerations of the planets as they orbit the sun, the acceleration of the moon in its path about the earth, and the acceleration of objects at the surface of the earth, all provided Newton with clues about the gravitational interaction. To study the nuclear interaction, one similarly examines how one nucleon is accelerated when it comes near another.

A typical experimental arrangement is shown in Figure 16–3. A proton (called the probe particle) is directed at a target—for example, a chamber filled with liquid hydrogen. Before the incident proton has come close enough to the target to have an appreciable interaction with the target nuclei, its velocity is constant.* After it leaves the target, it has another constant velocity. While it interacts with the target, the velocity of the proton generally changes—that is, it accelerates. If the target is sufficiently thin, there is an overwhelming chance that the incident proton will interact with the nucleus of no more than one of the target atoms. Consequently its change in velocity indicates the acceleration resulting from a single interaction between a proton and a nucleus, and hence furnishes direct information about the nuclear force. Such scattering experiments are the most important technique available for studying this force.

Figure 16–3. A typical scattering experiment. By observing the acceleration of a probe particle when it interacts with a nucleus of the target, the nuclear force can be investigated.

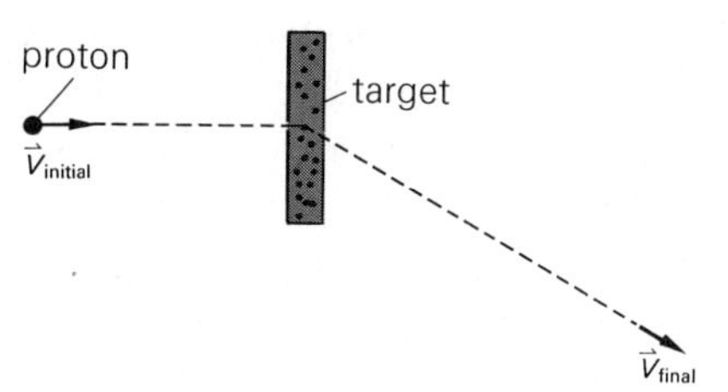

* Neglecting the gravitational force, since it is negligibly small compared with the nuclear force under investigation.

Particle Accelerators

To study the nuclear force in detail, it is necessary to accelerate probe particles to speeds very near that of light. A number of different kinds of particle accelerators have been developed to accomplish this. Most are based on the elementary idea illustrated in the basic accelerating device of Figure 16–4. With the aid of a battery or, more likely, some other means of producing a separation of electric charge, a net positive charge is placed on one metal plate (the positive electrode) and a net negative charge on another (the negative electrode). Suppose that the particle to be accelerated is a proton, which has a positive charge. A

Figure 16–4. The basic accelerating device. Protons emerging from the source are attracted by the negative electrode and repelled by the positive electrode.

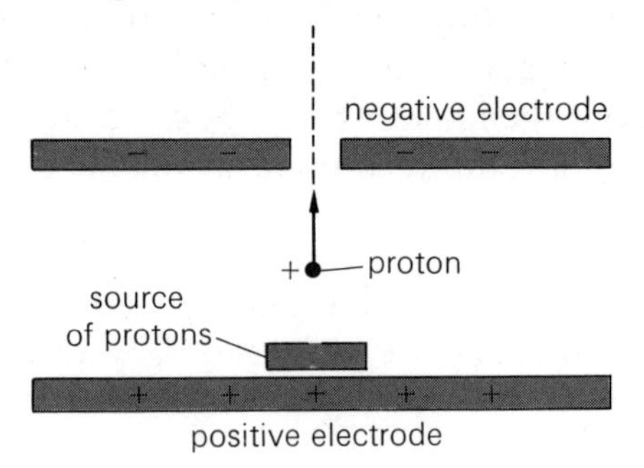

source of protons is located in the vicinity of the positive electrode. The positively charged proton experiences two electrostatic forces: it is repelled by the positive electrode and attracted by the negative electrode. Thus both forces act in concert to accelerate the proton from the source toward the negative electrode. If this electrode has a hole at its center, many of the protons will pass through, continuing with the speed they had acquired on reaching it.*

* It might seem that the proton, after passing through the hole in the negative electrode, would be attracted back toward that electrode. However, the electrodes can be designed so that a charged particle feels an electrical force only in the region between the electrodes.

In principle, one should be able to accelerate charged particles to very high speeds with such a device by placing enough charge on the electrodes. In practice, however, the largest amounts of charge one can achieve are too small to accelerate probe particles to the high speeds often desired. Fortunately, it is possible to employ the magnetic interaction between a moving charged particle and a magnetized piece of iron to steer such a particle into a circular orbit. Thus, even though a single passage through the basic accelerating device is not sufficient to produce extremely high speeds, one can design an apparatus in which the probe particle, moving in a circular path, passes through the basic accelerating device many times.

Figure 16–5 is a schematic illustration of this simple idea, on which the design of modern proton accelerators is based. Protons are injected into the ring, where magnets cause them to travel in a circular path. During each orbit they receive a "kick" from the basic accelerating device. After many orbits, very high speeds can be achieved. At this point a target within the ring is moved into the path of the protons, or, alternatively, the protons are deflected electromagnetically out of the ring and allowed to strike an external target. One then observes the deflection of the protons by the nuclei of the target.

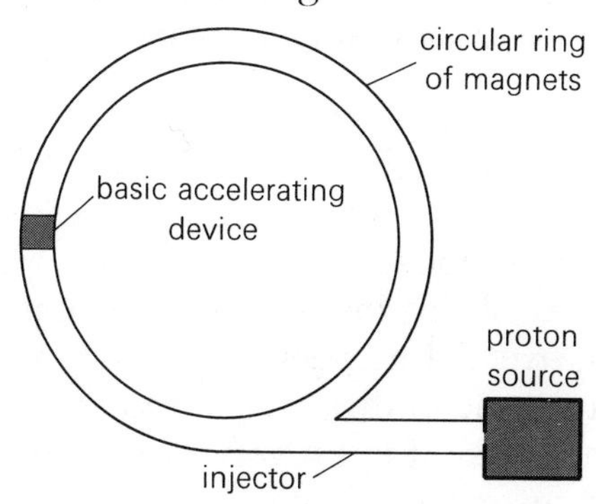

Figure 16–5. A schematic diagram of a proton accelerator. A ring of magnets steers the protons in a circular path, allowing them to pass through the basic accelerating device many times.

Figures 16–6 and 16–7 are photographs of two accelerators based on this principle—one of the earliest and one of the most recent. The striking contrast in their size and complexity reflects the growth of elementary particle physics in the past half-century.

Figure 16–6. An early cyclotron. The large coils of wire framed by the iron yoke are the magnet that steers the protons in a circular path. The cyclotron itself is barely visible, located between the two coils. (Lawrence Berkeley Laboratory of the University of California.)

Figure 16–6 shows the *cyclotron* constructed by Ernest Lawrence and Stanley Livingston at the University of California, Berkeley, in 1932. It measured less than a foot in diameter and could accelerate protons to about 5 percent of the speed of light. A modern descendant of this instrument is the huge *proton synchrotron* recently constructed at Batavia, Illinois, shown in Figure 16–7. More than a mile in diameter, it can accelerate protons to .999998 times the speed of light, very close to the limiting speed predicted by the theory of relativity.

The Nuclear Force Law

Studies of the nuclear force with particle accelerators like those described above have provided much information about the nuclear interaction. They indicate that a nucleon experiences a

Figure 16–7. The proton synchrotron at Batavia, Illinois, during its construction. The large circle, one kilometer in radius, marks the evacuated underground ring in which the protons move, surrounded by guiding magnets. The excavation at the bottom is for buildings to house the experimental areas where the protons, deflected out of the ring, interact with targets. (National Accelerator Laboratory.)

very large force, much larger than those of gravity or electromagnetism, when it comes close to another nucleon. This interaction, termed the *strong* nuclear force,* apparently does not distinguish between protons and neutrons. That is, if separated by the same distance, two protons, two neutrons, or a proton and a neutron interact with the same strong force.

There are striking differences, however, between the force law describing the strong nuclear interaction and the laws describing gravity and electromagnetism. The strong interaction has a very short range—about 10^{-15} m. Nucleons separated by more than this distance do not interact. Within this distance they

* There is also evidence of another nuclear force, the *weak* force. However, this interaction is much less important than the strong force in the binding of nucleons to form a nucleus.

experience an intense attraction—until the separation distance is less than about 10^{-16} m. Two nucleons separated by a distance of 10^{-16} m or less strongly repel one another.

Furthermore, in contrast to the appealing simplicity of the gravitational and electromagnetic force laws, the strong force law seems exceedingly complicated. Not only does the force have a short range and change from attractive to repulsive at small distances, but it also depends on other characteristics of the nucleons than their distance of separation—and in a very complex way.

For example, protons and neutrons have a property known as *spin*. Although this is a quantum-mechanical concept that cannot be represented adequately by a non-quantum model, we can loosely think of a proton or neutron as a tiny sphere spinning about an axis, as suggested in Figure 16–8(a). [Figure 16–8(b) indicates a convenient symbolism for the spin axis and the direction of rotation, which we shall use later in this chapter.] The strong nuclear force between two nucleons depends on the relative orientation of their spin axes as well as on the distance between the nucleons.

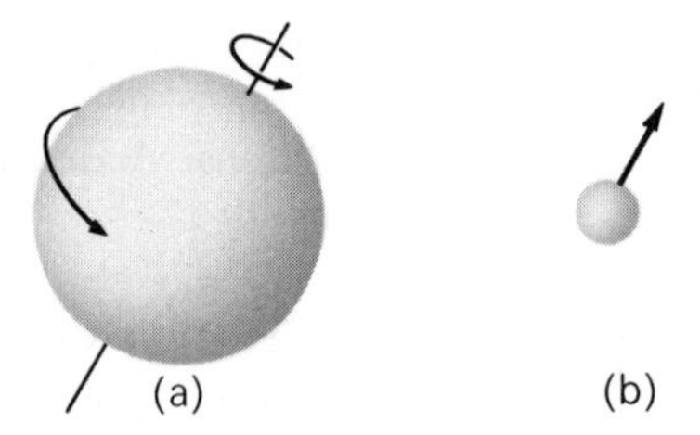

Figure 16–8. The spin of a nucleon. In (a) the particle is represented as a tiny sphere rotating about an axis. (b) is an abstract representation of this spin; the arrow is aligned in the direction of the rotation axis, and the point of the arrow indicates the sense of rotation about this axis. [A particle rotating in the opposite sense from that in (a) would be represented by an arrow pointing in the opposite direction.]

Knowing the strong nuclear force law, one can predict the results of scattering experiments and explain how nucleons are bound together in the nucleus of an atom. But the complexity of this force law is disquieting to many physicists. One might speculate that the Newtonian concept of force would not have been generally accepted had the gravitational force law been so complicated. By the same token, one might conclude that the Newtonian concept of force has proved inappropriate to deal with nuclear interactions; perhaps a new conceptual scheme is needed.

However, there is great reluctance among physicists to reject the force idea at the nuclear level—partly because of its success in describing the gravitational and electromagnetic interactions, and partly because no one has thought of a more fruitful alternative. Many believe that the complexity of the strong nuclear force law will eventually be explained in terms of simpler forces on the subnuclear level. For example, Victor Weisskopf, a prominent nuclear and elementary particle physicist, suggests:

> We are tempted to assume from the rather complicated form of the nuclear force that it is not as fundamental as, say, the electrostatic attraction. It may be an effect derived from a more basic phenomenon residing within the nucleons, a consequence of something much more powerful and simple, in the same way that the chemical force is a consequence of the simple electrostatic interaction.[3]

One intriguing speculation along these lines is that the nucleons are themselves composed of yet more basic constituent particles. Perhaps there is a simple force law describing the interaction between these particles. In this view, the apparent complexity of the force between nucleons might then be merely a consequence of their being composed of many of the constituent particles. Although these particles interact in a simple way, the combination of their forces might lead to the complicated interaction observed between two nucleons.

At present this idea is purely speculative, but it indicates the kind of intuition involved in developing an explanation of the nuclear force and a theory of elementary particles. Rather than being disturbed by the complexity of the nuclear force law, many physicists regard it as a welcome clue to the structure of nucleons. It suggests to some that a nucleon is a composite of yet more fundamental particles that interact according to a simple force law.

Pandora's Box

Since 1930, while particle accelerators evolved from cyclotrons that could be operated by a single man to huge complexes such as the proton synchrotron at Batavia, which requires a supporting staff of 2500, the focus of the experiments has changed as well. In the original scattering experiments, designed to investigate the nuclear force, a proton or other probe particle was directed at a nucleus, interacted with it, and emerged with a new velocity. As higher energies were attained, however, new particles, apparently created during the interaction between the probe particle and the target, began to be observed.

These particles had properties quite different from those of the electron, proton, and neutron. In particular, most were characterized by an extremely short lifetime: the μ meson, which we discussed in connection with time dilation, is one of the longest lived, achieving the ripe old age of 10^{-6} sec before it decays. The short lifetimes explain why these particles are not detected in ordinary matter. With the construction of very large accelerators, more and more particles were discovered and the emphasis in accelerator experiments shifted from the study of the nuclear force to the investigation of the particles produced.

Actually, the first discoveries of the short-lived particles occurred in investigations of cosmic rays in the 1930s. As discussed earlier, cosmic rays originate when high-energy protons bombard the earth's atmosphere much as high-energy protons in a particle accelerator bombard a target placed in their path. The interaction of the protons with the nuclei of the atoms in the air produces a wealth of particles that rains down on the earth. It was in studies of cosmic rays that the *anti-particle* of the electron, the positron, was discovered in 1934. This particle has the same mass and spin as the electron, but opposite charge. The μ meson, or muon, as it is sometimes called, was also found in the cosmic rays in 1936.

However, not until the construction of the large particle accelerators in the late 1950s and the 1960s could the dimensions of the elementary particle problem be seen. The trickle of new particles discovered in the cosmic rays in the 1930s and 1940s became a deluge in the 1960s. Table 16–1 lists some (though by no means all) of the more than one hundred particles known by 1971, along with a number of their properties.

This table presents a picture of relative chaos, although a few regularities can be discerned. The particles are tabulated in order of their masses, which are given in the second column in multiples of the electron mass, m_e. The charge of each particle is indicated in the third column; notice that in each case the charge is a simple integer multiple of the electron charge. The next column gives the particle's spin; notice that, like electric charge, the magnitude of the spin is quantized—it takes on only certain evenly spaced values: 0, 1/2, 1, and 3/2. The lifetimes of the particles are listed in the fifth column.

The sixth and seventh columns list two properties we have not yet mentioned. *Parity* is an esoteric property that is the analogue for particles of the everyday notion of right- or left-handedness. For our purposes it is sufficient to note that the parity of a particle is either positive or negative. *Hypercharge*, Y, is related to one of the regularities of the particles that can be seen in Table 16–1—that they appear in groups whose members are all called by the same name. Members of a group have nearly the same mass but different electric charge. Hypercharge is defined as twice the average electric charge of a group (in units of the electron charge, e). Thus the kaon group (K^+ and K^0) has an average charge of 1/2, which means that the group has a hypercharge of 1. The Σ group has an average charge of 0, giving a hypercharge of 0.

The discovery of so many different particles has led many physicists to feel that our present ideas of the structure of matter are inadequate. The simple world of the 1930s, in which there were only three fundamental particles—the proton, neutron, and electron—has given way to the complex world of the 1970s, which features more than a hundred particles. Just as ninety chemical elements seemed to be too many basic building blocks to nineteenth-century scientists, so today there is a feeling among physicists that one hundred "elementary" particles is too large a number. There is a widespread expectation among physicists that man's understanding of the structure of matter is due for another historic conceptual change.

Table 16–1. A partial listing of the particles known today. A complete tabulation would have over one hundred entries. In addition, for every particle listed, there is also an anti-particle with the same mass, spin, and lifetime, but opposite charge and hypercharge. (The masses listed have been rounded off to the nearest electron mass and have uncertainties of less than one electron mass except where noted.)

Particle (name and symbol)		Mass (in electron masses)	Charge	Spin	Lifetime (sec)	Parity	Hypercharge Y
Photon	γ	0	0	1	stable	*	*
Neutrino	ν	0	0	1/2	stable	*	*
Electron	e^-	1	$-e$	1/2	stable	*	*
Muon	μ^-	207	$-e$	1/2	2×10^{-6}	*	*
Pion	π^0	264	0	0	9×10^{-17}	−	0
	π^+	273	$+e$	0	3×10^{-8}	−	0
	π^-	273	$-e$	0	3×10^{-8}	−	0
Kaon	K^+	966	$+e$	0	1×10^{-8}	−	1
	K^0	974 ± 1	0	0	5×10^{-8}	−	0
Eta	η^0	1074 ± 1	0	0	about 10^{-23}	−	0
Nucleon	p^+	1836	$+e$	1/2	stable	+	1
	n^0	1839	0	1/2	900	+	1
Lambda	Λ^0	2183	0	1/2	3×10^{-10}	+	0
Sigma	Σ^+	2328	$+e$	1/2	8×10^{-11}	+	0
	Σ^0	2334	0	1/2	10^{-14}	+	0
	Σ^-	2343	$-e$	1/2	2×10^{-10}	+	0
Xi	Ξ^0	2573 ± 2	0	1/2	3×10^{-10}	+	−1
	Ξ^-	2586 ± 1	$-e$	1/2	2×10^{-10}	+	−1
Delta	Δ^{++}	2413 ± 6	$+2e$	3/2	about 10^{-23}	+	1
	Δ^+	?	$+e$	3/2	about 10^{-23}	+	1
	Δ^0	2415 ± 4	0	3/2	about 10^{-23}	+	1
	Δ^-	2429 ± 10	$-e$	3/2	about 10^{-23}	+	1
Sigma star	Σ^{*0}	2703 ± 8	0	3/2	about 10^{-23}	+	0
	Σ^{*+}	2706 ± 1	$+e$	3/2	about 10^{-23}	+	0
	Σ^{*-}	2712 ± 4	$-e$	3/2	about 10^{-23}	+	0
Xi star	Ξ^{*0}	2992 ± 2	0	3/2	about 10^{-23}	+	−1
	Ξ^{*-}	3002 ± 6	$-e$	3/2	about 10^{-23}	+	−1
Omega	Ω^-	3273 ± 1	$-e$	3/2	1×10^{-10}	+	−2

* Not defined.

SU(3): A Suggestive Grouping of Particles?

In the past, important conceptual changes in physics have often been preceded by the recognition of a particularly suggestive way to order the experimental data at hand. The heliocentric model and Kepler's three laws of planetary motion helped pave the way for Newton's theory of gravitation. Mendeleev's periodic table of the chemical elements set the stage for the nuclear model of the atom. In 1963, two theoretical physicists, Murray Gell-Mann in the United States and Yuval Ne'eman in England, independently proposed a new way to organize the particle data that many believe will play this role in the elementary particle problem. Gell-Mann and Ne'eman suggested that particles having the same spin and parity should be grouped together as a family. What is more, they noted, a graph of the hypercharge vs. charge for each such family produces characteristic patterns described by a mathematical scheme called SU(3).*

An example may clarify this idea. Consider the particles with spin 1/2 and positive parity. The SU(3) scheme predicts that there should be eight such particles (a so-called *octet*) and, indeed, that is precisely the number of particles with spin 1/2 and positive parity that have been discovered. They are the two nucleons, two Ξ particles, three Σ particles, and the Λ particle. (See Table 16–1.) Furthermore, the SU(3) scheme predicts that a graph of hypercharge vs. charge for these particles should yield the simple pattern shown in Figure 16–9.

Other families with eight particles have also been found. For example, Figure 16–10 illustrates the octet with spin 0 and negative parity. This family exhibits exactly the same SU(3) pattern as the family of spin-1/2, positive-parity particles in Figure 16–9.

The SU(3) scheme predicts a number of other families of

* The symbol SU(3), read "S U 3," is short for "*S*pecial *U*nitary group in *3* dimensions."

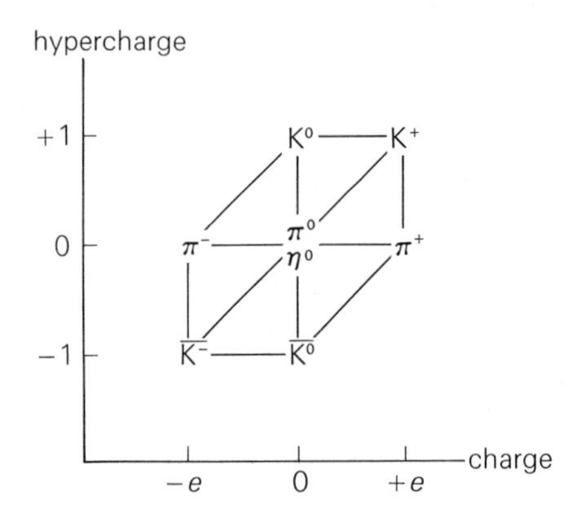

Figure 16–9. The octet of particles with spin $\frac{1}{2}$ and positive parity.

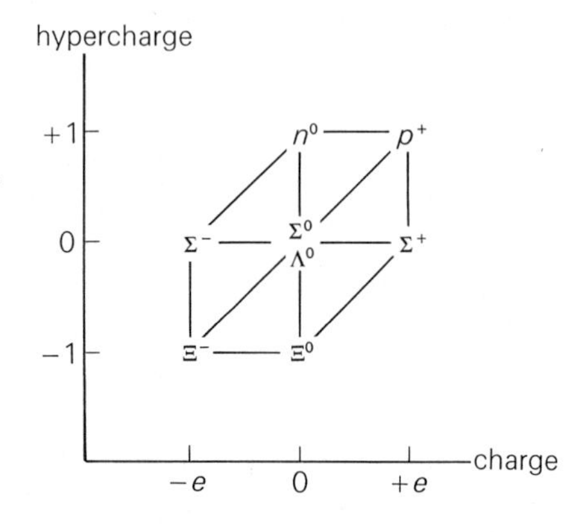

Figure 16–10. The octet of particles with spin 0 and positive parity.

particles and, in addition, indicates how the masses of particles within a family should be related. A most important further example is the family of ten particles (a *decuplet*) with spin 3/2 and positive parity. In 1963, when Gell-Mann and Ne'eman first proposed the SU(3) scheme, only nine such particles had been discovered in accelerator experiments, as illustrated in Figure 16–11. The SU(3) scheme predicted a tenth particle, located at the missing vertex of the triangular pattern.

Many properties of this missing particle, called the Ω^- ("omega minus") by Gell-Mann, were predicted by SU(3). From its position in the decuplet, one could conclude that its charge should be $-e$, its parity positive, and its spin 3/2. In addition, its mass could be predicted from a relationship be-

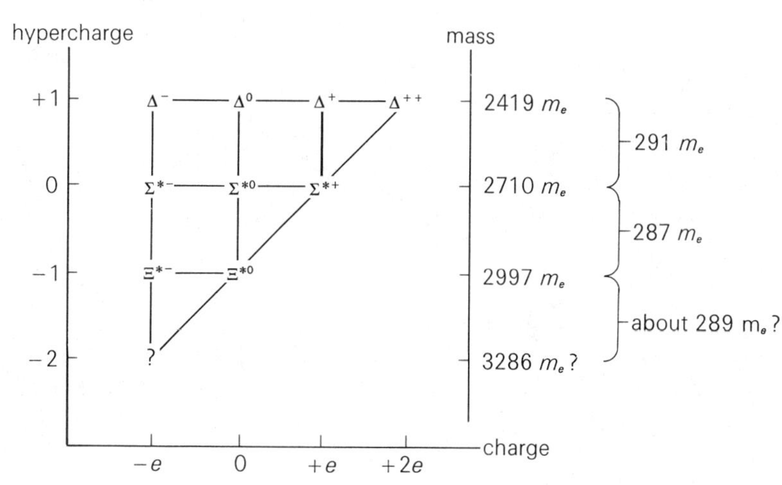

Figure 16–11. The nine particles with spin 3/2 and positive parity that were known in 1963. The question mark indicates a tenth particle predicted by the SU(3) scheme.

tween the masses of the particles within an SU(3) family. The precise form of this relationship depends on the family. It is particularly simple in the case of the spin-3/2, positive-parity decuplet we are considering. For every decrease in hypercharge by one unit, a constant increase in mass is predicted. As shown in Figure 16–11, the difference in mass between the Δ particles ($Y = 1$) and the Σ^* particles ($Y = 0$) is $291m_e$, while the difference in mass between the Σ^* particles and the Ξ^* particles ($Y = -1$) is $287m_e$, approximately the same. Hence, if the SU(3) scheme is appropriate, the missing Ω^- particle ($Y = -2$) was expected to have a mass about $289m_e$ (i.e., between $287m_e$ and $291m_e$) larger than the Ξ^* particle, or $3286m_e$.

The prediction of a new particle of specified charge, mass, spin, and parity presented a challenge to experimental physicists. At the Brookhaven proton accelerator on Long Island (then the largest in the world) an intensive search for the Ω^- was begun. High-energy K^- particles were allowed to bombard the liquid hydrogen in the large *bubble chamber* shown in Figure 16–12. In such a device, a track of bubbles forms in the liquid hydrogen along the path of a charged particle, rendering its trajectory

Figure 16–12. The liquid-hydrogen bubble chamber employed in the discovery of the Ω^- particle. The large iron yoke and the two round coils it encloses constitute the magnet that causes particles passing through the bubble chamber to bend, aiding the analysis of the tracks. The end of the bubble chamber itself can be seen between the coils; it extends for 80 inches into the magnet. (Brookhaven National Laboratory.)

visible. Photographs of the tracks are analyzed to determine the masses and charges of the particles.

Approximately 100,000 separate photographs of the bubble chamber were taken in the Brookhaven experiment. These photographs were then individually examined in an attempt to find at least one reaction in which an Ω^- was produced. Such a reaction was expected to be very rare, but it was predicted that in 100,000 photographs several Ω^-'s should be seen.

The tedious scanning of the photographs went on for weeks with no trace of an Ω^-. Apparently the reaction was even less likely than predicted—or perhaps the Ω^- did not exist. Rumors of these disappointing results spread through the expectant community of elementary particle physicists. But at last, after 50,000 pictures had failed to reveal the elusive particle, the

Figure 16–13. (a) The first bubble chamber photograph of an Ω^- particle. The track of the Ω^- is indicated by the arrow. (b) The interpretation of the photograph in (a). The solid lines represent tracks of charged particles, which may be seen in the photograph. Dotted lines represent the assumed tracks of neutral particles such as K^0, π^0, Λ^0, and γ, which leave no tracks in a bubble chamber. In this series of reactions, a K^- comes in from below, interacts with a p^+ (the nucleus of an atom of the liquid hydrogen in the chamber) to form a K^+, a K^0, and an Ω^-. The Ω^- travels only a short distance before decaying into a Ξ^0 and a π^-. The Ξ^0 decays into a Λ^0 and two γ's. The Λ^0 decays into a p^+ and a π^-, while the two γ's each decay into an electron-positron pair. (Brookhaven National Laboratory.)

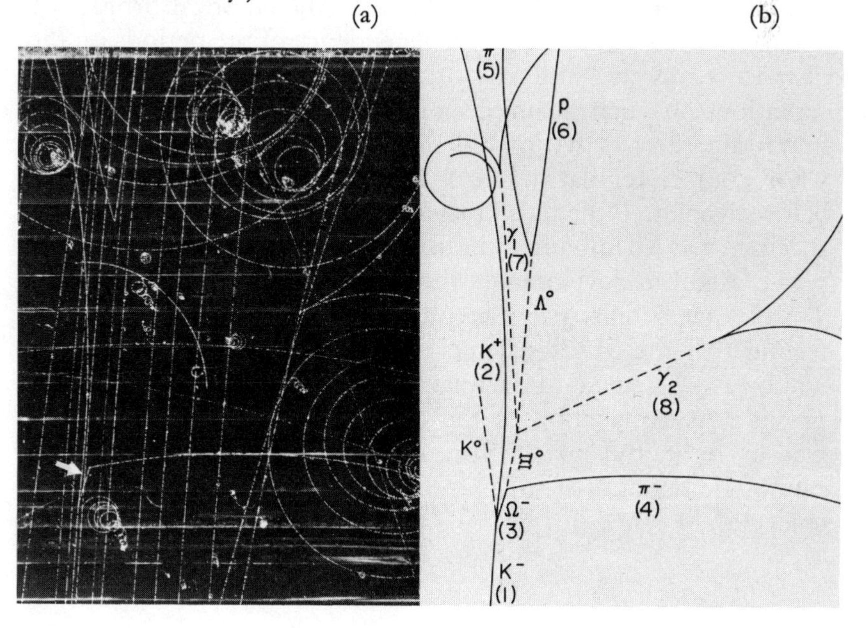

photograph shown in Figure 16–13(a) was examined and the tiny track indicated by the arrow was identified as an Ω^-. It had charge $-e$, spin 3/2, positive parity, and mass $3300 \pm 24m_e$, consistent with the SU(3) predictions.

Figure 16–13(b) indicates how the photograph was interpreted. The particle identifications were made from measurements of the length of the tracks, the density of the bubbles along each track, the curvature of the tracks (caused by the interaction of the charged particles with a magnetic field which was set up in the region of the bubble chamber), and the known energy of the incoming K^- meson.

Since the discovery of the Ω^-, the SU(3) scheme has achieved wide acceptance among particle physicists. It greatly simplifies the classification of particles. But by itself, it is not the theory of elementary particles people are seeking. Rather, it is widely regarded as a suggestive way to organize the particle data.

As we have noted, the isolation of regularities in observational data has often laid the groundwork for the development of a theory. The way the data are organized suggests what questions the theory must answer and often some of the concepts that must be employed in answering them. Kepler's three laws played such a role in the development of Newton's explanation of planetary motion. Kepler suggested that the features of the planets a theory should explain were the elliptical orbits, the equal-areas law, and the simple dependence of the periods on the distance from the sun. His organization of the planetary observations in such a simple and compelling fashion had the important effect of setting a definite problem to be solved.

Of course, regularities had been seen in planetary motion before Kepler. Ptolemy's epicycles were also an attractive and succinct way to organize the planetary data. Yet they did not prove useful in developing a theory of the planets.

Most elementary-particle physicists today have a strong feeling that the SU(3) scheme is of crucial importance—that it will be a key element in the imagination of the physicist who will provide an explanation of the elementary particles. They may well be right. But now, before such a theory is developed, one cannot be certain. SU(3) may be ellipses—or it just might be epicycles.

The Quark Model

The particle data, organized according to the SU(3) scheme, has in effect been presented to the theoretical physicists of the world with the charge, "Find a theory of the elementary particles." But what sort of theory? What questions about the particles should the theory answer? What concepts should it employ? Today, no one is really sure. Nevertheless, there are deeply felt and widely accepted prejudices about these matters—as, of course, there must be. Only a person with a strong conviction that he is on the right track will devote many years to the difficult and often tedious efforts associated with developing a physical theory.

One such prejudice, already mentioned, is that the SU(3) families should be a natural consequence of the theory. Another is that the particles observed are not really elementary. More than a hundred elementary particles seems too many—nature can't be that complicated. A third is that a theory should explain why we observe the particles we do with the particular properties they possess and no others.

Perhaps the leading candidate for a theory at the moment is the so-called *quark model.* This theory, proposed by Gell-Mann in 1964, assumes that there are only three truly elementary particles, called quarks,* and that all the observed particles are simply combinations of these quarks.

The three quarks postulated by Gell-Mann are illustrated in Figure 16–14. The p-quark has a positive charge whose magnitude is $(2/3)e$; it also has hypercharge 1/3 and spin 1/2. The n-quark has negative charge $-(1/3)e$, hypercharge 1/3, and spin 1/2. The s-quark has the same charge and spin as the n-quark, but its hypercharge is $-2/3$.

In Gell-Mann's theory, each elementary particle is assumed to be composed of a group of quarks. A few simple rules relate

* A word created by James Joyce in *Finnegans Wake*: "Three quarks for Muster Mark!"

symbol	+2/3	−1/3	−1/3
name	*p*	*n*	*s*
charge, *q*	+2/3 *e*	−1/3 *e*	−1/3 *e*
hypercharge, *Y*	1/3	1/3	−2/3
spin	1/2	1/2	1/2

Figure 16–14. The three quarks postulated by Gell-Mann.

the properties of the group to the properties of the constituent quarks. The charge and hypercharge of a combination of quarks are taken to be the sum of their charges and hypercharges respectively. The total spin of such a combination is the sum of the spins of the constituent quarks, with the proviso that spins in the same direction add, while spins in the opposite direction cancel. This is illustrated in Figure 16–15. A combination of two quarks, each with spin 1/2, whose spins are in the same direction [Figure 16–15(a)] gives a composite particle with spin 1. If the spins are oppositely directed [Figure 16–15(b)], the combination has spin 0. The remarkable consequence of these simple assumptions is that they lead to a quark theory that predicts precisely those families of particles described by SU(3) and observed in nature.

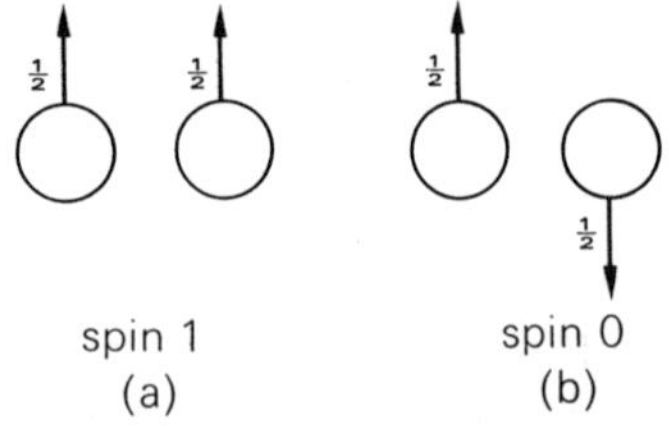

Figure 16–15. The addition of quark spins. (a) Two spin-$\frac{1}{2}$ quarks whose spins are in the same direction combine to give a particle with total spin 1. (b) Two spin-$\frac{1}{2}$ quarks can also combine with their spins oppositely directed, forming a particle with zero total spin.

For example, consider the particles with spin 3/2. Such a particle can be constructed from three quarks. Each quark has spin 1/2, and if the spins of the three quarks are in the same direction, the total spin will be 3/2. Figure 16–16 shows all

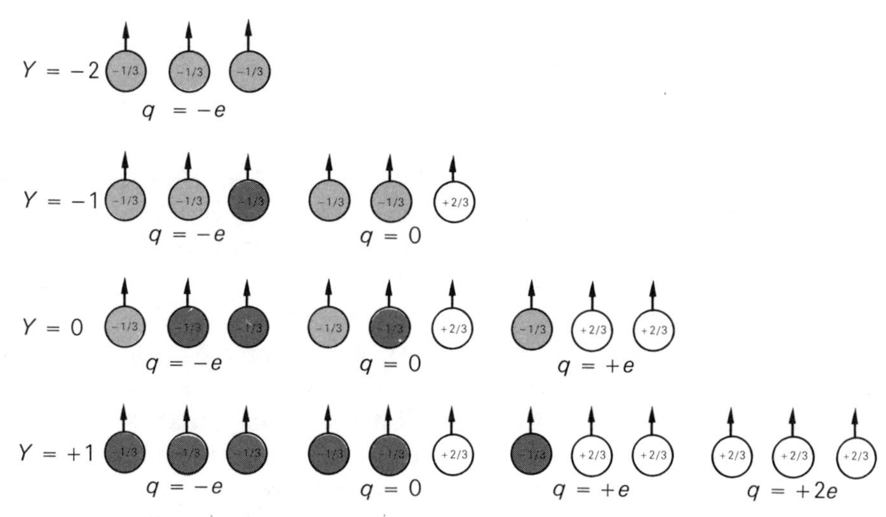

Figure 16–16. The ten possible combinations of three quarks whose spins are in the same direction.

possible groups of three quarks whose spins add to 3/2. There are ten such combinations, precisely the number of spin-3/2 particles that have been observed—the Ω^- decuplet. We are therefore tempted to associate each particle with a group of three quarks as shown in Figure 16–17.

The quark model also generates the other observed SU(3) families of particles.* This success provides strong support for

Ω^- *sss*			
Ξ^{*-} *ssn*	Ξ^{*0} *ssp*		
Σ^{*-} *snn*	Σ^{*0} *snp*	Σ^{*+} *spp*	
Δ^- *nnn*	Δ^0 *nnp*	Δ^+ *npp*	Δ^{++} *ppp*

Figure 16–17. Each particle in Figure 16–11 can be associated with one of the groups of quarks in Figure 16–16.

* The families of Figures 16–9 and 16–10 can be formed only by groups of quarks whose spins are not all in the same direction. In such a case, unfortunately, the choice of allowed combinations of quarks is somewhat more complicated than the example of Figure 16–16.

the model. For example, Laurie Brown, a particle physicist, writes that quarks "provide the simplest concrete way to understand fundamental particles."[4] However, Brown also notes one major difficulty with the quark idea:

> The one great objection to this approach is that no one has yet observed a quark. Perhaps they have not been searched for with sufficient seriousness. Perhaps they do not exist.[5]

In fact, ever since the quark model was first proposed, groups of physicists around the world have been searching for these particles. A distinctive characteristic of a quark is its fractional charge. Every other known particle has a charge that is some small integer (0, ± 1, ± 2) times the magnitude of the electron's charge, e. This should make it possible to recognize a quark experimentally. However, since very little else about quarks can be predicted with any assurance—their mass, for example, or their interaction with bulk matter—it is not clear where to look for quarks not already combined into nucleons or other particles. Attempts at the large particle accelerators to split known particles into their supposed quark constituents have failed, as have efforts to find quarks in naturally occurring matter such as sea water, ancient rocks, oyster shells, soil from under the ocean, and meteorites from outer space.

Only in the cosmic rays have there been any reports of evidence for quarks, and there the situation has been controversial. In 1969, a group of Australian cosmic-ray physicists announced that they had observed five particle tracks whose characteristics suggested that they were made by quarks, including one "for which the only explanation we can see is that it is produced by a fractionally charged particle."[6] This report was received initially with much acclaim, and then, in view of the far-reaching implications of such a discovery, it was subjected to critical scrutiny. Many physicists later expressed considerable skepticism. For example, one research group reported:

> Measurements which we have conducted, designed to detect quarks produced by cosmic rays under a variety of possible production mechanisms, have given negative results. A variety

> of measurements of comparable sensitivity by other observers have also failed to detect quarks . . . We suggest that their [the Australian group's] tracks . . . may be explained as statistical fluctuations in the density of shower tracks when the production of drops is correctly considered . . .[7]

Another group claimed:

> They [the Australians] did not, however, consider the effects of the relativistic rise of ionization, and we believe they underestimated the experimental errors. Consideration of the distribution of ionization expected due to known particles convinces us that their results can be explained without requiring fractionally charged particles.[8]

We see here yet another instance where the outcome of an experimental test of a theory is not clear-cut. Have quarks been detected? At this writing no one is certain, though the consensus seems to be that they have not.

In a sense the quark model would be a most unrevolutionary solution to the elementary particle problem. It is very much in the reductionist tradition, in which chemical properties have been explained by assuming matter to be composed of atoms, atomic properties explained by assuming the atom to be composed of electrons and a nucleus, and nuclear properties explained by assuming the nucleus to be composed of protons and neutrons. A model in which the properties of protons, neutrons and other elementary particles are explained by assuming them to be composed of quarks would not be a striking departure from this tradition. It would not ask new kinds of questions or radically alter the kinds of answers given to the old questions. And it raises an obvious further question—what are quarks made of, and where is the end of this reductionist sequence of a box within a box within a box?

The Bootstrap Model

Another proposal, called the *bootstrap model*, is more revolutionary. To appreciate it, we must first mention a result of Einstein's special theory of relativity we have ignored up to this point—the famous equation $E = mc^2$. While we shall not consider this relation in detail, its significance is easy to appreciate: Energy and mass are equivalent. When a physical system gains or loses energy, its mass is correspondingly increased or decreased.

Consider, for example, two magnets oriented so that they attract each other, as shown in Figure 16–18(a). If released on a relatively frictionless surface, they will accelerate toward each other, and just before they collide, the magnets will have acquired an energy of motion. During the collision, this motional energy is converted to heat energy, which is soon lost to the surroundings as the magnets cool off. According to the theory of relativity, this energy loss is accompanied by a corresponding loss of mass, so that the two magnets will have slightly less mass when joined and cooled off than they had originally when

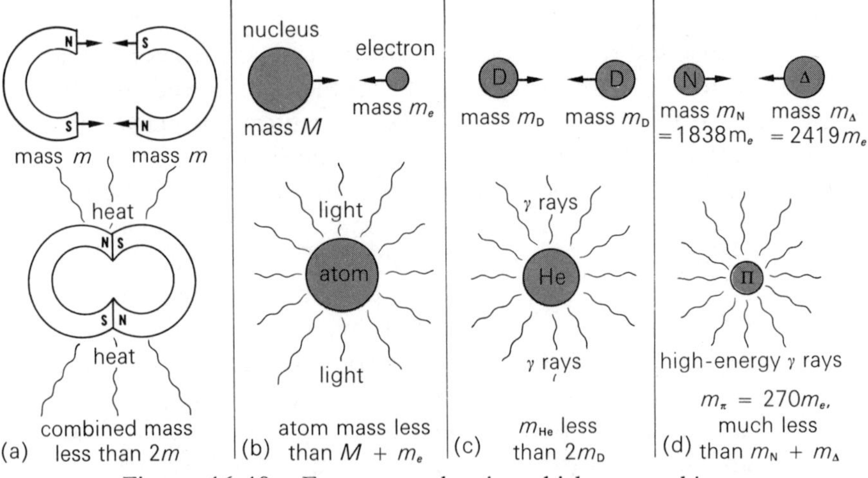

Figure 16–18. Four examples in which two objects combine to form one composite object with less mass than its separated constituents. (a) Two magnets combine to form a single object, losing energy in the form of heat. (b) An electron and a nucleus combine to form an atom, losing energy in the form of light. (c) Two deuterium nuclei combine to form a helium nucleus, losing energy in the form of γ rays. (d) A proton and a delta particle combine to form a pion, losing energy in the form of very high-energy γ rays.

separated. In this particular case, however, the magnetic forces are extremely weak, and the mass change is predicted to be extremely small—far too small to be measured in the laboratory.

Similarly, when an electron and a nucleus attract each other to form an atom, a small amount of energy is lost in the form of light. Consequently, the mass of the atom is slightly less than the sum of the masses of the separated electron and nucleus. As indicated in Figure 16–18(b), this mass difference is still very small, but it is a larger fraction of the mass of the combining objects than in the magnet example, since the forces involved are considerably stronger.

When protons and neutrons combine to form a nucleus the attractive forces are even stronger, leading to a still greater loss of mass. In this case, the nucleus has a mass about one percent less than the combined mass of the separated proton and neutron constituents. This phenomenon is responsible for the hydrogen bomb and for the energy release of the sun. In Figure 16–18(c), two deuterium (heavy hydrogen) nuclei brought together combine to form a helium nucleus whose mass is significantly less than the sum of the masses of the separated deuterium nuclei. The remainder of the mass is released as energy. If many such combinations can be made to take place over a very short period of time, the consequences can be devastating.

When two elementary particles combine, a peculiar thing can happen. The energy loss, in the form of γ rays, can be enormous, resulting in a mass loss which may be greater than the original mass of either of the particles alone. Thus it is in principle possible for two elementary particles to combine to form a particle with a smaller mass than either one of them. For example, as illustrated in Figure 16–18(d), it is possible for a nucleon (mass $1838m_e$) and a Δ (mass $2419m_e$) to combine to form a pion (mass $270m_e$). In this reaction the energy equivalent of $1838m_e + 2419m_e - 270m_e = 3987m_e$ is released. Or a nucleon might be formed from a pion and a Δ with the release of $851m_e$ of energy. It is also possible for a nucleon and a pion to produce a Δ when they collide, provided that they have enough energy of motion to supply the extra $311m_e$ needed to create the Δ.

In a sense a pion can therefore be considered to be composed of a nucleon and a Δ, a nucleon to be composed of a pion and a

Δ, and a Δ to be composed of a nucleon and a pion. This is the key to the bootstrap idea.

The bootstrap theory has a simple thesis: *All of the observed particles are equally "elementary" and they are all constituents of one another.* The elementary particles in effect lift themselves up by their own bootstraps. Geoffrey Chew, the theoretical physicist who originated the bootstrap concept, has described the goal of the theory:

> . . . each particle helps to generate other particles, which in turn generate it. In this circular and violently nonlinear situation it is possible to imagine that no free parameters appear and that the only self-consistent set of particles is the one we find in nature.[9]

In other words, it is hoped that one could show by calculation that the observed particles combine in a consistent manner to form each other. The ultimate triumph of the bootstrap theory would be to show that the observed particles constitute the only possible set of particles that could combine in this way. This would explain the existence of the particles found in nature.

Successful completion of this program would surely constitute a revolution in man's ideas about the structure of matter. It would mean that the reductionist chain ends with the observed elementary particles. Proponents of the bootstrap theory claim that it is plausible that the chain should end at this level, because only here do we encounter a situation where the mass losses of combining particles are comparable to the masses of the particles themselves.

Unfortunately, the calculations of the bootstrap theory are extremely difficult to perform. It is not hard to see why. In this theory a proton is not merely a combination of a π^+ and a Δ^0; it is also a π^0 and a Δ^+, or a π^- and a Δ^{++}, or many different combinations of other particles. This makes it very difficult to calculate the properties of a proton in the bootstrap model. It does appear that if certain assumptions are made about the way particles combine, the bootstrap model can be made to generate families of particles with the structure of SU(3). However, because of the complexity of the calculations, the program of explaining the observed elementary particles with this model has not yet been realized.

A Revolution in Progress?

This account of particle physics has of necessity been superficial and incomplete. Much more can be said of the huge accelerators that have produced the particles, the experiments that have detected them, and the properties they possess. Other regularities in the particle data in addition to the SU(3) scheme have been noted and other proposals for a theory of elementary particles besides the quark and bootstrap models are in the air.

As of today, the quark model would seem to be the most promising theoretical proposal, with the most support among elementary particle physicists. The model's success in generating the SU(3) families provides strong evidence in its favor. However, the failure of many years' search to uncover conclusive evidence for the existence of quarks is disconcerting. Perhaps we have not yet looked in the right places. Although it is the leading candidate, one could certainly not predict with confidence at this point that the quark model will ultimately be accepted as the theory of elementary particles.

In the midst of the search for a theory, intuition is important. For example, consider the following comment on the quark model by Geoffrey Chew:

> Physicists usually perceive their discipline's goal as the reduction of nature to fundamentals, and the high-energy arena has correspondingly been dominated by the search for "basic building blocks." Finding the quark is for the moment regarded by many as the ultimate prospective triumph; failure to find some such fundamental entity is equated with frustration. There exists, nonetheless, a 180-degree inverted point of view, which envisions the absence of fundamentals as the ultimate triumph; this is the bootstrap attitude . . . I would find it a crushing disappointment if in 1980 all of hadron* physics could be explained in terms of a few arbitrary entities. We should then be in essentially the same posture as in 1930,

* Hadron is the generic name for the elementary particles that interact through the strong nuclear force.

> when it seemed that neutrons and protons were the basic building blocks of nuclear matter. To have learned so little in half a century would to me be the ultimate frustration.[10]

This comment clearly focuses on philosophical differences between the quark and bootstrap theories rather than on differences in the predictions the theories make about experiments. Such considerations, commonly regarded as unscientific, have always been essential guides in the development of physical theories.

It is interesting to note in this regard that the quark and bootstrap models attempt to answer similar questions about experimental observations. Both interpret explaining the elementary particles as understanding why we observe the particles we do with the properties they have and no others. The systematic organization of the particles found in SU(3) makes this seem a reasonable question to ask and most physicists today believe we will be able to answer it. The situation may well be analogous to Mendeleev's empirical classification of the elements, which was explained by the atomic theory of matter. Or it might be like Kepler's attempt to explain why there are six and only six planets through his perfect solids model. The attempt to explain the periodic table turned out to be successful, but the attempt to explain the number of planets did not. We cannot know with certainty what questions it is useful to ask about elementary particles until a successful theory emerges.

It is not clear how the elementary particle problem will be resolved. In particular, it is not clear whether it will involve a conceptual revolution on the scale of those precipitated by Newton and Einstein or merely a straightforward extension of presently accepted ideas. What is generally believed is that the time is ripe for some kind of breakthrough. Many particles are known and regularities have been discerned. Abraham Pais, a particle theorist, has described the state of particle physics as like

> . . . a symphony hall a while before the start of the concert. On the podium one will see some but not yet all of the musicians.

> They are tuning up. Short brilliant passages are heard on some instruments; improvisations elsewhere; some wrong notes too. There is a sense of anticipation for the moment when the symphony starts.[11]

But will it be Beethoven or John Cage?

Suggested Reading

Holton, G. *Introduction to Concepts and Theories in Physical Science.* Reading, Mass.: Addison-Wesley, 1952. Chapter 19 presents a clear and authoritative account of the development of the atomic theory of matter.

Arons, A. *Development of Concepts of Physics.* Reading, Mass.: Addison-Wesley, 1965. Chapters 28 and 29 give a detailed exposition of the history of the atomic model.

Cooper, L. *An Introduction to the Meaning and Structure of Physics.* New York: Harper & Row, 1968. The final eight chapters are a good brief source for further information on elementary particles and their production, detection, and classification schemes.

Ford, K. W. *The World of Elementary Particles.* New York: Blaisdell, 1963. A well-written book requiring no advanced background. Somewhat dated, but still informative.

Fowler, W. B., and Samios, N. P. "The Omega-Minus Experiment." *Scientific American*, October 1964, pp. 36–45. An account of the experiment by two of the physicists who helped perform it.

"Strangeness Minus Three," BBC-TV film, Philip Daly producer, Robeck & Co. distributor, 230 Park Ave., New York 10017. An unusually good non-technical film, dealing with the discovery of the omega-minus and the SU(3) theory behind it. Includes interviews with Gell-Mann and Ne'eman, the SU(3) creators, Richard Feynman, another prominent particle theorist, and Nicholas Samios, leader of the group that found the omega-minus. We highly recommend this film as a supplement to this chapter.

Weisskopf, V. "The Three Spectroscopies." *Scientific American*, May 1968, pp. 15–29. A discussion of another point of view regarding a possible theory of elementary particles.

Questions

1. The [medieval] cathedrals were technological constructions of great beauty and significance that expressed the aspirations and spirituality of their age . . . I like to think that, six centuries hence, we will be as proud of our accelerators. I like to think that, even though they will only endure then as the Stonehenges of the future, the discoveries we make today will be part and parcel of the culture of that future.[12]

   —Robert R. Wilson
   Director, National Accelerator
   Laboratory (Batavia, Illinois)

   Comment on the notion that particle accelerators are the cathedrals of our time.

2. Consider the developments in the search for a solution to the elementary particle problem discussed in this chapter. To what extent do you find the traditional view of the scientific method discussed in Chapter 1 applicable to this search? In what respects does it seem least adequate?

3. In the passage quoted on page 451, Geoffrey Chew argues against the quark model on philosophical grounds. Suggest other such arguments that influenced the course of the revolutions discussed in the earlier chapters of this book.

4. If the bootstrap model is successful, it will mean the end of the reductionist chain in explanations of the structure of matter. In this case, could physicists rightly claim that they had finally found the ultimate solution to the problem of the structure of matter? Can physics ever find ultimate solutions, which make further questions unnecessary? Has the ultimate solution to the problem of motion, discussed in the first fifteen chapters, been found? If not, what would it be like?

5. It was suggested in the text that "SU(3) may be ellipses—or it just might be epicycles." What analogy do you see between SU(3) and ellipses or epicycles? What parallels can you see between the elementary particle and planetary problems?

6. Suppose the Ω^- had not been found in the 100,000 photographs taken at Brookhaven.
   (a) Would this have disproved the SU(3) scheme? Explain.
   (b) What would have been the likely reaction of experimental and theoretical physicists to this result?
7. (a) Speculate on what kinds of experimental or theoretical developments would disprove or cast grave doubts on the quark model of elementary particles.
   (b) Would the discovery of a quark in nature prove the quark model to be correct? Explain.
8. (a) A cube of copper metal, 1 cm on a side, has a mass of 9 gm. Using the data of Figure 16–2, and the mass of the electron and proton given in the text, calculate roughly how many atoms are incorporated in the cube of copper.
   (b) Use this result to calculate the approximate spacing between the atoms in the cube of copper. Compare with the size of an atom (i.e., of its electron cloud) given in the text.
9. In the Batavia accelerator, which has a radius of 1 km, protons travel around the ring for about four seconds before they are ejected and allowed to bombard a target. Assuming that the protons move with nearly the speed of light (3×10^8 m/sec), calculate how many revolutions around the accelerator ring a proton completes.
10. The effects of special relativity theory are constantly evident in elementary particle experiments. For example, consider the path of the Λ^0 particle in the bubble chamber photograph of Figure 16–13. The Λ^0 travels about 60 cm (as you can judge for yourself since the chamber is about 200 cm long) before decaying into a π^- and a p^+. It is well known, however, that the lifetime of a Λ^0 in its rest frame is about 2×10^{-10} sec.
    (a) If the Λ^0 lived 2×10^{-10} sec in the laboratory (where it is traveling with nearly the speed of light, 3×10^{10} cm/sec), how far would it travel?
    (b) How does special relativity theory explain the observation that the Λ^0 travels 60 cm before decaying?

(c) How near to the speed of light must you assume the Λ^0 to be traveling to explain the 60 cm path length?

11. Suppose a particle-accelerator research group reports the discovery of an Ω^0 particle, having spin 3/2, positive parity, mass $3273m_e$, and no charge. What implication would such a report have for our understanding of elementary particles? Speculate on how such a report would be received by particle physicists.
12. On Figure 16–11, enter the current best value for the omega-minus mass from Table 16–1, and calculate the difference between its mass and the listed Ξ^* mass. According to the SU(3) scheme, this mass difference and the other two mass differences listed in Figure 16–11 should be equal. Yet they appear to differ somewhat. What can you say about the significance of these differences?
13. The construction of the Batavia accelerator cost approximately \$260 million of taxpayers' money. Do you feel that this expenditure is worthwhile? (This sum might be compared with the \$80 billion defense budget, \$26 billion spent for medical and health care services, or the tobacco industry's income of about \$1.5 billion.)

APPENDICES

Appendix One: Straight-line Graphs

Consider a variable y whose value depends on a variable t. Perhaps the most elementary way in which y can depend on t is:

$$y = At + B \tag{A–1}$$

where A and B are constants. This is said to be a linear or straight-line relation because the graph of y vs. t is a straight line. The graph corresponding to equation (A–1) is shown in Figure A1–1. B is called the *y-intercept*. That is, B equals y_0, the value of

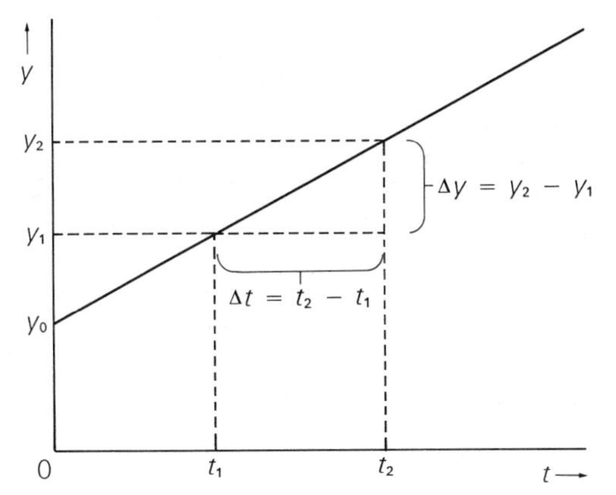

Figure A1–1. The straight-line graph of y vs. t corresponding to the equation $y = At + B$. The slope, $\Delta y/\Delta t$, is A, and the y-intercept, y_0, is B.

y corresponding to $t = 0$. This can be seen by substituting $t = 0$ in equation (A–1):

$$y_0 = A \cdot 0 + B$$

or

$$y_0 = B$$

The other constant, A, is the *slope* of the graph. Consider a time t_1 and a later time t_2. According to equation (A–1), the corresponding values of y are:

$$y_2 = At_2 + B$$
$$y_1 = At_1 + B$$

The slope of the graph is by definition the ratio of Δy, the change in y, to Δt, the corresponding change in t:

$$\text{slope} = \frac{\Delta y}{\Delta t} = \frac{y_2 - y_1}{t_2 - t_1}$$
$$= \frac{(At_2 + B) - (At_1 + B)}{t_2 - t_1} = \frac{A(t_2 - t_1)}{t_2 - t_1}$$
$$= A$$

Notice that if y_2 is less than y_1, as is the case in Figure A1–2, then $y_2 - y_1$ is negative. Since $t_2 - t_1$ is always positive (t_2 is always chosen to be a later time than t_1), this means that the slope of a graph like the one in Figure A1–2 is negative.

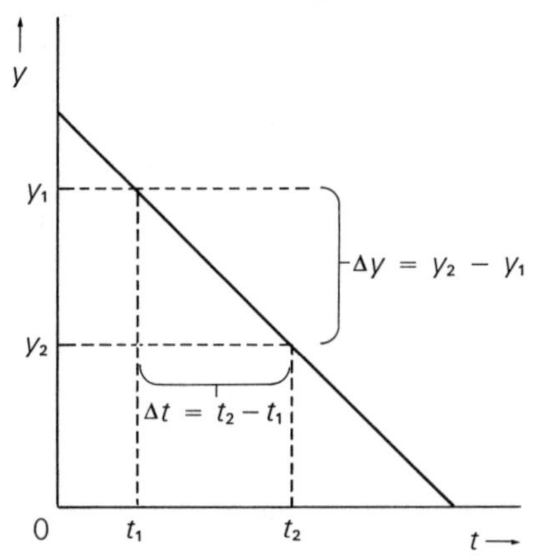

Figure A1–2. An example of a straight-line graph with negative slope. Since y_2 is less than y_1, Δy is negative, and hence the slope, $\Delta y/\Delta t$, is negative.

We encounter a number of straight-line graphs in this book. To understand their part in the discussion, one should be able to tell at a glance whether the slope of a particular graph is positive or negative and whether it is large or small. The following exercise will help develop this facility.

> Consider the straight-line graphs below, which have the same intercept but different slopes:
>
> (a) $y = 10t + 1$ (slope $= 10$)
> (b) $y = 2t + 1$ (slope $= 2$)
> (c) $y = 1$ (slope $= 0$)
> (d) $y = -2t + 1$ (slope $= -2$)
> (e) $y = -10t + 1$ (slope $= -10$)
>
> Sketch these graphs on the accompanying figure.

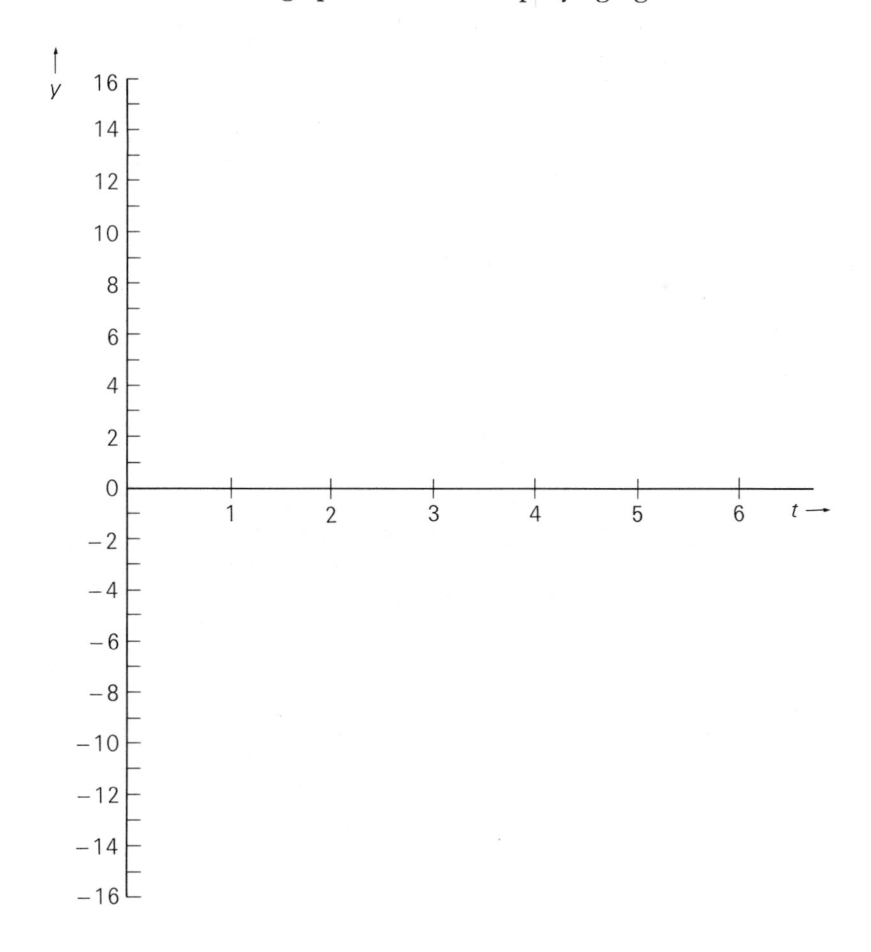

From this exercise it should be clear that a straight-line graph that looks like has a large positive slope, has a smaller positive slope, has zero slope, has a small negative slope and has a large negative slope.

Appendix Two: The Acceleration Vector in Uniform Circular Motion

Figure A2–1 represents the motion of an object traveling in a circular path with constant speed. At time t_i the object is at position A and has velocity $\vec{v}_i$; at a later time $t_f = t_i + \Delta t$ the object is at position B and has velocity $\vec{v}_f$. Since the object moves with constant speed, the velocity vectors are drawn with the same length.

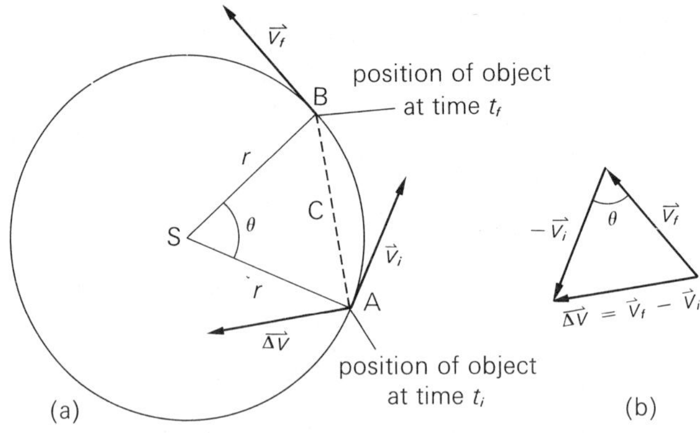

Figure A2–1. (a) During the time interval Δt between t_i and t_f the object moves from point A to point B. The velocity vectors, $\vec{v}_i$ and $\vec{v}_f$, have the same length and are tangent to the circle. The vector $\overrightarrow{\Delta v}$ is that obtained in (b). (b) $\overrightarrow{\Delta v}$, the change in the velocity vector during the time interval Δt, is obtained by adding $-\vec{v}_i$ to $\vec{v}_f$.

We wish to find the acceleration vector $\vec{a}$ at time t_i. To do this, we begin with the definition of the acceleration vector:

$$\vec{a} = \underset{\Delta t \to 0}{\text{limit}} \frac{\overrightarrow{\Delta v}}{\Delta t}$$

It is convenient to consider the direction and magnitude of $\vec{a}$ separately.

Direction: The definition implies that the direction of $\vec{a}$ is in the direction of $\overrightarrow{\Delta v}$ in the limit as Δt approaches zero. To find this limiting direction, we proceed in two steps: First we construct $\overrightarrow{\Delta v} = \vec{v}_f - \vec{v}_i$ and note its direction. Second, we take the limit as Δt approaches zero.

The first step is illustrated in Figure A2–1(b), where $-\vec{v}_i$ is added to $\vec{v}_f$ to give $\overrightarrow{\Delta v}$. The latter is drawn at point A in part (a) of the figure. In what direction does $\overrightarrow{\Delta v}$ point? We can find this by using a theorem from plane geometry: If each of the equal sides of one isosceles triangle is perpendicular to the corresponding side of another isosceles triangle, then the two triangles are similar and their third sides are also perpendicular.

There are two isosceles triangles in the figure. SAB is isosceles since the two sides marked r are both radii of the circle. The triangle in part (b) is isosceles because of the constant speed of the object; this means that the lengths of the vectors $-\vec{v}_i$ and $\vec{v}_f$ are equal. In addition, each of the equal sides of one triangle is perpendicular to the corresponding side of the other: SA is perpendicular to $-\vec{v}_i$ and SB is perpendicular to $\vec{v}_f$, since a radius of a circle is perpendicular to a tangent to the circle, and the velocity vectors are tangent to the circle. Consequently the theorem quoted above applies to these triangles, and allows us to conclude that the third sides of the triangles, $\overrightarrow{\Delta v}$ and the line C, are perpendicular. Thus $\overrightarrow{\Delta v}$ as redrawn in part (a) is perpendicular to the line C.

The second step, taking the limit as Δt approaches zero, gives the final result. As Δt becomes smaller, point B gets closer to point A. $\overrightarrow{\Delta v}$ remains perpendicular to C, whose direction approaches that of the tangent to the circle at A. Thus in the limit as Δt approaches zero, $\overrightarrow{\Delta v}$ is perpendicular to the tangent. This implies that $\overrightarrow{\Delta v}$ points from A toward the center of the circle. Thus $\vec{a}$ *is always directed toward the center of the circle from the position of the object.*

Magnitude: From the definition of $\vec{a}$ above, we may write a relation between a, the length of the vector $\vec{a}$, and Δv, the length of the vector $\overrightarrow{\Delta v}$:

$$a = \lim_{\Delta t \to 0} \frac{\Delta v}{\Delta t}$$

Notice that although the object is not speeding up, there *is* an acceleration; in our notation here, Δv does not mean "change in speed," but rather "magnitude of change in vector velocity." We can find Δv from the geometry of the figure. Since the velocity triangle in (b) was found to be similar to triangle SAB in (a), the corresponding sides are in the same proportion. Thus

$$\frac{\Delta v}{v} = \frac{C}{r}$$

or

$$\Delta v = \frac{Cv}{r}$$

where v is the speed—that is, the magnitude of the velocity vectors $\vec{v}_i$ and $\vec{v}_f$. Dividing the last equation by Δt, we have

$$\frac{\Delta v}{\Delta t} = \frac{C}{\Delta t}\frac{v}{r}$$

Then

$$a = \lim_{\Delta t \to 0} \frac{v}{r}\frac{C}{\Delta t} = \frac{v}{r} \lim_{\Delta t \to 0} \frac{C}{\Delta t}$$

since v and r are constant. But in the limit as Δt approaches zero, C becomes closer and closer to the length of the arc of the circle between A and B—that is, the distance traveled by the object during Δt. In other words

$$\lim_{\Delta t \to 0} \frac{C}{\Delta t} = v$$

the speed of the object. Thus we have finally,

$$a = \frac{v^2}{r}$$

Appendix Three: Central Forces and Kepler's Second Law

Newton showed in the *Principia* that for any object moving under the influence of a central force, (i.e., one directed toward a point—the center of force), the line between the object and the center of force sweeps out equal areas in equal intervals of time. We present here a proof of this statement very similar to that given by Newton:

Let us begin with a special case of a central force—no force at all. In Figure A3–1, the straight line *XYZAB* represents the path of an object that experiences no force and therefore moves

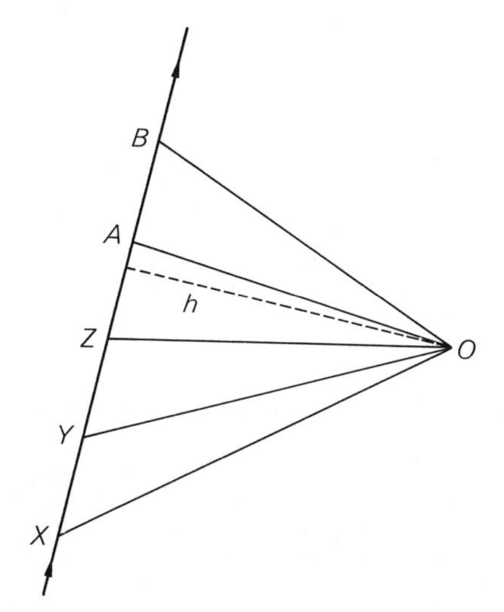

Figure A3–1. An object that experiences no force moves with a constant velocity along the line *XYZAB*. The triangles *XOY*, *YOZ*, *ZOA*, and *AOB* have equal areas since they have equal bases and the same altitude.

with a constant velocity. The distances $XY = YZ = ZA = AB$ are covered by the object in equal intervals of time. Consider some arbitrary point O, which later will be taken to be the center of force. A line from O to the object sweeps out the areas XOY, YOZ, ZOA, and AOB in equal intervals of time. We can easily see that these areas are equal: The area of a triangle is given by (1/2) × (base) × (altitude), where the base is a line between any two vertices and the altitude is the perpendicular line from the other vertex to the base. We can consider XY, YZ, ZA, and AB to be the equal bases of the triangles; the dashed line h is then the altitude of each triangle. Thus the areas are equal for the case of no forces.

Now suppose instead that when the object is at point A, it is struck a sharp blow directed toward O. Such a sudden force will bring about a sudden acceleration—that is, a sudden change in velocity $\overrightarrow{\Delta v}$—in the direction of the force, toward O. Before the blow, we can represent the initial velocity $\vec{v}_i$ by a vector drawn from A to B as shown in Figure A3–2. The direction of $\overrightarrow{\Delta v}$ is

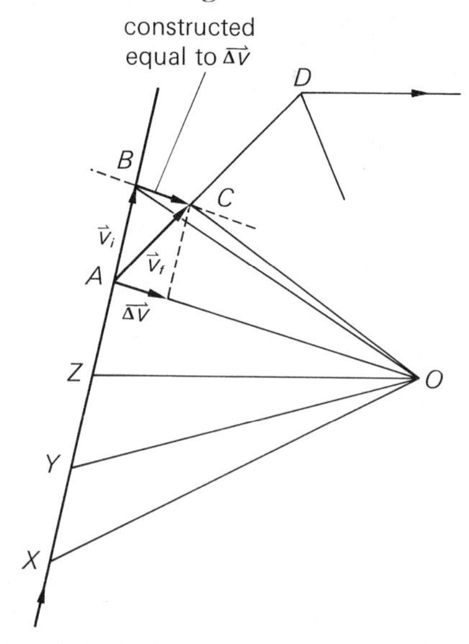

Figure A3–2. At point A, the object is struck a blow directed toward O, causing it to move from A to C during the next equal time interval. The triangles AOB and AOC have the same area since they have a common base, AO, and equal altitudes.

toward O; its magnitude depends on the strength of the blow. The velocity after the blow, $\vec{v}_f$, can then be constructed as shown by adding $\vec{v}_i + \overrightarrow{\Delta v}$. This means that in the time interval after the blow, area AOC is swept out instead of AOB. But we can show that these two areas are equal: Each triangle has the base AO in common; since the line BC was drawn parallel to AO, the altitudes of the two triangles, drawn from B and C perpendicular to AO, are equal.

Now the object continues to move along ACD. At some later time at an arbitrary point D we imagine it to be struck again with a blow toward O. Again, the motion will be diverted, and we can show by reasoning identical to that above that again equal areas are described by the object's motion in equal times before and after the blow. This will continue to be true no matter how many blows are applied, as long as they are directed toward the same point O. In particular, if we apply many small blows very closely spaced in time, we can approximate the effect of any continuous centrally directed force. In the limit as the time between blows approaches zero, the polygon-shaped path of the object becomes indistinguishable from a smooth curve. Thus we have proved that Kepler's Second Law is valid for any central force.

REFERENCES

INTRODUCTION

1. Bertrand Russell, *The Scientific Outlook* (New York: Norton, 1931), p. 14.

Chapter 1

1. Ibid., p. 57.

THE PROBLEM OF CELESTIAL MOTION

1. Stillman Drake, *Discoveries and Opinions of Galileo* (Garden City, N.Y.: Doubleday, 1957), p. 164. © 1967 by Stillman Drake.
2. *Encyclopedia Americana*, 1969 ed., s.v. "Copernicus."

Chapter 2

1. George Sarton, *A History of Science: Ancient Science Through the Golden Age of Science* (Cambridge, Mass.: Harvard University Press, 1952), p. 77.
2. Benjamin Farrington, *Greek Science, Its Meaning For Us* (Harmondsworth, England: Penguin, 1944), p. 42.
3. Aristotle, *De Caelo*, 290b, 15, in *The Origins of Scientific Thought*, trans. Giorgio de Santillana (Chicago: University of Chicago Press, 1961), p. 85. Used by permission of The New American Library.

Chapter 3

1. John Donne, "Upon the Translation of the Psalms by Sir Philip Sidney," in *The Poems of John Donne*, ed. H. J. C. Grierson (London: Oxford University Press, 1933), p. 349.

2. Ptolemy, *Almagest*, trans. A. J. Pomerans, in René Taton, *History of Science* (New York: Basic Books, 1963), vol. 1, p. 321.
3. J. L. E. Dreyer, *A History of Astronomy from Thales to Kepler*, 2nd ed. (New York: Dover, 1953), p. 137.

Chapter 4

1. Copernicus, "The Commentariolus," in *Three Copernican Treatises*, trans. Edward Rosen, 2nd ed. (New York: Dover, 1959), p. 57.
2. Thomas S. Kuhn, *The Copernican Revolution* (New York: Random House, 1959), p. 142.
3. Arthur Koestler, *The Sleepwalkers* (New York: Macmillan, 1959), p. 275.
4. Ibid., p. 322.
5. René Descartes, *Principia Philosophiae*, III, XXVIII, in René Dugas, *Mechanics in the Seventeenth Century*, trans. Freda Jacquot (Neuchatel, Switzerland: Griffon, 1958), p. 184.
6. René Descartes, *Lettres*, III, no. 103, p. 586, in J. F. Scott, *The Scientific Work of René Descartes* (London: Taylor & Francis, 1952), p. 169.
7. Figure from James B. Gerhart and Rudi H. Nussbaum, "Motion," paper prepared at the Conference on the New Instructional Materials in Physics, 1965, at the University of Washington.

THE PROBLEM OF TERRESTRIAL MOTION

1. Aristotle, *Physics*, 266b, 29, in Sir Thomas Heath, *Mathematics in Aristotle* (London: Oxford University Press, 1949), p. 155.
2. Marshall Clagett, *The Science of Mechanics in the Middle Ages* (Madison: University of Wisconsin Press, 1959), pp. 534–535.
3. Galileo Galilei, *Dialogue Concerning the Two Chief World Systems*, trans. Stillman Drake, 2nd ed. (Berkeley: University of California Press, 1967), p. 156. Originally published by the University of California Press; reprinted by permission of The Regents of the University of California.

Chapter 6

1. Aristotle, *Physics*, 194b, trans. Richard Hope (Lincoln: University of Nebraska Press, 1961), p. 28.
2. Aristotle, *Works*, ed. W. D. Ross, vol. II, *De Caelo*, trans. J. L. Stocks (Oxford: Clarendon Press, 1930), p. 277b.
3. Ibid., p. 309b.
4. Lane Cooper, *Aristotle, Galileo, and the Tower of Pisa* (Ithaca: Cornell University Press, 1935), p. 79. Copyright 1935 by Lane Cooper. Used by permission of Cornell University Press.

5. Galileo Galilei, *Dialogues Concerning Two New Sciences*, trans. Henry Crew and Alfonso de Salvio (New York: Macmillan, 1914), pp. 106–109.
6. Aristotle, *Physics*, p. 74.
7. Clagett, p. 263.
8. Galilei, *Two New Sciences*, pp. 212–213. (Italics added.)
9. Ibid., p. 202.

Chapter 7

1. Aristotle, *Physics*, p. 177.
2. Galilei, *Two World Systems*, pp. 145–147.
3. Ibid., p. 155.

THE NEWTONIAN SYNTHESIS

1. Richard Feynman, *The Character of Physical Law* (Cambridge, Mass.: M.I.T. Press, 1965), p. 18.

Chapter 8

1. Frank E. Manuel, *A Portrait of Isaac Newton* (Cambridge, Mass.: Harvard University Press, 1968), p. 39.
2. Isaac Newton, *Principia*, ed. Florian Cajori, trans. Andrew Motte (Berkeley: University of California Press, 1934), p. 13. Originally published by the University of California Press; reprinted by permission of The Regents of the University of California.
3. Ibid., pp. 13–14.

Chapter 9

1. William Stukeley, *Memoirs of Sir Isaac Newton's Life*, ed. A. Hastings White (London: Taylor & Francis, 1936; originally published in 1752), pp. 19–20.

Chapter 10

1. Newton, *Principia*, p. 421.
2. Ibid., p. 547.
3. Voltaire, *Lettres Philosophiques*, in Alexander Koyré, *Newtonian Studies* (Cambridge, Mass.: Harvard University Press, 1965), p. 55.
4. Alexander Pope, *Collected Works*, Cambridge Edition (Boston: Houghton Mifflin, 1903), p. 135.
5. Pierre Laplace, "Essai philosophique sur les probabilités," in Newton, *Principia*, p. 677.

Chapter 11

1. A. S. Eddington, *The Nature of the Physical World* (New York:

Macmillan, 1928), p. 124.
2. Newton, *Principia*, p. 6.
3. Galilei, *Two World Systems*, p. 154.
4. Newton, *Principia*, p. 20.
5. Ibid., p. 6.

THE THEORY OF RELATIVITY

1. Newton, *Principia*, p. 6.
2. A. Einstein, "Autobiographical Notes," in *Albert Einstein: Philosopher-Scientist*, ed. Paul A. Schilpp (New York: Tudor, 1949), p. 55. (Minor changes in translation are ours.)

Chapter 12

1. Albert A. Michelson and Edward W. Morley, "On the Relative Motion of the Earth and the Luminiferous Ether," *American Journal of Science*, Third Series, *34* (1887): 341.
2. Ibid.
3. R. S. Shankland, "Michelson-Morley Experiment," *American Journal of Physics*, *32* (1964): 52.
4. Alfred M. Bork, "The 'FitzGerald' Contraction," *Isis*, *57* (1966): 200.
5. G. F. FitzGerald, "The Ether and the Earth's Atmosphere," *Science*, *13* (1889): 390.
6. Bork, " The 'FitzGerald' Contraction," 201.
7. Charles Scribner, Jr., "Henri Poincaré and the Principle of Relativity," *American Journal of Physics*, *32* (1964): 673.
8. Ibid.
9. H. A. Lorentz, "Electromagnetic Phenomena in a System Moving with Any Velocity Less than That of Light," in *The Principle of Relativity*, trans. W. Perrett and G. B. Jeffery (London: Methuen, 1923), p. 13.

Chapter 13

1. Einstein, "Notes," p. 9.
2. Ibid., p. 53.
3. A. Einstein, "On the Electrodynamics of Moving Bodies," in *The Principle of Relativity*, pp. 37–38. (Italics added.)
4. William Francis Magie, "The Primary Concepts of Physics," *Physical Review*, *34* (1912): 125.
5. W. Wien, *Uber Elektronen*, 2nd ed. (Leipzig: Teubner, 1909), p. 32, in Gerald Holton, "Einstein, Michelson, and the 'Crucial' Experiment," *Isis*, *60* (1969): 140.

6. Richard C. Tolman, "The Second Postulate of Relativity," *Physical Review*, *31* (1910): 27.
7. Edward W. Morley and Dayton C. Miller, "Report of an Experiment to Detect the FitzGerald-Lorentz Effect," *Philosophical Magazine*, *9* (1905): 680.
8. "Conference on the Michelson-Morley Experiment," *Astrophysical Journal*, *68* (1928): 342.
9. W. Kaufmann, *Annalen der Physik*, *19* (1906): 495, in Gerald Holton, "On the Origins of the Special Theory of Relativity," *American Journal of Physics*, *28* (1960): 634.

Chapter 15

1. Newton, *Principia*, p. 10. (Translation slightly altered for clarity.)
2. Ibid., p. 11. (Italics added.)
3. Ernst Mach, *The Science of Mechanics*, trans. Thomas J. McCormack, 2nd ed. (Chicago: Open Court, 1907), p. 229.
4. Ibid., p. 232. (Italics added.)

EPILOGUE

1. Geoffrey F. Chew, Murray Gell-Mann, and Arthur H. Rosenfeld, "Strongly Interacting Particles," *Scientific American*, February 1964, p. 74. Copyright © 1964 by Scientific American, Inc. All rights reserved.

Chapter 16

1. Dmitri Mendeleev, "The Relation Between the Properties and Atomic Weights of the Elements," *Journal of the Russian Chemical Society*, *1* (1869): 60–77. Translation from Gerald Holton, *Introduction to Concepts and Theories in Physical Science* (Reading, Mass.: Addison-Wesley, 1952), p. 418.
2. Ibid. Translation from Henry M. Leicester and Herbert S. Klickstein, *A Source Book in Chemistry* (New York: McGraw-Hill, 1952), p. 440.
3. Victor F. Weisskopf, "Three Steps in the Structure of Matter," *Physics Today*, August 1970, p. 22.
4. Laurie M. Brown, "Quarkways to Particle Symmetry," *Physics Today*, February 1966, p. 44.
5. Ibid.
6. C. B. A. McCusker and I. Cairns, "Evidence of Quarks in Air-Shower Cores," *Physical Review Letters*, *23* (1969): 658.
7. R. K. Adair and H. Kasha, "Analysis of Some Results of Quark Searches," *Physical Review Letters*, *23* (1969): 1355–1358.

8. D. C. Rahn and R. I. Louttit, "Comments on 'Evidence of Quarks in Air-Shower Cores,' " *Physical Review Letters, 24* (1970): 279–280.
9. Geoffrey F. Chew, "Elementary Particles?" *Physics Today*, April 1964, p. 34.
10. Geoffrey F. Chew, "Hadron Bootstrap: Triumph or Frustration?" *Physics Today*, October 1970, pp. 23–25.
11. Abraham Pais, "Particles," *Physics Today*, May 1968, p. 28.
12. Robert Rathbun Wilson, "Particles, Accelerators, and Society," *American Journal of Physics, 36* (1968): 492.

INDEX